Dissipation and Diffusion of Quantum Light Field

Hongyi Fan · Liyun Hu

Dissipation and Diffusion of Quantum Light Field

中国科学技术大学出版社
UNIVERSITY OF SCIENCE AND TECHNOLOGY OF CHINA PRESS

Springer

Hongyi Fan
University of Science and Technology
of China
Hefei, Anhui, China

Liyun Hu
Jiangxi Normal University
Nanchang, China

Translated by
Xiangguo Meng
Shandong Second Medical University
Weifang, Shandong, China

ISBN 978-981-95-1527-1 ISBN 978-981-95-1528-8 (eBook)
https://doi.org/10.1007/978-981-95-1528-8

Jointly published with University of Science and Technology of China Press
The print edition is not for sale in China (Mainland). Customers from China (Mainland) please order the print book from: University of Science and Technology of China Press.
ISBN of the Co-Publisher's edition: 978-7-312-04909-5

Translation from the Chinese Simplified language edition: "Dissipation and Diffusion of Quantum Light Field" by Hongyi Fan and Liyun Hu, © The University of Science and Technology of China Press 2020. Published by The University of Science and Technology of China Press. All Rights Reserved.

The original submitted manuscript has been translated into English. The translation was done using artificial intelligence. A subsequent revision was performed by the author(s) to further refine the work and to ensure that the translation is appropriate concerning content and scientific correctness. It may, however, read stylistically different from a conventional translation.

This Springer imprint is published by the registered company Springer Nature Singapore Pte Ltd.
The registered company address is: 152 Beach Road, #21-01/04 Gateway East, Singapore 189721, Singapore

Preface

The nature of light is still not fully understood. Einstein classified the radiation transitions into three categories in 1917: stimulated absorption, spontaneous emission, and stimulated emission. He wrote in 1917, "I will reflect on what light is for the rest of my life." After 34 years, he said, fifty years of conscious contemplation have not brought me close to the answer to this question. Of course, anyone who thinks they know the answer now is actually deceiving themselves.

The scientific exploration of light began with Galileo, who was the first person to create a telescope and used it to observe planets and their satellites in the solar system. Later, Fermat proposed the law of light walking, i.e., the principle of shortest path. And then, Huygens and Newton debated whether the light is a wave or a particle. Yang's double slit experiment, Marius' polarization phenomenon of light, Fresnel's diffraction theory, and Maxwell's electrodynamics support the wave theory of light. Surprisingly, the study of light actually promoted the birth of the theory of special relativity. Michelson's aberration experiment showed that the rotation of the earth does not affect the speed of light, which led Einstein to abandon the concept of "ether". Michael Sun loved optical experiments and said, if a poet is also a physicist, he may be able to express to others the joy, enjoyment, and almost reverent feelings that this discipline inspires. I admit that the aesthetic content of this discipline is not lacking in appeal to me, especially in the field of studying light where I feel its charm.

The study of the radiation theory of light began with Kirchhoff, and this theory along with Boltzmann's entropy theory helped Planck discover the quantum. Einstein proposed the concept of photons from the perspective of the photoelectric effect, and the wave particle duality of light promoted the birth of de Broglie's theory of matter waves.

The study of atomic optical radiation using quantum mechanics began with Bohr (early 20th century), and later Dirac deepened Einstein's stimulated and spontaneous radiation using quantum mechanics and quantum field theory methods. Although the wave particle duality of light promoted the development of photon theory, quantum optics truly became a discipline due to the birth of lasers in the laboratory in the 1960s, resulting in the Jaynes Cummings model of single-mode optical fields and two-level atoms, as well as coherent and squeezed state optical field theories. The non-classical properties of squeezed lights have also promoted the development of quantum statistical mechanics. Quantum optics is a discipline that studies the non-classical properties of lights (i.e. properties that cannot be explained by classical optics and classical electrodynamics), the light-atom interactions, and the generation, propagation, and detection of lights based on the quantum theory of radiation. The lights described by the coherent states and the squeezed states have their unique properties, which enlightens people to further discover the new light fields to profoundly reveal the nature of light.

Discovering new quantum states of light fields and studying their non-classical properties is an important topic in quantum optics. Our research indicates that the dissipation and diffusion can lead to the emergence of new light fields, which is why we wrote this book. We use our own invented the integration theory within ordered operators (IWOP) and the entangled state representations to study the dissipation and diffusion processes of certain light fields. By arranging the operators appropriately and ordering the density operators that describe the evolutions of the light fields, we obtain new light fields that can be expressed by special functions. The effectiveness of this method lies in its ordered arrangement, which maximizes the entropies of light fields and increases the chances of forming new light fields. On the contrary, if they are not ordered, their true faces cannot be seen.

The IWOP theory is a supplement and development to the Bohr Heisenberg's matrix theory and the Dirac's sign method. It is beneficial for the operator ordering, and the diversification of representation and the transformation theory in quantum mechanics, as well as for solving the master equations of density operators.

Hefei, China — Hongyi Fan
Nanchang, China — Liyun Hu

Contents

Chapter 1
Research on Quantum Mechanical Representations via the IWOP Method

To understand the nature of light from the perspective of quantum mechanics, there must be a theory that describes quantum light fields, one of which is the representation theory. Observable quantities are represented by Hermite operators in quantum mechanics, and their complete eigen-states can form the representations, meaning that each representation has its own functional basis vectors [1–3]. Commonly used functional basis vectors have single- and two-variable Hermite polynomials (closely related to Laguerre polynomials proved by Fan [4], which can be derived from Hermite polynomials), Legendre polynomials and Bessel functions.

The representation theory was introduced into quantum theory by Dirac in a symbolic way [5], which is abstract and difficult to understand, like a pile of building blocks that are unattainable. In the 1970s, the integration within ordered operators (IWOP) invented by Fan made the symbolic method more flexible and useful [6–8]. In other words, converting the operators composed of Ket-Bra into their ordering arrangement, in which the original non-commutative components can be interchanged (without losing the nature of the operator), making it possible to directly integrate them. In this chapter, the completeness relations of the representations are expressed in pure Gaussian integral forms using the integration methods within the normal ordering, anti-normal ordering, and Weyl ordering. This has a unique effect on the development of the classical-quantum correspondence theory and the operator theory for quantum light fields.

Physical theories need to be appreciated, and only those who can appreciate them can discuss and study them.

H. Fan and L. Hu, *Dissipation and Diffusion of Quantum Light Field*,
https://doi.org/10.1007/978-981-95-1528-8_1

1.1 Integration Within Ordered Products

1.1.1 Integration Within Normal Ordering

For an operator function with the bose creation and annihilation operators $a^\dagger$, a, we can always move all the creation operators $a^\dagger$ to the left of a by using the identity $\left[a, a^\dagger\right] = 1$, and thus this operator function is arranged in the normal ordering form, marked with the symbol $::$. Within $::$, a and $a^\dagger$ can be interchanged (because no matter how they are interchanged internally, and when removing $::$, all the creation operators $a^\dagger$ must be placed to the left of a, any interchange between a and $a^\dagger$ within $::$ will not change its final result). Now, we list other properties of the normal ordering products [6]:

(1) The normal ordering operator $:F\left(a^\dagger, a\right):$ obeys

$$\langle z| : F\left(a^\dagger, a\right) : \left|z'\right\rangle = F\left(z^*, z'\right) \langle z| z'\rangle . \tag{1.1}$$

(2) The operators $a, a^\dagger$ are commutative within the symbol $: \, :$, that is

$$: a^\dagger a: \; =: aa^\dagger : \; = a^\dagger a. \tag{1.2}$$

(3) c numbers can freely enter and exit the normal ordering symbol $: \, :$, and it is possible to perform the integral or differential operations on c numbers within $: \, :$, and the former requires the integral to convergent.

(4) The symbol $: \, :$ within normal ordering product can be canceled, i.e.,

$$: f(a^\dagger, a) \colon g(a^\dagger, a) \colon : \; =: f(a^\dagger, a) g(a^\dagger, a) : . \tag{1.3}$$

(5) The sum of two normal ordering products reads

$$: f(a^\dagger, a) : + : g(a^\dagger, a) : \; =: \left[f(a^\dagger, a) + g(a^\dagger, a)\right] : . \tag{1.4}$$

(6) Hermite conjugate operation can freely enter and exit the symbol $: \, :$, i.e.,

$$: (W \cdots V) :^\dagger =: (W \cdots V)^\dagger : . \tag{1.5}$$

(7) Within the symbol $: \, :$, the operator $f(a, a^\dagger)$ satisfies the following identities:

$$: \frac{\partial}{\partial a} f(a, a^\dagger) : \; = \left[: f(a, a^\dagger) : , a^\dagger\right], \tag{1.6}$$

$$: \frac{\partial}{\partial a^\dagger} f(a, a^\dagger) : \; = \left[: f(a, a^\dagger) : , a\right]. \tag{1.7}$$

For multi-mode cases, the above two equations can be generalized as

$$
\begin{aligned}
&:\frac{\partial}{\partial a_i}\frac{\partial}{\partial a_j} f(a_i, a_i^\dagger, a_j, a_j^\dagger): = \left[\left[: f(a_i, a_i^\dagger, a_j, a_j^\dagger): , a_j^\dagger\right], a_i^\dagger\right], \\
&:\frac{\partial}{\partial a_i^\dagger}\frac{\partial}{\partial a_j^\dagger} f(a_i, a_i^\dagger, a_j, a_j^\dagger): = \left[\left[: f(a_i, a_i^\dagger, a_j, a_j^\dagger): , a_j\right], a_i\right].
\end{aligned}
\tag{1.8}
$$

One application of the IWOP method is to study the displaced Fock states that are obtained by acting the displaced operator $D(z)$ on the number states $|m\rangle$:

$$
D(z)|m\rangle = \frac{1}{\sqrt{m!}}\left(a^\dagger - z^*\right)^m D(z)|0\rangle = \frac{1}{\sqrt{m!}}\left(a^\dagger - z^*\right)^m |z\rangle . \tag{1.9}
$$

Using the IWOP method to prove its completeness relation, that is

$$
\begin{aligned}
&\int \frac{d^2 z}{\pi} D(z)|m\rangle\langle n| D^{-1}(z) \\
&= \int \frac{d^2 z}{\pi}\frac{1}{\sqrt{m!n!}}\left(a^\dagger - z^*\right)^m |z\rangle\langle z| (a - z)^n \\
&= \int \frac{d^2 z}{\pi}\frac{1}{\sqrt{m!n!}} : \left(a^\dagger - z^*\right)^m (a - z)^n \exp\left[-\left(a^\dagger - z^*\right)(a - z)\right] : \\
&= \frac{(-1)^{m+n}}{\sqrt{m!n!}}\int \frac{d^2 z}{\pi} z^{*m} z^n e^{-|z|^2} = \delta_{m,n}.
\end{aligned}
\tag{1.10}
$$

1.1.2 Integration Within Anti-Normal Ordering

In a monomial consisting of the operators a and $a^\dagger$, if all the annihilation operators a are arranged to the left of all the creation $a^\dagger$, it called an anti-normal ordering product, marked with $\vdots\ \vdots$. The main properties of the anti-normal ordering operators are [9]:

(1) The operators $a, a^\dagger$ are commutative within the anti-normal ordering symbol $\vdots\ \vdots$, i.e., $\vdots a^\dagger a \vdots = \vdots a a^\dagger \vdots = a a^\dagger$.

(2) c numbers can freely enter and exit the symbol $\vdots\ \vdots$, and it is possible to perform the integral or differential operations on c numbers within the symbol $\vdots\ \vdots$, and the former requires the integral to convergent.

(3) In the coherent state representation, the anti-normal ordering operator $\vdots g\left(a, a^\dagger\right)\vdots$ is represented as [2]

$$
\vdots g\left(a, a^\dagger\right)\vdots = \int \frac{d^2 z}{\pi} g\left(z, z^*\right)|z\rangle\langle z| . \tag{1.11}
$$

For example, we use Eq. (1.11) to prove that

$$\begin{aligned}
\mathrm{e}^{-\lambda}{:}\mathrm{e}^{\left(1-\mathrm{e}^{-\lambda}\right)a^{\dagger}a}{:} &= \mathrm{e}^{-\lambda}{:}\mathrm{e}^{\left(1-\mathrm{e}^{-\lambda}\right)a^{\dagger}a}{:}\int\frac{\mathrm{d}^2z}{\pi}\left|z\right\rangle\left\langle z\right| \\
&= \mathrm{e}^{-\lambda}\int\frac{\mathrm{d}^2z}{\pi}\mathrm{e}^{\left(1-\mathrm{e}^{-\lambda}\right)|z|^2}\left|z\right\rangle\left\langle z\right| \\
&= \mathrm{e}^{-\lambda}\int\frac{\mathrm{d}^2z}{\pi}:\exp\left(-\mathrm{e}^{-\lambda}|z|^2+za^{\dagger}+z^*a-a^{\dagger}a\right): \\
&= :\exp\left[\left(\mathrm{e}^{\lambda}-1\right)a^{\dagger}a\right]: = \mathrm{e}^{\lambda a^{\dagger}a},
\end{aligned} \tag{1.12}$$

which leads to

$$\vdots\mathrm{e}^{\lambda aa^{\dagger}}\vdots = (1-\lambda)^{-1}:\exp\left(\frac{-\lambda a^{\dagger}a}{\lambda-1}\right):. \tag{1.13}$$

Using the generating function of the Laguerre polynomial

$$(1-z)^{-1}\exp\left(\frac{zx}{z-1}\right)=\sum_{l=0}^{\infty}\mathrm{L}_n\left(x\right)z^n, \tag{1.14}$$

Equation (1.13) becomes

$$\vdots\mathrm{e}^{\lambda aa^{\dagger}}\vdots =: \sum_{l=0}^{\infty}\lambda^l\mathrm{L}_l\left(-a^{\dagger}a\right):, \tag{1.15}$$

which is a new operator identity.

1.1.3 Integration Within Weyl Ordering

When the classical function $\mathrm{e}^{z\alpha^*}\mathrm{e}^{-z^*\alpha}$ is quantized into an operator, there are multiple choices, either $\mathrm{e}^{za^{\dagger}}\mathrm{e}^{-z^*a}$ (normal ordering), $\mathrm{e}^{-z^*a}\mathrm{e}^{za^{\dagger}}$ (anti-normal ordering), or $\mathrm{e}^{za^{\dagger}-z^*a}$ (Weyl ordering). In the operator $\mathrm{e}^{za^{\dagger}-z^*a}$, $za^{\dagger}$ and z^*a can add and subtract on the exponent, which itself is called Weyl ordering [10], marked with the symbol $\begin{smallmatrix}:&:\\:&:\end{smallmatrix}$, that is

$$\mathrm{e}^{za^{\dagger}-z^*a} = \vdots\,\mathrm{e}^{za^{\dagger}-z^*a}\,\vdots \equiv D\left(z\right). \tag{1.16}$$

Within the symbol $\begin{smallmatrix}:&:\\:&:\end{smallmatrix}$, $a^{\dagger}$ is interchangeable with a, so the IWOP method can be used. So what is the Weyl ordering form of the vacuum projection operator $\left|0\right\rangle\left\langle 0\right|$? Using the IWOP method, we can write the normal ordering product

$$|0\rangle\langle 0| =: \mathrm{e}^{-a^\dagger a}: \tag{1.17}$$

as the following integral form

$$|0\rangle\langle 0| = \int \frac{\mathrm{d}^2 z}{\pi} : \mathrm{e}^{-|z|^2 + z a^\dagger - z^* a} : = \int \frac{\mathrm{d}^2 z}{\pi} D(z)\, \mathrm{e}^{-|z|^2/2}. \tag{1.18}$$

Using Eq. (1.16), we have

$$|0\rangle\langle 0| = \int \frac{\mathrm{d}^2 z}{\pi} \vdots\, \mathrm{e}^{z a^\dagger - z^* a} \,\vdots\, \mathrm{e}^{-|z|^2/2} = 2 \,\vdots\, \mathrm{e}^{-2a^\dagger a} \,\vdots, \tag{1.19}$$

which is the Weyl ordering form of the vacuum projector $|0\rangle\langle 0|$. Similarly, we derive the Weyl ordering product of the coherent state $|z\rangle\langle z|$, that is

$$\begin{aligned} |z\rangle\langle z| &= \int \frac{\mathrm{d}^2 z'}{\pi} : \mathrm{e}^{-|z'|^2 + z'(a^\dagger - z^*) + z'^*(z-a)} : \\ &= \int \frac{\mathrm{d}^2 z'}{\pi} \vdots\, \mathrm{e}^{z'(a^\dagger - z^*) + z'^*(z-a)} \,\vdots\, \mathrm{e}^{-|z'|^2/2} \\ &= 2 \,\vdots\, \mathrm{e}^{-2(a^\dagger - z^*)(a-z)} \,\vdots, \end{aligned} \tag{1.20}$$

which leads to the following completeness relation:

$$\int \frac{\mathrm{d}^2 z}{\pi} |z\rangle\langle z| = 2 \int \frac{\mathrm{d}^2 z}{\pi} \vdots\, \mathrm{e}^{-2(a^\dagger - z^*)(a-z)} \,\vdots = 1. \tag{1.21}$$

1.2 Several Commonly Used Quantum Mechanical Representations

Representation is a set of complete basis vectors that can represent any operator as a ordinary number, first introduced by Dirac. French sculptor Rodin said, "A master is someone who uses their own eyes to see what others have seen, and can discover beauty in things that others are accustomed to". Another person said, "The only point of genius is that what no one can express, or is difficult to express, he can express it". Dirac was such a genius. In the embryonic stage of quantum mechanics, since the lack of mathematical symbols that could express its essence, it makes Heisenberg and others very difficult. Therefore, Dirac invented the symbol method, which not only unified the Heisenberg's matrix mechanics and the Schrödinger's wave mechanics, but also reflected the de Broglie's wave particle duality. With the Dirac's Ket-Bra symbol, the quantum mechanical representations and transformation theory can be established, which ultimately became the language of quantum mechanics. So, in his later years, Dirac said, "Symbolism is my beloved, I won't exchange it for anything".

Dirac's symbolic method has the characteristic of simplicity, without any traces of carving, which is difficult for later generations to imitate. Just as Tao Yuanming's simplicity comes from nature, learning from it would be far from it. So how to integrate the Ket-Bra operator? Dirac did not think of this question, while Fan put forward a solution (the IWOP method), and use it to greatly develop the representation and transformation theory of quantum mechanics.

1.2.1 Photon Number State Representation

Viewing light as photons originated from Einstein's explanation of the photoelectric effect, where each light field mode corresponds to an oscillator of the electromagnetic field. Later, Dirac regarded the electromagnetic radiation as a transition of atomic energy states caused by acting the external perturbations on the atomic systems, during which quantum energy can be absorbed or emitted. This explained the dynamic mechanism of the stimulated radiation of light proposed by Einstein from a quantum mechanics perspective, making the theory more substantial. It is worth noting that the Einstein coefficients for the stimulated radiation involve a very weak effect, which was initially proposed with little hope of being observed. But later, people found the ways to enhance this effect, thus pioneering the theory of lasers. It can be seen that in order to understand the quantum nature of light, the first step is to have a representation that describes the creation and annihilation of photons. Classical optics mainly discusses the behaviors of light propagation in a medium, such as interference, diffraction, and polarization. Maxwell's classical electrodynamics took a step forward by recognizing light as an electromagnetic field, viewing light as composed of electromagnetic waves, and treating each wave as a oscillator, reflecting the wave theory of light. But the creation and annihilation mechanism of light was not discussed. Just as we see lightning and thunder happening in the vast sky, the explanation of the creation and annihilation of light also requires a theoretical "space" that people have envisioned, which is the photon number state representation. The so-called representation is an operator that can be described numerically in the set of its own eigen-states. The set of state vectors with completeness relation forms a representation, and the representation that describes the creation and annihilation mechanism of light is called the photon number state representation or Fock state representation.

After recognizing Bohr's view that the creation and annihilation of light are related to the energy level transitions of atoms, the genius physicist Heisenberg started his research from the frequency and intensity of observable spectral lines, and together with Bohr et al., discovered $[Q, P] = \mathrm{i}\hbar$. Taking his thinking mode as a model, we believe that, given that experiments can observe the creation and annihilation of light, it is more acceptable to intuitively introduce the photon number state representation to readers when taking the quantization of harmonic oscillators (the mechanism of quantum creation and annihilation) as an example. This is because it is easy to transition from the classical vibrational eigen-mode of the harmonic oscillator to a quantum energy level. In classical mechanics, string vibration is a typical harmonic oscillator motion, and the two ends of a fixed string are called nodes. When the length of the string fixed at both ends is L, the string length must be an integer multiple of the half wavelength of the oscillating wave. Only in this way, the entire chord length is precisely embedded in an integer number of half wavelengths. In addition, the vibration of strings has a fundamental frequency and a overtone frequency, so the quantization of harmonic oscillators can maintain similar characteristics to classical cases, while also conforming to the characteristics of de Broglie waves. Although there is no theory of the creation and annihilation of light in classical optics, the vibration of a harmonic oscillator can generate waves. If we compare this with de Broglie's wave particle duality, the generation of light waves corresponds to the creation of photons (or, to put it far fetched, particles accompanied by a wave). Therefore, in order to enable the theory to describe the creation and annihilation of photons, we need to compare the various eigen-vibration modes of the harmonic oscillator to a "photon reservoir". When considering the following quantization condition

$$[Q, P] = \mathrm{i}\hbar, \tag{1.22}$$

the Hamiltonian $h = p^2/(2m) + m\omega^2 q^2/2$ of the classical harmonic oscillator can be transformed into the following operator form:

$$H = \frac{1}{2m}P^2 + \frac{1}{2}m\omega^2 Q^2. \tag{1.23}$$

Using the combination of the operators Q and P, the operators a and $a^\dagger$ are defined as

$$a = \frac{1}{\sqrt{2}}\left(\sqrt{\frac{m\omega}{\hbar}}Q + \mathrm{i}\frac{P}{\sqrt{m\hbar\omega}}\right), \tag{1.24}$$

$$a^\dagger = \frac{1}{\sqrt{2}}\left(\sqrt{\frac{m\omega}{\hbar}}Q - \mathrm{i}\frac{P}{\sqrt{m\hbar\omega}}\right). \tag{1.25}$$

According to Eq. (1.22), we easily obtain

$$\left[a, a^\dagger\right] = 1, \tag{1.26}$$

thus Eq. (1.23) becomes

$$H = \left(N + \frac{1}{2}\right)\hbar\omega, \; N = a^\dagger a. \tag{1.27}$$

Given that the classical harmonic oscillators have their own eigen-vibration modes, labeled as integers, so the quantum harmonic oscillators should also have their own eigen-vibration modes, denoted as $|n\rangle$ $[|\rangle$(Ket), which was invented by Dirac and is called the "right vector", while $\langle|$(Bra) is called the "left vector"], n is a zero or positive integer, and the set of $|n\rangle$ is the "quantum reservoir" of the harmonic oscillator. When $n = 0$, there still exists $\hbar\omega/2$, which is called zero point energy and can't occur in classical mechanics. Record the fact that $|n\rangle$ is the eigen-mode (eigen-state) of H as

$$N\,|n\rangle = n\,|n\rangle\,. \tag{1.28}$$

For example, if $|n\rangle$ is seen as a pocket containing n yuan of money, $a^\dagger a$ represents the operation of "counting" money (operator). Specifically, for $|n\rangle$ with the action of a, it means taking out 1 yuan from the pocket, $n \to n-1$, and then putting this 1 yuan back into the pocket (this operation is represented by the action of $a^\dagger$ on $|n-1\rangle$), and the money in the pocket returns to n yuan. It can be seen that taking it out and putting it back is equivalent to "counting" money, and the pocket still contains n yuan. So, a is the annihilation operator, $a^\dagger$ is the creation operator, and when there is no money in the pocket (represented by $|0\rangle$, called vacuum), no more money can be withdrawn from it, so

$$a\,|0\rangle = 0. \tag{1.29}$$

It follows from the equation $a^\dagger a\,|n\rangle = n\,|n\rangle$ that

$$a\,|n\rangle = \sqrt{n}\,|n-1\rangle\,, \tag{1.30}$$

$$a^\dagger\,|n\rangle = \sqrt{n+1}\,|n+1\rangle\,. \tag{1.31}$$

Clearly, the state $|n\rangle$ is the result of applying the creation operator $a^\dagger$ on $|0\rangle$ n times, i.e.,

$$|n\rangle = \frac{a^{\dagger n}}{\sqrt{n!}}\,|0\rangle\,, \tag{1.32}$$

which is easy to understand because it satisfies Eqs. (1.30) and (1.31). Further, using the identities $\langle n|\,a^\dagger a\,|n\rangle = n$ and

$$\langle n|\,a^\dagger a\,|n\rangle = \langle n|\,a^\dagger\sqrt{n}\,|n-1\rangle\,, \tag{1.33}$$

we get

$$(a\,|n\rangle)^\dagger = \langle n|\,a^\dagger = \sqrt{n}\,\langle n-1|\,, \tag{1.34}$$

where $\langle n-1|$ is the conjugate imaginary quantity of $|n-1\rangle$, so the conjugate imaginary quantity of $a\,|n\rangle$ "in the mirror" is $\langle n|\,a^\dagger$, denoted as

$$(a\,|n\rangle)^\dagger = \left(\sqrt{n}\,|n-1\rangle\right)^\dagger = \sqrt{n}\,\langle n-1| = \langle n|\,a^\dagger, \tag{1.35}$$

where the operation "†" is called Hermite conjugation. The set of the eigen-states of a quantum harmonic oscillator is complete, requiring

$$\sum_{n=0}^{\infty} |n\rangle \langle n| = 1, \tag{1.36}$$

we call it the completeness relation, that is, the set of eigenvalues $n = 0, 1, 2...$ is complete. Later, we can further prove the validity of Eq. (1.36). The set $|n\rangle$, as the eigenfunction set of the operator $a^{\dagger}a$, is a set of basis vectors in an abstract space, forming the photon number state representation. The characteristic of this representation is that the photon number operator can be described by positive integers in its own eigen-state representation. Equation (1.36) is also known as the decomposition of unit 1, where $|n\rangle \langle n|$ is a pure state. Besides, we also introduce the mixed state decomposition of unit 1 in the following content.

Next, we discuss the vacuum measurement operator $|0\rangle \langle 0|$. In the eigen-state set of $a^{\dagger}a$, we first need to consider vacuum (the state without photons), $|0\rangle \langle 0|$ is the vacuum measurement operator. However, vacuum is not absolute nothingness. The operator $|0\rangle \langle 0|$ is a function of the number operator N, where $|0\rangle \langle 0| \, |0 \langle = |0\rangle$, and 0^0 is an indefinite form, so we can speculate that the operator $|0\rangle \langle 0|$ was written as

$$\begin{aligned} |0\rangle \langle 0| &= 0^N = (1-1)^N = \sum_{l=0}^{\infty} C_N^l (-1)^l \\ &= 1 - N + \frac{1}{2!}N(N-1) - \frac{1}{3!}N(N-1)(N-2) + \cdots \\ &= \sum_{l=0}^{\infty} \frac{(-1)^l}{n!} N(N-1)\cdots(N-l+1). \end{aligned} \tag{1.37}$$

From Eqs. (1.30) and (1.36), we can obtain

$$N(N-1)\cdots(N-l+1) = \sum_{m=0}^{\infty} a^{\dagger l} |m\rangle \langle m| a^l = a^{\dagger l} a^l. \tag{1.38}$$

It follows from Eqs. (1.37) and (1.38) that

$$|0\rangle \langle 0| = \sum_{l=0}^{\infty} \frac{(-1)^l}{l!} a^{\dagger l} a^l. \tag{1.39}$$

Since $a^{\dagger l}$ is placed to the left of a^l in Eq. (1.39), it is called the normal ordering product, so $a^{\dagger l} a^l =: a^{\dagger l} a^l :$, and

$$|0\rangle \langle 0| = \sum_{l=0}^{\infty} \frac{(-1)^l}{l!} : a^{\dagger l} a^l : \; =: \mathrm{e}^{-a^{\dagger}a} : . \tag{1.40}$$

Thus we obtain the normal ordering form of the operator $|0\rangle \langle 0|$ in a concise manner.

In a monomial consisting of a and $a^\dagger$ function variables, when all annihilation operators a are placed to the left of $a^\dagger$, it is said to have been arranged as an anti-normal ordering product, marked with $\vdots\,\vdots$. So, what is the anti-normal ordering product of $|0\rangle\langle 0|$? Writing Eq. (1.40) as its integral form

$$|0\rangle\langle 0| =: \mathrm{e}^{-a^\dagger a}: = \int \frac{\mathrm{d}^2\xi}{\pi} : \mathrm{e}^{\mathrm{i}\xi^* a^\dagger + \mathrm{i}\xi a - |\xi|^2} : . \tag{1.41}$$

When integrating $\mathrm{d}^2\xi$, a and $a^\dagger$ within :: can be regarded as integration parameters. Using the Baker-Hausdorff formula commonly used in quantum theory and the Fourier transform of the delta function, we have

$$\begin{aligned} |0\rangle\langle 0| &= \int \frac{\mathrm{d}^2\xi}{\pi} \mathrm{e}^{\mathrm{i}\xi^* a^\dagger} \mathrm{e}^{\mathrm{i}\xi a - |\xi|^2} = \int \frac{\mathrm{d}^2\xi}{\pi} \mathrm{e}^{\mathrm{i}\xi a} \mathrm{e}^{\mathrm{i}\xi^* a^\dagger} \\ &= \pi \delta(a)\, \delta\left(a^\dagger\right) = \pi \vdots \delta(a)\, \delta\left(a^\dagger\right) \vdots, \end{aligned} \tag{1.42}$$

which is just the anti-normal ordering product of the operator $|0\rangle\langle 0|$ because $\delta(a)$ is placed to the left of $\delta\left(a^\dagger\right)$, indicating that it first creates and then annihilates (commonly known as self generation and self destruction, rather than self destruction and self generation). This is consistent with the intuitive meaning of vacuum, that is, where photons are generated [represented by $\delta\left(a^\dagger\right)$], they annihilated [represented by $\delta(a)$], indicating the importance of paying attention to the ordering of operators in Fock space at all times.

Using the normal ordering product of the operator $|0\rangle\langle 0|$, we can verify the completeness relation

$$\sum_{n=0}^{\infty} |n\rangle\langle n| = \sum_{n=0}^{\infty} \frac{a^{\dagger n}}{\sqrt{n!}} |0\rangle\langle 0| \frac{a^n}{\sqrt{n!}} = \sum_{n=0}^{\infty} \frac{1}{n!} : \left(a^\dagger a\right)^n \mathrm{e}^{-a^\dagger a} : = 1, \tag{1.43}$$

and obtain an operator identity

$$\begin{aligned} \mathrm{e}^{\lambda a^\dagger a} &= \sum_{n=0}^{\infty} \mathrm{e}^{\lambda n} |n\rangle\langle n| = \sum_{n=0}^{\infty} \mathrm{e}^{\lambda n} \frac{a^{\dagger n}}{\sqrt{n!}} |0\rangle\langle 0| \frac{a^n}{\sqrt{n!}} \\ &= \sum_{n=0}^{\infty} : \frac{1}{n!} \left(\mathrm{e}^{\lambda} a^\dagger a\right)^n \mathrm{e}^{a^\dagger a} : =: \exp\left[\left(\mathrm{e}^{\lambda} - 1\right) a^\dagger a\right] : , \end{aligned} \tag{1.44}$$

which is very useful for removing the symbol : :, for example,

$$: \exp\left(\lambda a^\dagger a\right) : = \exp\left[a^\dagger a \ln(\lambda + 1)\right]. \tag{1.45}$$

In Chap. 5, we will point out that the completeness relation in Eq. (1.43) can also be composed of other optical field states. Obviously,

$$: \exp\left(-2a^{\dagger}a\right): \; = \exp\left[a^{\dagger}a \ln(-1)\right] = \exp\left(\mathrm{i}\pi a^{\dagger}a\right) = (-1)^{N} \tag{1.46}$$

is the parity operator, which leads to

$$(-1)^{N} |n\rangle = (-1)^{n} |n\rangle , \quad (-1)^{N} a (-1)^{N} = -a. \tag{1.47}$$

1.2.2 Coherent State Representation

This section introduces the question of what quantum states are used to represent lasers (with good monochromaticity). Introducing the displaced operator $D(z) = \mathrm{e}^{za^{\dagger}-z^{*}a}$ and using the Bogoliubov transformation

$$\begin{aligned}\mathrm{e}^{A} B \mathrm{e}^{-A} &= \left(1 + A + \frac{A^2}{2!} + \frac{A^3}{3!} + \dots\right) B \left(1 - A + \frac{A^2}{2!} - \frac{A^3}{3!} + \cdots\right) \\ &= B + [A, B] + \frac{1}{2!}[A, [A, B]] + \frac{1}{3!}[A, [A, [A, B]]] + \cdots ,\end{aligned} \tag{1.48}$$

we give

$$D(z) a D^{-1}(z) = a - z. \tag{1.49}$$

Using Eq. (1.42) and the normal ordering of vacuum projector, i.e., $|0\rangle\langle 0| =: \mathrm{e}^{-a^{\dagger}a}:$, we obtain

$$\begin{aligned}D(z) |0\rangle \langle 0| D^{\dagger}(z) &= \pi \delta(a - z) \delta\left(a^{\dagger} - z^{*}\right) \\ &= \int \frac{\mathrm{d}^2\xi}{\pi} \mathrm{e}^{\mathrm{i}\xi(a-z)} \mathrm{e}^{\mathrm{i}\xi^{*}(a^{\dagger}-z^{*})} \\ &= \int \frac{\mathrm{d}^2\xi}{\pi} \mathrm{e}^{\mathrm{i}\xi^{*}(a^{\dagger}-z^{*})} \mathrm{e}^{\mathrm{i}\xi(a-z)-|\xi|^2} \\ &= \int \frac{\mathrm{d}^2\xi}{\pi} : \mathrm{e}^{\mathrm{i}\xi^{*}(a^{\dagger}-z^{*})+\mathrm{i}\xi(a-z)-|\xi|^2} : \\ &= : \mathrm{e}^{-(a^{\dagger}-z^{*})(a-z)} : \; = |z\rangle \langle z| ,\end{aligned} \tag{1.50}$$

where

$$|z\rangle = \mathrm{e}^{-\frac{|z|^2}{2} + za^{\dagger}} |0\rangle = D(z) |0\rangle \tag{1.51}$$

is the eigenstate of the operator a, that is

$$a\left|z\right\rangle = \mathrm{e}^{-\frac{1}{2}|z|^2}\left[a, \mathrm{e}^{za^\dagger}\right]\left|z\right\rangle = z\left|z\right\rangle, \; z = |z|\,\mathrm{e}^{\mathrm{i}\theta}. \tag{1.52}$$

Obviously, the state $|z\rangle$ is called a coherent state, which is composed of the superposition of different number states, i.e.,

$$|z\rangle = \mathrm{e}^{-\frac{1}{2}|z|^2}\sum_{n=0}^{\infty}\frac{|z|^n\,\mathrm{e}^{\mathrm{i}n\theta}}{\sqrt{n!}}\,|n\rangle\,. \tag{1.53}$$

Using Eq. (1.51), we can calculate the inner product $\langle z'\,|z\rangle$ as

$$\langle z'\,|z\rangle = \mathrm{e}^{-\frac{1}{2}\left(|z|^2+|z'|^2\right)}\,\langle 0|\,\mathrm{e}^{z'^*a}\mathrm{e}^{za^\dagger}\,|0\rangle = \mathrm{e}^{-\frac{1}{2}\left(|z|^2+|z'|^2\right)+z'^*z}, \tag{1.54}$$

which means that the coherent states are non-orthogonal. Using Eq. (1.40) and the IWOP method, we give

$$\begin{aligned}\int\frac{\mathrm{d}^2z}{\pi}\,|z\rangle\,\langle z| &= \int\frac{\mathrm{d}^2z}{\pi}\mathrm{e}^{-|z|^2}\mathrm{e}^{za^\dagger}\,|0\rangle\,\langle 0|\,\mathrm{e}^{z^*a}\\ &= \int\frac{\mathrm{d}^2z}{\pi}\colon\,\exp\left(-|z|^2+za^\dagger+z^*a-a^\dagger a\right)\colon\\ &= \int\frac{\mathrm{d}^2z}{\pi}\colon\mathrm{e}^{-\left(z^*-a^\dagger\right)(z-a)}\colon\; = \int\frac{\mathrm{d}^2z}{\pi}\mathrm{e}^{-z^*z} = 1,\end{aligned} \tag{1.55}$$

which shows that the set of coherent states $|z\rangle$ yields the completeness relation, so any optical field operator can be expanded using the coherent state representation. Using Eq. (1.55), we also verify the operator identity $\mathrm{e}^A\mathrm{e}^B = \mathrm{e}^A\mathrm{e}^B\mathrm{e}^{[A,\,B]}$, which holds for $[[A, B], A] = [[A, B], B] = 0$. For example,

$$\begin{aligned}\mathrm{e}^{fa}\mathrm{e}^{ga^\dagger} &= \int\frac{\mathrm{d}^2z}{\pi}\mathrm{e}^{fa}\,|z\rangle\,\langle z|\,\mathrm{e}^{ga^\dagger}\\ &= \int\frac{\mathrm{d}^2z}{\pi}\colon\,\exp\left(fz+gz^*-|z|^2+za^\dagger+z^*a-a^\dagger a\right)\colon\\ &= \colon\,\exp\left[\left(a^\dagger+f\right)(a+g)-a^\dagger a\right]\colon\; = \mathrm{e}^{ga^\dagger}\mathrm{e}^{fa}\mathrm{e}^{[fa,ga^\dagger]}.\end{aligned} \tag{1.56}$$

Further, using the following integration formula

$$\begin{aligned}&\int\frac{\mathrm{d}^2z}{\pi}\exp\left(\zeta|z|^2+\xi z+\eta z^*+w'z^2+v'z^{*2}\right)\\ &= \frac{1}{\sqrt{\zeta^2-4w'v'}}\exp\left(\frac{-\zeta\xi\eta+\xi^2v'+\eta^2w'}{\zeta^2-4w'v'}\right)\end{aligned} \tag{1.57}$$

with the convergence condition $\mathrm{Re}\left(\frac{\zeta^2-4\zeta w'v'}{\xi^2\pm w'\pm v'}\right) < 0$, $\mathrm{Re}(\xi^2 \pm w' \pm v') < 0$, we also obtain the following operator identity:

$$
\begin{aligned}
e^{fa^2}e^{ga^{\dagger 2}} &= \int \frac{d^2 z}{\pi} e^{fa^2} |z\rangle \langle z| e^{ga^{\dagger 2}} \\
&= \int \frac{d^2 z}{\pi} : \exp\left(-|z|^2 + za^\dagger + z^* a + fz^2 + gz^{*2} - a^\dagger a\right) : \\
&= \frac{1}{\sqrt{1-4fg}} e^{\frac{ga^{\dagger 2}}{1-4fg}} e^{-a^\dagger a \ln(1-4fg)} e^{\frac{fa^2}{1-4fg}}.
\end{aligned} \tag{1.58}
$$

On the other hand, using Eq. (1.53), we obtain

$$
|\langle n\ |z\rangle|^2 = e^{-|z|^2} \frac{|z|^{2n}}{n!},
$$

which is the probability of finding n photons in a coherent state, a Poisson distribution. Owing to $\langle N\rangle = \langle z| N |z\rangle = |z|^2$ and $\langle N^2\rangle = \langle z| N^2 |z\rangle = |z|^2 + |z|^4$, the photon number fluctuation in a coherent state is

$$
\Delta N = \sqrt{\langle N^2\rangle - \langle N\rangle^2} = |z|, \quad \frac{\Delta N}{\langle N\rangle} = \frac{1}{|z|}. \tag{1.59}
$$

It means that the fluctuation $\Delta N / \langle N\rangle$ is smaller when the average photon number $\langle N\rangle$ is larger (i.e., $|z|$ is large), which is approach to the classical light field. Mandel once introduced a parameter, denoted as

$$
\mathfrak{M} = \frac{\langle N^2\rangle - \langle N\rangle^2}{\langle N\rangle} - 1. \tag{1.60}
$$

For a coherent state, $\mathfrak{M} = 0$. If a certain light field with $\mathfrak{M} > 0$, its photon number distribution is called a super-Poisson distribution, otherwise it is called a sub-Poisson distribution, belonging to a non-classical effect.

Experiments have found that under the high excitation conditions, the photon statistics of lasers tends to follow a Poisson distribution. Therefore, coherent states are the quantum states that describe lasers, and their importance is evident. For a very small $|z|$, we give

$$
|z\rangle \to \frac{1}{\sqrt{1+|z|^2}} (|0\rangle + z\,|1\rangle) = |\mathfrak{Z}\rangle, \tag{1.61}
$$

which is normalized. The probability of measuring a single photon state in the state $|\mathfrak{Z}\rangle$ is

$$\langle 3| a^\dagger a |3\rangle = \frac{|z|^2}{1+|z|^2} = \frac{\langle N\rangle}{1+\langle N\rangle}. \tag{1.62}$$

This indicates that when $|z|$ is very small, the probability of obtaining the single photon state $|1\rangle$ is very low, so it seems difficult to obtain a stable single photon source through weak laser decay.

Next, we shall calculate the quantum fluctuations of the conjugate orthogonal components $X_1 = \left(a^\dagger + a\right)/2$ and $X_2 = \left(a - a^\dagger\right)/(2\mathrm{i})$ in the coherent light field. Noting the average values

$$\langle z| X_1 |z\rangle = \frac{1}{2}\left(z + z^*\right), \ \langle z| X_2 |z\rangle = \frac{1}{2\mathrm{i}}\left(z - z^*\right), \tag{1.63}$$

$$\begin{aligned} \langle z| X_1^2 |z\rangle &= \frac{1}{4}\left(z^2 + z^{*2} + 2|z|^2 + 1\right), \\ \langle z| X_2^2 |z\rangle &= \frac{1}{4}\left(z^2 + z^{*2} - 2|z|^2 - 1\right), \end{aligned} \tag{1.64}$$

thus the mean square error is

$$\begin{aligned} (\Delta X_1)^2 &= \langle z| X_1^2 |z\rangle - (\langle z| X_1 |z\rangle)^2 = \frac{1}{4}, \\ (\Delta X_2)^2 &= \langle z| X_2^2 |z\rangle - (\langle z| X_2 |z\rangle)^2 = \frac{1}{4}. \end{aligned} \tag{1.65}$$

So, we obtain

$$\Delta X_1 \Delta X_2 = \frac{1}{4}. \tag{1.66}$$

Noting that $X_1 = X/\sqrt{2}$, $X_2 = P/\sqrt{2}$, $[X, P] = \mathrm{i}$, thus in the coherent state, we have

$$\Delta X \Delta P = \frac{1}{2}. \tag{1.67}$$

So, the coherent state $|z\rangle$ is a state that minimizes the coordinate momentum uncertainty relation. Letting $z = (x + \mathrm{i}p)/\sqrt{2}$, $\langle z| X |z\rangle = x$, $\langle z| P |z\rangle = p$, thus, in the x-p phase space, the coherent state is represented as a small circle with an area of $1/2$ and a center at point (x, p), rather than a point. Therefore, the motion theory describing classical phase points (classical Liuville theorem) also needs to be modified accordingly.

Finally, we point out that the dynamic Hamiltonians for generating coherent states of harmonic oscillators are

$$H_0 = \omega a^\dagger a + \mathrm{i} f a - \mathrm{i} f^* a^\dagger, \tag{1.68}$$

please ask readers to prove this point.

1.2.3 Coordinate Representation and Momentum Representation

In quantum mechanics, the most commonly used representation is the coordinate representation, represented by $|x\rangle$. In this representation, the coordinate operator X is represented as an ordinary number, which corresponds to the following eigen-state equation:

$$X|x\rangle = x|x\rangle. \tag{1.69}$$

Also, $|x\rangle\langle x|$ formed by $|x\rangle$ and $\langle x|$ is an operator that functions as a projection operator and can also be considered as an operator for measuring the coordinate value x . Therefore, it can also be represented by Dirac's delta δ function, that is

$$|x\rangle\langle x| = \delta(x - X), \quad X = \frac{a + a^{\dagger}}{\sqrt{2}}. \tag{1.70}$$

Using the Fourier transform of the delta δ function, that is

$$\delta(x - X) = \frac{1}{2\pi}\int_{-\infty}^{\infty} \mathrm{d}p e^{\mathrm{i}p(x-X)} = \frac{1}{2\pi}\int_{-\infty}^{\infty} \mathrm{d}p e^{\mathrm{i}p\left(x - \frac{a+a^{\dagger}}{\sqrt{2}}\right)} \tag{1.71}$$

and the IWOP method, we find

$$\begin{aligned}\delta(x - X) &= \frac{1}{2\pi}\int_{-\infty}^{\infty} \mathrm{d}p \colon e^{-\frac{p^2}{4} + \mathrm{i}p\left(x - \frac{a^{\dagger}}{\sqrt{2}}\right) - \mathrm{i}p\frac{a}{\sqrt{2}}} \colon \\ &= \frac{1}{\sqrt{\pi}} \colon \exp\left[-\left(x - \frac{a + a^{\dagger}}{\sqrt{2}}\right)^2\right] \colon . \end{aligned} \tag{1.72}$$

It should be emphasized that due to the commutation between a and $a^{\dagger}$ within : : , they can be considered as the integral parameters when calculating the integration (1.72). Using the operator identity $|0\rangle\langle 0| = \colon e^{-a^{\dagger}a} \colon$, we can decompose Eq. (1.72) into

$$\begin{aligned}\delta(x - X) &= |x\rangle\langle x| \\ &= \frac{1}{\pi^{1/2}} e^{-\frac{x^2}{2} + \sqrt{2}xa^{\dagger} - \frac{a^{\dagger 2}}{2}} |0\rangle\langle 0| e^{-\frac{x^2}{2} + \sqrt{2}xa - \frac{a^2}{2}}, \end{aligned} \tag{1.73}$$

thus we can obtain the specific expression of the state $|x\rangle$ in the Fock space, i.e.,

$$|x\rangle = \pi^{-1/4} \exp\left(-\frac{x^2}{2} + \sqrt{2}xa^{\dagger} - \frac{a^{\dagger 2}}{2}\right)|0\rangle. \tag{1.74}$$

So, the completeness of the coordinate representation can be written in the form of a pure Gaussian integral:

$$\int_{-\infty}^{\infty} \mathrm{d}x \, |x\rangle \langle x| = \int_{-\infty}^{\infty} \frac{\mathrm{d}x}{\sqrt{\pi}} : \mathrm{e}^{-(x-X)^2} : \; = 1. \tag{1.75}$$

Similarly, we can also calculate the integral $\int_{-\infty}^{\infty} \frac{\mathrm{d}x}{\sqrt{\mu}} \left| \frac{x}{\mu} \right\rangle \langle x|$ as

$$\begin{aligned} &\int_{-\infty}^{\infty} \frac{\mathrm{d}x}{\sqrt{\mu}} \left| \frac{x}{\mu} \right\rangle \langle x| \\ =& \frac{1}{\pi^{1/2}} \int_{-\infty}^{\infty} \mathrm{e}^{-\frac{x^2}{2\mu^2} + \sqrt{2}xa^\dagger/\mu - \frac{a^{\dagger 2}}{2}} |0\rangle \langle 0| \, \mathrm{e}^{-\frac{x^2}{2} + \sqrt{2}xa - \frac{a^2}{2}} \\ =& \mathrm{e}^{-\frac{a^{\dagger 2}}{2} \tanh \lambda} \mathrm{e}^{(a^\dagger a + \frac{1}{2}) \ln \operatorname{sech} \lambda} \mathrm{e}^{\frac{a^2}{2} \tanh \lambda}, \quad \mu = \mathrm{e}^\lambda \end{aligned} \tag{1.76}$$

which is called the single-mode squeezed operator (the quantum correspondence of the classical scale transformation $x \longrightarrow x/\mu$), where a^2, $\left(a^\dagger a + \frac{1}{2}\right)$ and $a^{\dagger 2}$ form the SU (1, 1) Lie algebra. Therefore, the state

$$\operatorname{sech}^{\frac{1}{2}} \lambda \mathrm{e}^{-\frac{a^{\dagger 2}}{2} \tanh \lambda} |0\rangle \tag{1.77}$$

is called the squeezed vacuum state. Using Eq. (1.75), we also derive the following operator identity:

$$\begin{aligned} \mathrm{e}^{fX^2} &= \int_{-\infty}^{\infty} \mathrm{d}x \mathrm{e}^{fx^2} |x\rangle \langle x| = \int_{-\infty}^{\infty} \frac{\mathrm{d}x}{\sqrt{\pi}} : \mathrm{e}^{-(x-X)^2 + fx^2} : \\ &= \frac{1}{\sqrt{1-f}} : \mathrm{e}^{-\frac{f}{f-1} X^2} : . \end{aligned} \tag{1.78}$$

Note that the state $|p\rangle$ is the eigen-state of the momentum operator $P = \left(a - a^\dagger\right) / \left(\mathrm{i}\sqrt{2}\right)$ yielding the eigen-equation $P|p\rangle = p|p\rangle$, similar to obtaining Eqs. (1.71)–(1.75), we have

$$|p\rangle \langle p| = \delta(p - P) = \frac{1}{\sqrt{\pi}} : \exp\left[-\left(p - \frac{a - a^\dagger}{\mathrm{i}\sqrt{2}} \right)^2 \right] : , \tag{1.79}$$

using the operator identity $|0\rangle \langle 0| = \, : \mathrm{e}^{-a^\dagger a} :$ to split the above equation, thus the expression of the state $|p\rangle$ is obtained as

$$|p\rangle = \pi^{-1/4} \exp\left(-\frac{p^2}{2} + \mathrm{i}\sqrt{2} p a^\dagger + \frac{a^{\dagger 2}}{2} \right) |0\rangle , \tag{1.80}$$

whose Gaussian completeness relation is

$$\int_{-\infty}^{\infty} \mathrm{d}p \left|p\right\rangle \left\langle p\right| = \int_{-\infty}^{\infty} \frac{\mathrm{d}p}{\sqrt{\pi}} : \mathrm{e}^{-(p-P)^2} : = 1. \tag{1.81}$$

From the above analysis, we reveal the profound beauty of Dirac symbols, for example, the appropriate ordering products of $|\rangle\langle|$ introduced by Dirac possess the Gaussian forms. This is in line with his prophecy written in "Principles of Quantum Mechanics": "The method of symbols, which directly deals with quantities of fundamental importance in an abstract way,……, however it seems to more deeply understand the essence of things. It enables people to express physical laws in a concise and clear way, and in the future, as people's understanding of it deepens and its own mathematics develops, it may have increasing applications".

1.2.4 Mixed State Representation

Combining Eqs. (1.75) and (1.81), we obtain

$$\frac{1}{\pi} \int\int_{-\infty}^{\infty} \mathrm{d}x \mathrm{d}p : \mathrm{e}^{-(x-X)^2-(p-P)^2} : = 1, \tag{1.82}$$

which provides the completeness relation in the x-p phase space, and

$$\frac{1}{\pi} : \mathrm{e}^{-(x-X)^2-(p-P)^2} : \equiv \Delta\left(x, p\right). \tag{1.83}$$

Since $\Delta\left(x, p\right)$ cannot be written as the form of a pure state $|\rangle\langle|$, it belongs to the mixed state representation [11, 12]. However, the marginal integrals of $\Delta\left(x, p\right)$ lead to two pure states, i.e., $\int_{-\infty}^{\infty} \mathrm{d}x \Delta\left(x, p\right) \to \frac{1}{\sqrt{\pi}} : \mathrm{e}^{-(p-P)^2} : = \left|p\right\rangle \left\langle p\right|$ and $\int_{-\infty}^{\infty} \mathrm{d}p \Delta\left(x, p\right) \to \frac{1}{\sqrt{\pi}} : \mathrm{e}^{-(x-X)^2} := \left|x\right\rangle \left\langle x\right|$. Noting the completeness relation in Eq. (1.82), and using $\Delta\left(x, p\right)$, any operator $H\left(X, P\right)$ can be expanded as

$$H\left(X, P\right) = \int\int_{-\infty}^{\infty} \mathrm{d}x \mathrm{d}p \Delta\left(x, p\right) h\left(x, p\right), \tag{1.84}$$

where the expansion function $h\left(x, p\right)$ is a classical correspondence of $H\left(X, P\right)$, i.e., the Weyl-Wigner correspondence. Due to historical reasons, $\Delta\left(x, p\right)$ is called the Wigner operator, but it was not previously introduced in the form of a normal ordering product in Eq. (1.83). Letting

$$a = \frac{X + \mathrm{i}P}{\sqrt{2}},\ a^{\dagger} = \frac{X - \mathrm{i}P}{\sqrt{2}}, \tag{1.85}$$

thus we have

$$\Delta(x,p) \longrightarrow \Delta\left(\alpha,\alpha^*\right) = \frac{1}{\pi} : \mathrm{e}^{-2(a-\alpha)\left(a^\dagger-\alpha^*\right)} : \\ = \frac{1}{\pi}\mathrm{e}^{2\alpha a^\dagger} : \mathrm{e}^{-2a^\dagger a} : \mathrm{e}^{2\alpha^* a - 2|\alpha|^2}, \tag{1.86}$$

where

$$: \mathrm{e}^{-2a^\dagger a} : = (-1)^{a^\dagger a} \tag{1.87}$$

is the parity operator. $\langle\psi|\,\Delta(x,p)\,|\psi\rangle$ is called the Wigner function for the state $|\psi\rangle$. Since $(-1)^{a^\dagger a}|n\rangle = (-1)^n|n\rangle$, so the Wigner function is not always positive definite.

Next, we introduce the coherent state representation of the Wigner operator. Using the IWOP method and $|0\rangle\langle 0| =: \mathrm{e}^{-a^\dagger a} :$, we can rewrite Eq. (1.86) as

$$\begin{aligned}\Delta(\alpha,\alpha^*) &= \frac{1}{\pi} : \mathrm{e}^{-2(a-\alpha)\left(a^\dagger-\alpha^*\right)} : \\ &= \int \frac{\mathrm{d}^2 z}{\pi^2} \exp\left[-|z|^2 + z\left(a^\dagger-\alpha^*\right) + (\alpha-a)z^* \right. \\ &\quad \left. + \alpha a^\dagger + \alpha^*\alpha - a^\dagger a - |\alpha|^2\right] : \\ &= \int \frac{\mathrm{d}^2 z}{\pi^2} \exp\left[-|z|^2 + (\alpha+z)a^\dagger + \left(\alpha^*-z^*\right)a \right. \\ &\quad \left. - a^\dagger a + \alpha z^* - z\alpha^* - |\alpha|^2\right] : \mathrm{e}^{\alpha z^* - z\alpha^*} \\ &= \int \frac{\mathrm{d}^2 z}{\pi^2} \mathrm{e}^{-\frac{|\alpha+z|^2}{2} + (\alpha+z)a^\dagger} |0\rangle\langle 0| \, \mathrm{e}^{-\frac{|\alpha-z|^2}{2} + (\alpha^*-z^*)a} \mathrm{e}^{\alpha z^* - z\alpha^*}.\end{aligned} \tag{1.88}$$

Further, using the definition of the coherent states, we thus have

$$\Delta(\alpha,\alpha^*) = \int \frac{\mathrm{d}^2 z}{\pi^2} |\alpha+z\rangle\langle\alpha-z| \, \mathrm{e}^{\alpha z^* - z\alpha^*}, \tag{1.89}$$

which is the coherent state representation of the Wigner operator. Thus, Eq. (1.84) can be written as

$$H\left(a^\dagger,a\right) = 2\int \mathrm{d}^2\alpha \Delta(\alpha,\alpha^*) h\left(\alpha,\alpha^*\right). \tag{1.90}$$

Using Eqs. (1.54) and (1.86), we obtain

$$
\begin{aligned}
&\mathtt{tr}\left[\Delta\left(\alpha,\alpha^{*}\right)\Delta\left(\alpha',\alpha'^{*}\right)\right] \\
&=\mathtt{tr}\left[\int\frac{d^{2}z}{\pi^{2}}\left|\alpha+z\right\rangle\left\langle\alpha-z\right|e^{\alpha z^{*}-z\alpha^{*}}\Delta\left(\alpha',\alpha'^{*}\right)\right] \\
&=\int\frac{d^{2}z}{\pi^{2}}\left\langle\alpha-z\right|\Delta\left(\alpha',\alpha'^{*}\right)\left|\alpha+z\right\rangle e^{\alpha z^{*}-z\alpha^{*}} \\
&=\int\frac{d^{2}z}{\pi^{3}}\left\langle\alpha-z\right|:e^{-2(\alpha-\alpha')(a^{\dagger}-\alpha'^{*})}:\left|\alpha+z\right\rangle e^{\alpha z^{*}-z\alpha^{*}} \\
&=\frac{1}{4\pi}\delta\left(\alpha-\alpha'\right)\delta\left(\alpha^{*}-\alpha'^{*}\right),
\end{aligned}
\tag{1.91}
$$

thus, using Eq. (1.90) again, we derive

$$
\begin{aligned}
&2\pi\mathtt{tr}\left[\Delta\left(\alpha,\alpha^{*}\right)H\left(a^{\dagger},a\right)\right] \\
&=4\pi\int d^{2}\alpha'\mathtt{tr}\left[\Delta\left(\alpha,\alpha^{*}\right)\Delta\left(\alpha',\alpha'^{*}\right)\right]h\left(\alpha',\alpha'^{*}\right) \\
&=\int d^{2}\alpha'\delta\left(\alpha-\alpha'\right)\delta\left(\alpha^{*}-\alpha'^{*}\right)h\left(\alpha',\alpha'^{*}\right) \\
&=h\left(\alpha,\alpha^{*}\right),
\end{aligned}
\tag{1.92}
$$

which is the formula for calculating the classical Weyl correspondence function $h(\alpha, \alpha^*)$ of the operator $H\left(a^{\dagger}, a\right)$ [13, 14].

References

1. Louisell WH (1973) Quantum statistical properties of radiation. Wiley, New York
2. Klauder JR, Sudarshan ECG (1968) Fundamentals of quantum optics. W. A. Benjamin Inc, New York
3. Mehta CL (1967) Diagonal coherent-state representation of quantum operators. Phys Rev Lett 18(18):752–754
4. Fan HY, Zhou J (2012) Coherent state and normal ordering method for transiting Hermite polynomials to Laguerre polynomials. Sci China Phys Mech Astron 55(4):605–608
5. Dirac PAM (1958) Principles of quantum mechanics. Oxford University Press, New York
6. Fan HY (2004) Entangled states, squeezed states gained via the route of developing Dirac's symbolic method and their applications. Int J Mod Phys B 18(10–11):1387–1455
7. Fan HY, Hu LY (2008) Operator-sum representation of density operators as solution to master equations obtained via the entangled states approach. Mod Phys Lett B 22(25):2435–2468
8. Fan HY, Lu HL, Fan Y (2006) Newton-Leibniz integration for ket-bra operators in quantum mechanics and derivation of entangled state representations. Ann Phys 321(2):480–494
9. Fan HY (2007) Newton-Leibniz integration for ket-bra operators (II)-application in deriving density operator and generalized partition function formula. Ann Phys 322(4):866–885
10. Fan HY (2008) Newton-Leibniz integration for ket-bra operators in quantum mechanics (IV)-integrations within Weyl ordered product of operators and their applications. Ann Phys 323(2):500–526
11. Wang JS, Fan HY, Meng XG (2012) A generalized Weyl-Wigner quantization scheme unifying P-Q and Q-P ordering and Weyl ordering of operators. Chin Phys B 21(6):064204

12. Wang JS, Meng XG, Fan HY (2017) New relationship between quantum state's tomogram and its wave function. J Mod Opt 64(14):1398–1403
13. Fan HY, Ruan TN (1983) Coherent state formulation of the Weyl correspondence and the Wigner function. Commun Theor Phys 2(6):1563–1574
14. Fan HY, Zaidi HR (1987) Application of IWOP technique to the generalized Weyl correspondence. Phys Lett A 124(6–7):303–307

Chapter 2
Normal and Anti-Normal Ordering Method of Field Operators

When discussing the properties of light fields, the proper ordering of field operators can reveal the true faces of the fields [1, 2]. When calculating the expected value of an operator in the coherent state, it is often necessary to use the normal ordering product of this operator. In this chapter, we first discuss the normal ordering of the function $f(X)$ of the orthogonal component operator X of light field, which can be achieved by virtue of a differential operation [3]. And then, we discuss the differential operation of turning the density operator ρ within anti-normal ordering into its normal ordering, and provide two important operator identities. Besides, we extend the normal ordering problem to two-mode case and present the anti-normal ordering formula [3].

2.1 Differential Operation of Converting $f(X)$ into $:f(X):$

The orthogonal component operator X of light field is $X = \left(a + a^{\dagger}\right)/\sqrt{2}$, how to achieve $f(X) \rightarrow : f(X) :$? It should be noted that $f(X)$ can be expanded as the power series of X^n. For this, we first consider the following differential-integral relation

$$\begin{aligned}
\exp\left(\tau\frac{\partial^2}{\partial x^2}\right) f(x) &= \sum_{n=0}^{\infty}\frac{\tau^n}{n!}\int_{-\infty}^{\infty} \mathrm{d}s f(s)\left(\frac{\partial^2}{\partial x^2}\right)^n \delta(x-s) \\
&= \sum_{n=0}^{\infty}\frac{\tau^n}{n!}\frac{1}{2\pi}\int_{-\infty}^{\infty}\mathrm{d}s f(s)\int_{-\infty}^{\infty}\mathrm{d}t\left(-t^2\right)^n \mathrm{e}^{-\mathrm{i}(x-s)t} \\
&= \frac{1}{2\pi}\int_{-\infty}^{\infty}\mathrm{d}s f(s)\int_{-\infty}^{\infty}\mathrm{d}t \exp\left(-\tau t^2 - \mathrm{i}xt + \mathrm{i}st\right) \\
&= \sqrt{\frac{1}{4\pi\tau}}\int_{-\infty}^{\infty}\mathrm{d}s f(s)\exp\left[-\frac{(x-s)^2}{4\tau}\right].
\end{aligned} \tag{2.1}$$

H. Fan and L. Hu, *Dissipation and Diffusion of Quantum Light Field*,
https://doi.org/10.1007/978-981-95-1528-8_2

Especially, for $\tau = 1/4$, Eq. (2.1) becomes

$$\exp\left(\frac{1}{4}\frac{\partial^2}{\partial x^2}\right) f(x) = \sqrt{\frac{1}{\pi}}\int_{-\infty}^{\infty} \mathrm{d}s f(s) \exp\left[-(x-s)^2\right]. \tag{2.2}$$

Further, letting $f(x) = \mathrm{H}_n(x)$ in Eq. (2.2), thus we have

$$\begin{aligned}\exp\left(\frac{1}{4}\frac{\partial^2}{\partial x^2}\right) \mathrm{H}_n(x) &= \sqrt{\frac{1}{\pi}}\int_{-\infty}^{\infty} \mathrm{d}s \mathrm{H}_n(s) \exp\left[-(x-s)^2\right] \\ &= \sqrt{\frac{1}{\pi}}\int_{-\infty}^{\infty} \mathrm{d}s \exp\left[-(x-s)^2\right] \left.\frac{\mathrm{d}^n}{\mathrm{d}t^n}\mathrm{e}^{-t^2+2ts}\right|_{t=0} \\ &= 2^n x^n,\end{aligned} \tag{2.3}$$

it follows that

$$\exp\left(-\frac{1}{4}\frac{\partial^2}{\partial x^2}\right) x^n = 2^{-n}\mathrm{H}_n(x). \tag{2.4}$$

In fact, Eq. (2.4) can also be calculated via direct differentiation. The proof is as follows:

$$\begin{aligned}\exp\left(-\frac{1}{4}\frac{\partial^2}{\partial x^2}\right) x^n &= \sum_{k=0}^{\infty}\frac{1}{k!}\left(-\frac{1}{4}\frac{\partial^2}{\partial x^2}\right)^k x^n \\ &= \sum_{k=0}^{[n/2]}\frac{(-1)^k n!}{2^{2k}k!(n-2k)!}x^{n-2k} \\ &= 2^{-n}\mathrm{H}_n(x),\end{aligned} \tag{2.5}$$

which also leads to the series expansions of single-variable Hermite polynomials. Now, replacing x with the operator X, we get

$$\exp\left(-\frac{1}{4}\frac{\partial^2}{\partial X^2}\right) X^n = 2^{-n}\mathrm{H}_n(X). \tag{2.6}$$

In terms of $\mathrm{H}_n(X) = 2^n : X^n :$, so we get

$$\exp\left(-\frac{1}{4}\frac{\partial^2}{\partial X^2}\right) X^n =: X^n : . \tag{2.7}$$

For $f(X)$ that can be expanded as a power series of X^n, we obtain

$$\exp\left(-\frac{1}{4}\frac{\partial^2}{\partial X^2}\right) f(X) =: f(X) : , \tag{2.8}$$

which is a shortcut to achieving $f(X) \to : f(X) :$. Further, we have

$$\begin{aligned} \exp\left(-\frac{1}{4}\frac{\partial^2}{\partial X^2}\right) X^n &= : X^n : = 2^{-n}\mathrm{H}_n(X) \\ &= \sum_{k=0}^{[n/2]} \frac{(-1)^k n!}{2^{2k}k!(n-2k)!} X^{n-2k} \\ &= n! \sum_{k=0}^{[n/2]} \frac{1}{k!(n-2k)!} 2^{-n} : \mathrm{H}_{n-2k}(X) : . \end{aligned} \tag{2.9}$$

It follows that

$$: X^n : = n! \sum_{k=0}^{[n/2]} \frac{1}{k!(n-2k)!} 2^{-n} : \mathrm{H}_{n-2k}(X) : . \tag{2.10}$$

For example, using Eq. (2.8), we have

$$\exp\left(-\frac{1}{4}\frac{\partial^2}{\partial X^2}\right) \mathrm{e}^{-(x-X)^2} =: \mathrm{e}^{-(x-X)^2} : . \tag{2.11}$$

Noting that

$$\frac{1}{\sqrt{\pi}} : \mathrm{e}^{-(x-X)^2} : = |x\rangle\langle x| = \delta(x-X), \tag{2.12}$$

thus Eq. (2.11) becomes

$$\frac{1}{\sqrt{\pi}} \exp\left(-\frac{1}{4}\frac{\partial^2}{\partial X^2}\right) \mathrm{e}^{-(x-X)^2} = \delta(x-X). \tag{2.13}$$

Another example is using Eq. (2.8) to obtain

$$\exp\left(-\frac{1}{4}\frac{\partial^2}{\partial X^2}\right) \mathrm{e}^{-\lambda X^2} =: \mathrm{e}^{-\lambda X^2} : . \tag{2.14}$$

On the other hand, using the completeness relation of coordinate states and the IWOP method [3, 4], we have

$$\begin{aligned} \mathrm{e}^{-\lambda X^2} &= \int_{-\infty}^{\infty} \mathrm{d}x\, |x\rangle\langle x| \mathrm{e}^{-\lambda x^2} = \frac{1}{\sqrt{\pi}} \int_{-\infty}^{\infty} \mathrm{d}x : \mathrm{e}^{-(x-X)^2 - \lambda x^2} : \\ &= \frac{1}{\sqrt{1+\lambda}} : \exp\left(\frac{-\lambda}{1+\lambda} X^2\right) : . \end{aligned} \tag{2.15}$$

Letting $t = \lambda/(1+\lambda)$, Eq. (2.15) becomes

$$:\exp\left(-tX^2\right): = \frac{1}{\sqrt{1-t}} e^{-\frac{t}{1-t}X^2}, \tag{2.16}$$

so we have

$$\exp\left(-\frac{1}{4}\frac{\partial^2}{\partial X^2}\right) e^{-tX^2} =: \exp\left(-tX^2\right): = \frac{1}{\sqrt{1-t}} e^{-\frac{t}{1-t}X^2}, \tag{2.17}$$

or

$$\exp\left(-\frac{1}{4}\frac{\partial^2}{\partial x^2}\right) e^{-tx^2} = \frac{1}{\sqrt{1-t}} e^{-\frac{t}{1-t}x^2}. \tag{2.18}$$

Besides, using Eq. (2.15) to obtain

$$e^{-\lambda X^2} = \frac{1}{\sqrt{1+\lambda}} : \sum_{n=0}^{\infty} \frac{1}{n!}\left(\frac{-\lambda}{1+\lambda}\right)^n X^{2n} :, \tag{2.19}$$

which means that

$$e^{-\lambda x^2} = \frac{1}{\sqrt{1+\lambda}} \sum_{n=0}^{\infty} \frac{1}{2^{2n} n!}\left(\frac{-\lambda}{1+\lambda}\right)^n \mathrm{H}_{2n}(x). \tag{2.20}$$

2.2 Differential Operation for Converting Anti-Normal Ordering of Density Operator into Its Normal Ordering Product

Noting that the differential operation $\exp\left(-\frac{1}{4}\frac{\partial^2}{\partial X^2}\right) f(X) =: f(X):$ that turns $f(X)$ into $: f(X):$, we can obtain the differential operation for transforming the anti-normal ordering of the density operator ρ into its normal ordering product, that is

$$\rho =: \exp\left(\frac{\partial^2}{\partial z \partial z^*}\right) \mathrm{P}\left(z, z^*\right)\Big|_{z^* \to a^\dagger, z \to a} :, \tag{2.21}$$

where $\mathrm{P}(z, z^*)$ is the P-representation of the density operator ρ in the coherent state representation [5], i.e.,

$$\rho = \int \frac{d^2 z}{\pi} \mathrm{P}\left(z, z^*\right) |z\rangle \langle z|. \tag{2.22}$$

Proof. Using the IWOP method to rewrite the operator $|z\rangle\langle z|$ as the following integral form

$$|z\rangle\langle z| = \int \frac{d^2\beta}{\pi} : \exp\left[-|\beta|^2 + i\beta\left(a^\dagger - z^*\right) + i\beta^*(a-z)\right] : . \tag{2.23}$$

Owing to $\partial z/\partial z^* = 0$, thus

$$\exp\left(-|\beta|^2 - iz^*\beta - i\beta^* z\right) = \exp\left(\frac{\partial^2}{\partial z \partial z^*}\right)\exp\left(-iz^*\beta - i\beta^* z\right), \tag{2.24}$$

so the operator $|z\rangle\langle z|$ can be written as

$$\begin{aligned} |z\rangle\langle z| &= \exp\left(\frac{\partial^2}{\partial z \partial z^*}\right)\int \frac{d^2\beta}{\pi} : \exp\left[i\beta\left(a^\dagger - z^*\right) + i\beta^*(a-z)\right] : \\ &= \exp\left(\frac{\partial^2}{\partial z \partial z^*}\right)\delta\left(a^\dagger - z^*\right)\delta^*(a-z). \end{aligned} \tag{2.25}$$

So, substituting Eq. (2.25) into Eq. (2.22) leads to Eq. (2.21). Especially, for the operator $\rho = a^n a^{\dagger m}$, thus $P(z, z^*) = z^n z^{*m}$, using Eq. (2.21), we can directly obtain

$$\begin{aligned} a^n a^{\dagger m} &= : \exp\left(\frac{\partial^2}{\partial z \partial z^*}\right) z^n z^{*m}\bigg|_{z^* \to a^\dagger, z \to a} : \\ &= : \sum_{l=0}^{\min(m,n)} \frac{1}{l!}\left(\frac{\partial^2}{\partial z \partial z^*}\right)^l z^n z^{*m}\bigg|_{z^* \to a^\dagger, z \to a} : \\ &= \sum_{l=0}^{\min(m,n)} \frac{m!n!}{l!(m-l)!(n-l)!} a^{\dagger m-l} a^{n-l}. \end{aligned} \tag{2.26}$$

According to the series expansions of two-variable Hermite polynomials, we can see

$$a^n a^{\dagger m} =: \exp\left(\frac{\partial^2}{\partial z \partial z^*}\right) z^n z^{*m}\bigg|_{z^* \to a^\dagger, z \to a} : = (-i)^{m+n} : H_{m,n}\left(ia^\dagger, ia\right) : , \tag{2.27}$$

which is the normal ordering product of the anti-normal ordering operator $a^n a^{\dagger m}$. As an alternative, using the generating function for the two-variable Hermite polynomial, we get

$$\begin{aligned}\sum_{n,m=0}^{\infty}\frac{\tau^n t^m}{n!m!}a^n a^{\dagger m} &= \mathrm{e}^{\tau a}\mathrm{e}^{t a^\dagger} =: \exp\left(\tau a + t a^\dagger + \tau t\right): \\ &= :\exp\left[-\left(-\mathrm{i}\tau\right)\left(-\mathrm{i}t\right) + \left(-\mathrm{i}t\right)\left(\mathrm{i}a^\dagger\right) + \left(-\mathrm{i}\tau\right)\left(\mathrm{i}a\right)\right]: \\ &= \sum_{n,m=0}^{\infty}\frac{\left(-\mathrm{i}\tau\right)^n\left(-\mathrm{i}t\right)^m}{n!m!}: \mathrm{H}_{n,m}\left(\mathrm{i}a, \mathrm{i}a^\dagger\right):,\end{aligned} \tag{2.28}$$

which leads to

$$a^n a^{\dagger m} = \left(-\mathrm{i}\right)^{m+n}: \mathrm{H}_{m,n}\left(\mathrm{i}a^\dagger, \mathrm{i}a\right):. \tag{2.29}$$

On the other hand, using the identity

$$\begin{aligned}a^{\dagger m}a^n &= \left.\frac{\partial^{n+m}}{\partial t^m\partial\tau^n}\mathrm{e}^{t a^\dagger}\mathrm{e}^{\tau a}\right|_{t,\tau=0} = \left.\frac{\partial^{n+m}}{\partial t^m\partial\tau^n}\mathrm{e}^{\tau a}\mathrm{e}^{t a^\dagger}\mathrm{e}^{-t\tau}\right|_{t,\tau=0} \\ &= \left.\frac{\partial^{n+m}}{\partial t^m\partial\tau^n}\vdots\mathrm{e}^{t a^\dagger+\tau a-t\tau}\vdots\right|_{t,\tau=0} = \vdots\mathrm{H}_{m,n}\left(a^\dagger, a\right)\vdots \\ &= \vdots\mathrm{H}_{n,m}\left(a, a^\dagger\right)\vdots,\end{aligned} \tag{2.30}$$

we obtain the anti-normal ordering of the normal ordering operator $a^{\dagger m}a^n$, i.e.,

$$a^{\dagger m}a^n = \vdots\mathrm{H}_{n,m}\left(a, a^\dagger\right)\vdots. \tag{2.31}$$

Using Eq. (2.31), we can verify Eq. (1.42). Using the operator identities

$$|0\rangle\langle 0| = \sum_{m=0}^{\infty}\frac{(-1)^m}{m!}: a^{\dagger m}a^m: = \sum_{m=0}^{\infty}\frac{(-1)^m}{m!}a^{\dagger m}a^m \tag{2.32}$$

and

$$a^{\dagger m}a^m = \vdots\mathrm{H}_{m,m}\left(a^\dagger, a\right)\vdots = m!\left(-1\right)^m\vdots\mathrm{L}_m\left(aa^\dagger\right)\vdots, \tag{2.33}$$

as well as the generating function for the Laguerre polynomial, i.e.,

$$(1-z)^{-1}\mathrm{e}^{\frac{zx}{z-1}} = \sum_{m=0}^{\infty}\mathrm{L}_m(x)\,z^m, \tag{2.34}$$

we can obtain the anti-normal ordering product of vacuum projector as

$$\begin{aligned}|0\rangle\langle 0| &= \sum_{m=0}^{\infty}\vdots\mathrm{L}_m\left(aa^\dagger\right)\vdots 1^m = \lim_{z\to 1}\vdots(1-z)^{-1}\mathrm{e}^{\frac{z}{z-1}aa^\dagger}\vdots \\ &= \vdots\pi\delta\left(aa^\dagger\right)\vdots = \pi\delta\left(a\right)\delta\left(a^\dagger\right).\end{aligned} \tag{2.35}$$

2.3 Differential Operation for Converting $G(a+b^\dagger, a^\dagger+b)$ into $:G(a+b^\dagger, a^\dagger+b):$

This section extends the above discussion to the two-mode case, introducing the annihilation and creation operators $b, b^\dagger$ of the second mode, $\left[b, b^\dagger\right]=1$, and

$$\left[a+b^\dagger, a^\dagger+b\right]=0, \tag{2.36}$$

which means that the operators $a+b^\dagger$ and $a^\dagger+b$ have the common eigen-states. Also, the eigen-states can be easily obtained using the IWOP method. According to the completeness relation of coherent states [5], i.e.,

$$\begin{aligned}\int \frac{\mathrm{d}^2 z}{\pi}|z\rangle\langle z| &= \int \frac{\mathrm{d}^2 z}{\pi}: \exp\left(-|z|^2+za^\dagger+z^*a-a^\dagger a\right): \\ &= \int \frac{\mathrm{d}^2 z}{\pi}: \mathrm{e}^{-\left(z^*-a^\dagger\right)(z-a)}: =1,\end{aligned} \tag{2.37}$$

we conclude that 1 has the following integration representation

$$\int \frac{\mathrm{d}^2 \xi}{\pi}: \mathrm{e}^{-\left[\xi^*-\left(a^\dagger+b\right)\right]\left[\xi-\left(a+b^\dagger\right)\right]}: =1. \tag{2.38}$$

Using the normal ordering product of two-mode vacuum projector, i.e.,

$$|00\rangle\langle 00| =: \mathrm{e}^{-a^\dagger a-b^\dagger b}:, \tag{2.39}$$

Equation (2.38) can be decomposed into

$$1=\int \frac{\mathrm{d}^2 \xi}{\pi}: \mathrm{e}^{-\left[\xi^*-\left(a^\dagger+b\right)\right]\left[\xi-\left(a+b^\dagger\right)\right]}: =\int \frac{\mathrm{d}^2 \xi}{\pi}|\xi\rangle\langle\xi|, \tag{2.40}$$

where the states

$$|\xi\rangle=\exp\left(-\frac{1}{2}|\xi|^2+\xi a^\dagger+\xi^* b^\dagger-a^\dagger b^\dagger\right)|00\rangle \tag{2.41}$$

yield the completeness relation $\pi^{-1}\int \mathrm{d}^2\xi\,|\xi\rangle\langle\xi|=1$ and the following eigen-equations

$$\left(a^\dagger+b\right)|\xi\rangle=\xi^*|\xi\rangle, \quad \left(a+b^\dagger\right)|\xi\rangle=\xi|\xi\rangle, \tag{2.42}$$

so the set of the states $|\xi\rangle$ can form a new representation, called the entangled state representation. Using the IWOP method, the operator $|\xi\rangle\langle\xi|$ can be rewritten as

$$|\xi\rangle\langle\xi| = \int\frac{d^2\beta}{\pi}: \exp\left[-|\beta|^2 + i\beta\left(a^{\dagger} + b - \xi^*\right) + i\beta^*\left(a + b^{\dagger} - \xi\right)\right]: . \quad (2.43)$$

Owing to $\partial\xi/\partial\xi^* = 0$, thus

$$\exp\left(-|\beta|^2 - i\xi^*\beta - i\beta^*\xi\right) = \exp\left(\frac{\partial^2}{\partial\xi\partial\xi^*}\right)\exp -i\xi^*\beta - i\beta^*\xi), \quad (2.44)$$

so Eq. (2.43) becomes

$$\begin{aligned}|\xi\rangle\langle\xi| &= \exp\left(\frac{\partial^2}{\partial\xi\partial\xi^*}\right)\int\frac{d^2\beta}{\pi}: \exp\left[i\beta\left(a^{\dagger} + b - \xi^*\right) + i\beta^*\left(a + b^{\dagger} - \xi\right)\right]: \\ &= \exp\left(\frac{\partial^2}{\partial\xi\partial\xi^*}\right): \delta\left(a^{\dagger} + b - \xi^*\right)\delta^*\left(a + b^{\dagger} - \xi\right): . \quad (2.45)\end{aligned}$$

Using the completeness relation of the entangled state $|\xi\rangle$ and Eq. (2.45), thus we expand the operator $G\left(a^{\dagger} + b, a + b^{\dagger}\right)$ as

$$\begin{aligned}G\left(a^{\dagger} + b, a + b^{\dagger}\right) &= \int\frac{d^2\xi}{\pi}G\left(\xi, \xi^*\right)|\xi\rangle\langle\xi| \\ &= \int\frac{d^2\xi}{\pi}G\left(\xi, \xi^*\right)\exp\left(\frac{\partial^2}{\partial\xi\partial\xi^*}\right) \\ &\quad : \delta\left(a^{\dagger} + b - \xi^*\right)\delta\left(a + b^{\dagger} - \xi\right): \\ &= : \exp\left(\frac{\partial^2}{\partial\xi\partial\xi^*}\right)G\left(\xi, \xi^*\right)\Big|_{\xi\mapsto a+b^{\dagger}, \xi^*\mapsto a^{\dagger}+b} : . \quad (2.46)\end{aligned}$$

In particular, when $G = \left(a + b^{\dagger}\right)^n\left(a^{\dagger} + b\right)^m$, $G\left(\xi, \xi^*\right) = \xi^n\xi^{*m}$, we thus have

$$\begin{aligned}\left(a + b^{\dagger}\right)^n\left(a^{\dagger} + b\right)^m &= : \exp\left(\frac{\partial^2}{\partial\xi\partial\xi^*}\right)\xi^n\xi^{*m}\Big|_{a+b^{\dagger}\to\xi, a^{\dagger}+b\to\xi^*} : \\ &= (-i)^{m+n} : H_{n,m}\left[i\left(a + b^{\dagger}\right), i\left(a^{\dagger} + b\right)\right]: , \quad (2.47)\end{aligned}$$

which can be verified by the following proof. Using the generating function for the two-mode Hermite polynomial, we have

$$\begin{aligned}
&\sum_{m,n=0}^{\infty}\frac{t^{n}t'^{m}}{n!m!}\left(a+b^{\dagger}\right)^{n}\left(a^{\dagger}+b\right)^{m}\\
&=\mathrm{e}^{t\left(a+b^{\dagger}\right)}\mathrm{e}^{t'\left(a^{\dagger}+b\right)}\\
&=\mathrm{e}^{tt'}:\mathrm{e}^{tb^{\dagger}}\mathrm{e}^{t'a^{\dagger}}\mathrm{e}^{ta}\mathrm{e}^{t'b}:\\
&=\mathrm{e}^{-(-\mathrm{i}t)(-\mathrm{i}t')}:\mathrm{e}^{-\mathrm{i}t\left[\mathrm{i}\left(a+b^{\dagger}\right)\right]}\mathrm{e}^{-\mathrm{i}t'\left[\mathrm{i}\left(a^{\dagger}+b\right)\right]}:\\
&=\sum_{m,n=0}^{\infty}\frac{(-\mathrm{i}t)^{n}\left(-\mathrm{i}t'\right)^{m}}{n!m!}:\mathrm{H}_{n,m}[\mathrm{i}\left(a+b^{\dagger}\right),\mathrm{i}\left(a^{\dagger}+b\right)]:,
\end{aligned}\tag{2.48}$$

which leads to the identity

$$\left(a+b^{\dagger}\right)^{n}\left(a^{\dagger}+b\right)^{m}=(-\mathrm{i})^{m+n}:\mathrm{H}_{n,m}[\mathrm{i}\left(a+b^{\dagger}\right),\mathrm{i}\left(a^{\dagger}+b\right)]:.\tag{2.49}$$

As another way to verify Eq. (2.47), we use the completeness relation of the entangled states $|\xi\rangle$ and the eigen-equations in Eq. (2.42) to obtain

$$\begin{aligned}
\left(a+b^{\dagger}\right)^{n}\left(a^{\dagger}+b\right)^{m}&=\int\frac{\mathrm{d}^{2}\xi}{\pi}\left(a+b^{\dagger}\right)^{n}|\xi\rangle\langle\xi|\left(a^{\dagger}+b\right)^{m}\\
&=\int\frac{\mathrm{d}^{2}\xi}{\pi}\xi^{n}\xi^{*m}:\mathrm{e}^{-\left(\xi^{*}-a^{\dagger}-b\right)\left(\xi-a-b^{\dagger}\right)}:.
\end{aligned}\tag{2.50}$$

Integrating Eq. (2.50) via the IWOP method yields the normal ordering product on the right side of Eq. (2.49).

Combining Eqs. (2.49) and (2.50) leads to the following integral formula

$$\mathrm{e}^{-zz^{*}}\int\frac{\mathrm{d}^{2}\xi}{\pi}\xi^{n}\xi^{*m}\mathrm{e}^{-|\xi|^{2}+\xi^{*}z+z^{*}\xi}=(-\mathrm{i})^{m+n}\,\mathrm{H}_{n,m}\left(\mathrm{i}z,\mathrm{i}z^{*}\right),\tag{2.51}$$

or

$$\mathrm{H}_{m,n}\left(\xi,\xi^{*}\right)=\mathrm{i}^{m+n}\mathrm{e}^{|\xi|^{2}}\int\frac{\mathrm{d}^{2}z}{\pi}z^{n}z^{*m}\exp\left(-|z|^{2}-\mathrm{i}\xi z-\mathrm{i}\xi^{*}z^{*}\right).\tag{2.52}$$

Using Eqs. (2.40) and (2.42), we have

$$\mathrm{H}_{m,n}\left(a+b^{\dagger},a^{\dagger}+b\right)=\int\frac{\mathrm{d}^{2}\xi}{\pi}\mathrm{H}_{m,n}\left(\xi,\xi^{*}\right)|\xi\rangle\langle\xi|,\tag{2.53}$$

thus using the integral formula in Eq. (2.52), we can calculate Eq. (2.53) as

$$\mathrm{H}_{m,n}\left(a+b^{\dagger},a^{\dagger}+b\right)=:\left(a+b^{\dagger}\right)^{m}\left(a^{\dagger}+b\right)^{n}:.\tag{2.54}$$

Similar to deriving Eq. (2.27), we obtain

$$\exp\left[\frac{\partial^2}{\partial\left(a+b^{\dagger}\right)\partial\left(a^{\dagger}+b\right)}\right]\left(a+b^{\dagger}\right)^n\left(a^{\dagger}+b\right)^m$$
$$=(-\mathrm{i})^{m+n}\,\mathrm{H}_{n,m}\left[\mathrm{i}\left(a+b^{\dagger}\right),\mathrm{i}\left(a^{\dagger}+b\right)\right], \tag{2.55}$$

thus we get

$$\begin{aligned}
&\exp\left[\frac{\partial^2}{\partial\left(a+b^{\dagger}\right)\partial\left(a^{\dagger}+b\right)}\right]\left(a+b^{\dagger}\right)^n\left(a^{\dagger}+b\right)^m\\
&=(-\mathrm{i})^{m+n}\int\frac{\mathrm{d}^2\xi}{\pi}\mathrm{H}_{n,m}\left(\mathrm{i}\xi,\mathrm{i}\xi^*\right)|\xi\rangle\langle\xi|\\
&=(-\mathrm{i})^{m+n}\int\frac{\mathrm{d}^2\xi}{\pi}\mathrm{H}_{n,m}\left(\mathrm{i}\xi,\mathrm{i}\xi^*\right):\exp\{-\left[\xi^*-\left(a^{\dagger}+b\right)\right]\left[\xi-\left(a+b^{\dagger}\right)\right]\}:\\
&=:\left(a+b^{\dagger}\right)^n\left(a^{\dagger}+b\right)^m:,
\end{aligned} \tag{2.56}$$

which leads to a general formula

$$\exp\left[\frac{\partial^2}{\partial\left(a+b^{\dagger}\right)\partial\left(a^{\dagger}+b\right)}\right]G\left(a+b^{\dagger},a^{\dagger}+b\right)=:G\left(a+b^{\dagger},a^{\dagger}+b\right):. \tag{2.57}$$

For example,

$$\exp\left[\frac{\partial^2}{\partial\left(a+b^{\dagger}\right)\partial\left(a^{\dagger}+b\right)}\right]\mathrm{H}_{m,n}\left(a+b^{\dagger},a^{\dagger}+b\right)=:\mathrm{H}_{m,n}\left(a+b^{\dagger},a^{\dagger}+b\right): \tag{2.58}$$

and

$$\begin{aligned}
&\exp\left[\frac{\partial^2}{\partial\left(a+b^{\dagger}\right)\partial\left(a^{\dagger}+b\right)}\right]\exp\left\{-\left[\xi^*-\left(a^{\dagger}+b\right)\right]\left[\xi-\left(a+b^{\dagger}\right)\right]\right\}\\
&=:\exp\left\{-\left[\xi^*-\left(a^{\dagger}+b\right)\right]\left[\xi-\left(a+b^{\dagger}\right)\right]\right\}:=|\xi\rangle\langle\xi|\\
&=\delta\left[\xi^*-\left(a^{\dagger}+b\right)\right]\delta\left[\xi-\left(a+b^{\dagger}\right)\right],
\end{aligned} \tag{2.59}$$

as well as

$$\exp\left[\frac{\partial^2}{\partial\left(a+b^{\dagger}\right)\partial\left(a^{\dagger}+b\right)}\right]\mathrm{e}^{-\lambda\left(a+b^{\dagger}\right)\left(a^{\dagger}+b\right)}=:\mathrm{e}^{-\lambda\left(a+b^{\dagger}\right)\left(a^{\dagger}+b\right)}:. \tag{2.60}$$

On the other hand, using the completeness relation of the entangled states $|\xi\rangle$ and the IWOP method, we obtain

$$
\begin{aligned}
e^{-\lambda(a+b^\dagger)(a^\dagger+b)} &= \int \frac{d^2\xi}{\pi} e^{-\lambda|\xi|^2} |\xi\rangle\langle\xi| \\
&= \int \frac{d^2\xi}{\pi} : e^{-(1+\lambda)|\xi|^2+\xi^*(a+b^\dagger)+\xi(a^\dagger+b)-(a+b^\dagger)(a^\dagger+b)} : \\
&= \frac{1}{1+\lambda} : \exp\left[\frac{-\lambda}{1+\lambda}\left(a+b^\dagger\right)\left(a^\dagger+b\right)\right] : . \qquad (2.61)
\end{aligned}
$$

Letting $\lambda/(1+\lambda)=t$, i.e., $\lambda=t/(1-t)$, thus Eq. (2.61) becomes

$$
: \exp\left[-t\left(a+b^\dagger\right)\left(a^\dagger+b\right)\right] : = (1-t)\, e^{-\frac{t}{1-t}(a+b^\dagger)(a^\dagger+b)}. \qquad (2.62)
$$

It follows from Eq. (2.61) that

$$
\begin{aligned}
&\exp\left[\frac{\partial^2}{\partial\left(a+b^\dagger\right)\partial\left(a^\dagger+b\right)}\right] e^{-\lambda(a+b^\dagger)(a^\dagger+b)} \\
&= : e^{-\lambda(a+b^\dagger)(a^\dagger+b)} : \\
&= (1-\lambda)\, e^{-\frac{\lambda}{1-\lambda}(a+b^\dagger)(a^\dagger+b)}. \qquad (2.63)
\end{aligned}
$$

Similarly, we have

$$
\exp\left[\frac{\partial^2}{\partial\left(a+b^\dagger\right)\partial\left(a^\dagger+b\right)}\right] G\left(a+b^\dagger, a^\dagger+b\right) =: G\left(a+b^\dagger, a^\dagger+b\right) : , \qquad (2.64)
$$

which is the generalization of single-mode case in Eq. (2.8). Using Eq. (2.54), we can also obtain

$$
\begin{aligned}
&: \left(a+b^\dagger\right)^n \left(a^\dagger+b\right)^m : \\
&= H_{n,m}\left(a+b^\dagger, a^\dagger+b\right) \\
&= \sum_{l=0}^{\min(m,n)} \frac{n!m!(-1)^l}{l!(m-l)!(n-l)!} \left(a+b^\dagger\right)^{n-l} \left(a^\dagger+b\right)^{m-l} \\
&= (-i)^{m+n} \sum_{l=0}^{\min(m,n)} \frac{n!m!}{l!(m-l)!(n-l)!} : H_{n-l,m-l}\left[i\left(a+b^\dagger\right), i\left(a^\dagger+b\right)\right] : . \qquad (2.65)
\end{aligned}
$$

Combining Eqs. (2.49) and (2.65), we can obtain the following expansion

$$
x^n y^m = (-i)^{m+n} \sum_{l=0}^{\min(m,n)} \frac{n!m!}{l!(m-l)!(n-l)!} H_{n-l,m-l}(ix, iy), \qquad (2.66)
$$

which is the inverse expansion of

$$\mathrm{H}_{m,n}(x,y)=\sum_{l=0}^{\min(m,n)}\frac{n!m!(-1)^l}{l!\,(m-l)!\,(n-l)!}x^{m-l}y^{n-l}. \tag{2.67}$$

2.4 Anti-Normal Ordering Formula

We know that using the completeness relation of coherent states, any single-mode operator ρ can be expanded as

$$\rho=\int\frac{d^2z}{\pi}\mathrm{P}(z)\,|z\rangle\langle z|=\int\frac{d^2z}{\pi}\mathrm{P}(z):\exp\left[-(z^*-a^\dagger)(z-a)\right]:, \tag{2.68}$$

where the expansion function P(z) is called the P-representation of the density operator ρ. Especially, for $\rho=\pi\delta(a)\,\delta\left(a^\dagger\right)$, we have

$$\begin{aligned}\pi\delta(a)\,\delta\left(a^\dagger\right)&=\pi\int\frac{d^2z}{\pi}\delta(a)\,|z\rangle\langle z|\,\delta\left(a^\dagger\right)\\&=\int d^2z\delta(z)\,|z\rangle\langle z|\,\delta\left(z^*\right)=|0\rangle\langle 0|,\end{aligned} \tag{2.69}$$

it follows that

$$|0\rangle\langle 0|=\pi\delta(a)\,\delta\left(a^\dagger\right), \tag{2.70}$$

which leads to the anti-normal ordering product of the density operator $|z\rangle\langle z|$, that is

$$|z\rangle\langle z|=D(z)\,|0\rangle\langle 0|\,D^\dagger(z)=\pi\delta(a-z)\,\delta\left(a^\dagger-z^*\right). \tag{2.71}$$

Noting that the inner product of two different coherent states reads

$$\langle z|\,\beta\rangle=\exp\left[-\frac{1}{2}\left(|z|^2+|\beta|^2\right)+z^*\beta\right], \tag{2.72}$$

thus we have

$$\begin{aligned}\langle-\beta|\,\rho(a,a^\dagger)\,|\beta\rangle&=\int\frac{d^2z}{\pi}\mathrm{P}(z)\,\langle-\beta|z\rangle\,\langle z|\beta\rangle\\&=\int\frac{d^2z}{\pi}\mathrm{P}(z)\exp\left(-|z|^2+\beta^*z-\beta z^*-|\beta|^2\right),\end{aligned} \tag{2.73}$$

where $\exp\left(\beta^*z-\beta z^*\right)$ is a Fourier integral transform kernel, so its inverse transform is given by

$$\mathrm{P}(z) = \mathrm{e}^{|z|^2} \int \frac{\mathrm{d}^2\beta}{\pi} \langle -\beta| \rho(a, a^\dagger) |\beta\rangle \exp\left(|\beta|^2 + \beta^* z - \beta z^*\right). \tag{2.74}$$

Owing to

$$|z\rangle \langle z| =: \exp\left(-|z|^2 + z a^\dagger + a z^* - a^\dagger a\right) : = \pi \vdots \delta(z - a)\, \delta\left(z^* - a^\dagger\right) \vdots, \tag{2.75}$$

substituting Eqs. (2.74) and (2.75) into Eq. (2.68) leads to

$$\begin{aligned} \rho(a, a^\dagger) &= \int \mathrm{d}^2 z \mathrm{e}^{|z|^2} \int \frac{\mathrm{d}^2\beta}{\pi} \langle -\beta| \rho(a, a^\dagger) |\beta\rangle \exp\left(|\beta|^2 + \beta^* z - \beta z^*\right) \\ &\quad \times \vdots \delta(z - a)\, \delta\left(z^* - a^\dagger\right) \vdots \\ &= \int \frac{\mathrm{d}^2\beta}{\pi} \vdots \langle -\beta| \rho(a, a^\dagger) |\beta\rangle \exp\left(|\beta|^2 + \beta^* a - \beta a^\dagger + a^\dagger a\right) \vdots, \end{aligned} \tag{2.76}$$

which is the formula for calculating the anti-normal ordering product of any operator $\rho(a, a^\dagger)$ [6, 7], $|\beta\rangle$ is a coherent state, $a|\beta\rangle = \beta|\beta\rangle$. In particular, if $\rho = 1$, $\langle -\beta| \beta\rangle = \mathrm{e}^{-2|\beta|^2}$, thus using Eq. (2.76) to obtain

$$\int \frac{\mathrm{d}^2\beta}{\pi} \vdots \exp\left(-|\beta|^2 + \beta^* a - \beta a^\dagger + a^\dagger a\right) \vdots = 1, \tag{2.77}$$

which also demonstrates the correctness of Eq. (2.76). When $\rho = a^{\dagger m} a^n$, using Eq. (2.76), we immediately obtain

$$\begin{aligned} a^{\dagger m} a^n &= \int \frac{\mathrm{d}^2\beta}{\pi} \vdots \langle -\beta| a^{\dagger m} a^n |\beta\rangle \exp\left(|\beta|^2 + \beta^* a - \beta a^\dagger + a^\dagger a\right) \vdots \\ &= \vdots \int \frac{\mathrm{d}^2\beta}{\pi} (-1)^m z^{*m} z^n \exp\left(-|\beta|^2 + \beta^* a - \beta a^\dagger + a^\dagger a\right) \vdots \\ &= \vdots \mathrm{H}_{m,n}\left(a^\dagger, a\right) \vdots = \vdots \mathrm{H}_{n,m}\left(a, a^\dagger\right) \vdots, \end{aligned} \tag{2.78}$$

which is the anti-normal ordering product of the operator $a^{\dagger m} a^n$, a two-variable Hermite polynomial.

Now, we present the anti-normal ordering products of the parity operator and the Wigner operator. In Chap. 1, we already know the normal ordering product of the parity operator, that is

$$(-1)^N = \sum_{n=0}^{\infty} (-1)^n |n\rangle \langle n| =: \mathrm{e}^{-2a^\dagger a} : . \tag{2.79}$$

Substituting Eq. (2.79) into Eq. (2.76), there will be the problem of integral divergence. To avoid this problem, we expand the parity operator according to the normal ordering product, and use Eq. (2.78), thus the operator $(-1)^N$ becomes

$$(-1)^N = \sum_{n=0}^{\infty} \frac{(-2)^n a^{\dagger n} a^n}{n!} = \sum_{n=0}^{\infty} \frac{(-2)^n}{n!} \vdots \mathrm{H}_{n,n}\left(a, a^\dagger\right) \vdots. \tag{2.80}$$

Noting the relations between Laguerre polynomials $\mathrm{L}_n(\cdot)$ and two-variable Hermite polynomials $\mathrm{H}_{n,n}(\cdot, \cdot)$, i.e.,

$$\mathrm{L}_n(xy) = \frac{(-1)^n}{n!} \mathrm{H}_{n,n}(x, y), \tag{2.81}$$

and using the series expansions of Laguerre polynomials in Eq. (2.34), thus Eq. (2.80) becomes

$$(-1)^N = \sum_{n=0}^{\infty} 2^n \vdots \mathrm{L}_n\left(aa^\dagger\right) \vdots = -\vdots \mathrm{e}^{2aa^\dagger} \vdots, \tag{2.82}$$

which is the anti-normal ordering product of the parity operator $(-1)^N$.

Using Eq. (2.82), we can directly get the anti-normal ordering product of the Wigner operator. Noting that the Wigner operator $\Delta(\alpha, \alpha^*) = D(\alpha)(-1)^N D^\dagger(\alpha)$ and the operator identity

$$(-1)^N a = -a(-1)^N, \tag{2.83}$$

we obtain

$$\begin{aligned} \Delta\left(\alpha, \alpha^*\right) &= \frac{1}{\pi} \mathrm{e}^{2|\alpha|^2} \mathrm{e}^{-2\alpha^* a} (-1)^N \mathrm{e}^{-2\alpha a^\dagger} \\ &= -\frac{1}{\pi} \mathrm{e}^{2|\alpha|^2} \mathrm{e}^{-2\alpha^* a} \vdots \mathrm{e}^{2aa^\dagger} \vdots \mathrm{e}^{-2\alpha a^\dagger} \\ &= -\frac{1}{\pi} \vdots \mathrm{e}^{2(\alpha^* - a^\dagger)(\alpha - a)} \vdots, \end{aligned} \tag{2.84}$$

which is the anti-normal ordering product of the Wigner operator $\Delta(\alpha, \alpha^*)$ [8].

References

1. Xu SM, Zhang YH, Xu XL, Li HQ, Wang JS (2016) New approach for anti-normally and normally ordering bosonic-operator functions in quantum optics. Chin Phys B 25(12):120301
2. Meng XG, Li KC, Wang JS, Yang ZS, Zhang XY, Zhang ZT, Liang BL (2020) Multi-variable special polynomials using an operator ordering method. Front Phys 15(5):52501
3. Fan HY, Lu HL, Fan Y (2006) Newton-Leibniz integration for ket-bra operators in quantum mechanics and derivation of entangled state representations. Ann Phys 321(2):480–494

4. Fan HY (2007) Newton-Leibniz integration for ket-bra operators (II)-application in deriving density operator and generalized partition function formula. Ann Phys 322(4):866–885
5. Fan HY (2004) Entangled states, squeezed states gained via the route of developing Dirac's symbolic method and their applications. Int J Mod Phys B 18(10–11):1387–1455
6. Hu LY, Fan HY, Zhang ZM (2013) New formulas for normalizing photon-added (-subtracted) two-mode squeezed thermal states. Chin Phys B 22(3):034202
7. Meng XG, Wang JS, Liang BL (2023) Entangled state representations in quantum optics. Springer, Singapore
8. Jia F, Zhang HL, Hu LY, Fan HY (2015) From Weyl expansion of operator to ordering of operator and its new applications. Int J Theor Phys 54:1283–1291

Chapter 3
Thermal Vacuum Pure States Corresponding to Mixed States

Given that the light field is a multiple-degree-of-freedom system, it is generally a mixed state represented by a density operator. One problem is to find the corresponding thermal vacuum pure state $|\psi(\beta)\rangle$ for the density operator ρ (also known as the mixed state), so that solving the ensemble average of the mixed state ρ can be transformed into solving the average value in the pure state $|\psi(\beta)\rangle$ [1–3]. In other words, what conditions should the state $|\psi(\beta)\rangle$ meet? Let's start with the chaotic light field.

3.1 Thermal Vacuum State Corresponding to Chaotic Light Field

The density operator of the chaotic light field is defined as $\rho_c = \left(1 - \mathrm{e}^{-\frac{\hbar\omega}{k_B T}}\right)\mathrm{e}^{-\frac{\hbar\omega}{k_B T}a^\dagger a}$, or denoted as

$$\rho_c = \gamma(1-\gamma)^{a^\dagger a}, \gamma = 1 - \mathrm{e}^{-\frac{\hbar\omega}{k_B T}}, \quad \mathtt{tr}\rho_c = 1, \tag{3.1}$$

taking $\tanh\theta = \exp\left(-\frac{\hbar\omega}{2k_B T}\right)$, thus the density operator ρ_c becomes

$$\rho_c = \mathrm{sech}^2\theta \mathrm{e}^{a^\dagger a \ln\tanh^2\theta} = \mathrm{sech}^2\theta \colon \mathrm{e}^{-a^\dagger a \mathrm{sech}^2\theta} \colon, \tag{3.2}$$

whose second-order coherence is $g^{(2)} = 2$. Generally speaking, the density operator of the boson system is defined as $\rho = \mathrm{e}^{-\beta H}/Z$, $Z = \mathtt{tr}\mathrm{e}^{-\beta H}$ is the partition function, where $\beta = 1/(k_B T)$, k_B is the Boltzmann constant.

Can we convert the chaotic state ρ_c into its corresponding pure state $|\psi(\beta)\rangle$? We need to introduce the fictitious space as a companion to the free degree of the real system, where $|(\beta)\rangle$ is defined in the expanded space. And, $\mathtt{Tr} = \mathtt{tr}\widetilde{\mathtt{tr}}$ is defined as a sign that traces both real and fictitious spaces, where $\mathtt{tr}$ refers to tracing only

H. Fan and L. Hu, *Dissipation and Diffusion of Quantum Light Field*,
https://doi.org/10.1007/978-981-95-1528-8_3

the real space and $\widetilde{\mathtt{tr}}$ represents tracing only the fictitious space. Therefore, for the real mechanical quantity A, we have

$$\langle A\rangle = \langle\psi(\beta)|\,A\,|\psi(\beta)\rangle = \mathtt{Tr}\,(A\,|\psi(\beta)\rangle\langle\psi(\beta)|) = \mathtt{tr}\left\{A\left[\widetilde{\mathtt{tr}}\,(|\psi(\beta)\rangle\langle\psi(\beta)|)\right]\right\}. \quad (3.3)$$

Since $|\psi(\beta)\rangle$ involves the real and fictitious spaces, so we need to pay attention to the following equation:

$$\widetilde{\mathtt{tr}}\,(|\psi(\beta)\rangle\langle\psi(\beta)|) \neq \langle\psi(\beta)|\,\psi(\beta)\rangle\,. \quad (3.4)$$

In terms of the ensemble average $\langle A\rangle = \mathtt{tr}\,(A\rho)$ and Eq. (3.3), the state $|\psi(\beta)\rangle$ to be solved shall satisfy the following identity:

$$\widetilde{\mathtt{tr}}\,(|\psi(\beta)\rangle\langle\psi(\beta)|) = \rho. \quad (3.5)$$

For example, the chaotic light field ρ_c in Eq. (3.1) can be rewritten as

$$\begin{aligned}\rho_c &= \sum_{m=0}^{\infty}\gamma\,(1-\gamma)^m\,|m\rangle\langle m| \\ &= \sum_{m'=0}^{\infty}\sqrt{\gamma\,(1-\gamma)^{m'}}\sum_{m=0}^{\infty}\sqrt{\gamma\,(1-\gamma)^m}\,|m'\rangle\langle m|\,\langle\tilde{m}|\,\tilde{m}'\rangle \\ &= \widetilde{\mathtt{tr}}\left(\sum_{m'=0}^{\infty}\sqrt{\gamma\,(1-\gamma)^{m'}}\,|m'\rangle\,|\tilde{m}'\rangle\sum_{m=0}^{\infty}\sqrt{\gamma\,(1-\gamma)^m}\,\langle m|\,\langle\tilde{m}|\right), \end{aligned} \quad (3.6)$$

where $\sqrt{m!}\,|\tilde{m}\rangle = \tilde{a}^{\dagger m}\left|\tilde{0}\right\rangle$, $\left|\tilde{0}\right\rangle$ is annihilated by $\tilde{a}$. Comparing Eq. (3.6) and the trace $\widetilde{\mathtt{tr}}\,(|0(\beta)\rangle\langle 0(\beta)|)$, we get

$$|0(\beta)\rangle = \sum_{m=0}^{\infty}\sqrt{\gamma\,(1-\gamma)^m}\,|m,\tilde{m}\rangle\,, \quad (3.7)$$

which is called a binomial thermal vacuum state according to its mathematical form [1]. Or,

$$|0(\beta)\rangle = \operatorname{sech}\theta\exp\left(a^{\dagger}\tilde{a}^{\dagger}\tanh\theta\right)\left|0,\tilde{0}\right\rangle, \quad (3.8)$$

so we say that the state $|0(\beta)\rangle$ is a pure state corresponding to the chaotic light field in the expanded Fock space. Thus, the average value of the number operator $a^{\dagger}a$ in the state $|0(\beta)\rangle$ is

$$\begin{aligned}\langle 0(\beta)|\, a^{\dagger}a\, |0(\beta)\rangle &= \left(1-\mathrm{e}^{-\frac{\hbar\omega}{k_BT}}\right)\sum_{m=0}\mathrm{e}^{-\frac{\hbar\omega}{k_BT}}\langle m|\, a^{\dagger}a\,|m\rangle \\ &= \frac{\mathrm{e}^{-\frac{\hbar\omega}{k_BT}}}{1-\mathrm{e}^{-\frac{\hbar\omega}{k_BT}}} = \frac{1-\gamma}{\gamma} \equiv n_c,\end{aligned} \tag{3.9}$$

which is the Planck's photon distribution formula.

Another way to obtain the thermal vacuum state of the chaotic light field is to use the IWOP method. Using the IWOP method and the coherent states $|z\rangle$, Eq. (3.2) can be rewritten as

$$\begin{aligned}\rho_c &= \mathrm{sech}^2\theta \int \frac{\mathrm{d}^2 z}{\pi} : \mathrm{e}^{-|z|^2 + a^{\dagger}z^* \tanh\theta + az\tanh\theta - a^{\dagger}a} : \\ &= \mathrm{sech}^2\theta \int \frac{\mathrm{d}^2 z}{\pi} \mathrm{e}^{a^{\dagger}z^*\tanh\theta}\,|0\rangle\,\langle 0|\,\mathrm{e}^{az\tanh\theta}\mathrm{e}^{-|z|^2}.\end{aligned} \tag{3.10}$$

Using the coherent states $|\tilde{z}\rangle$ of fictitious mode, which yield the inner product

$$\mathrm{e}^{-|z|^2/2} = \left\langle \tilde{0}\middle|\,\tilde{z}\right\rangle \tag{3.11}$$

and the completeness relation

$$\int \frac{\mathrm{d}^2 z}{\pi}\,|\tilde{z}\rangle\,\langle\tilde{z}| = 1,\ \tilde{a}\,|\tilde{z}\rangle = z\,|\tilde{z}\rangle\,, \tag{3.12}$$

we can write Eq. (3.10) as

$$\begin{aligned}\rho_c &= \mathrm{sech}^2\theta \int \frac{\mathrm{d}^2 z}{\pi}\,\langle\tilde{z}|\,\mathrm{e}^{a^{\dagger}z^*\tanh\theta}\left|0,\tilde{0}\right\rangle\left\langle 0,\tilde{0}\right|\mathrm{e}^{az\tanh\theta}\,|\tilde{z}\rangle \\ &= \mathrm{sech}^2\theta \int \frac{\mathrm{d}^2 z}{\pi}\,\langle\tilde{z}|\,\mathrm{e}^{a^{\dagger}\tilde{a}^{\dagger}\tanh\theta}\left|0,\tilde{0}\right\rangle\left\langle 0,\tilde{0}\right|\mathrm{e}^{a\tilde{a}\tanh\theta}\,|\tilde{z}\rangle\,.\end{aligned} \tag{3.13}$$

Letting

$$|0\,(\beta)\rangle = \mathrm{sech}\theta\mathrm{e}^{a^{\dagger}\tilde{a}^{\dagger}\tanh\theta}\left|0,\tilde{0}\right\rangle, \tag{3.14}$$

thus Eq. (3.13) is changed into

$$\begin{aligned}\rho_c &= \int \frac{\mathrm{d}^2 z}{\pi}\,\langle\tilde{z}|\,0\,(\beta)\rangle\,\langle 0\,(\beta)|\,\tilde{z}\rangle \\ &= \widetilde{\mathtt{tr}}\left(\int \frac{\mathrm{d}^2 z}{\pi}\,|\tilde{z}\rangle\,\langle\tilde{z}|\,0\,(\beta)\rangle\,\langle 0\,(\beta)|\right) \\ &= \widetilde{\mathtt{tr}}\,(|0\,(\beta)\rangle\,\langle 0\,(\beta)|)\,,\end{aligned} \tag{3.15}$$

which obeys Eq. (3.5), and its formal solution is Eq. (3.14). Also, using $\mathtt{tr}\rho_c = 1$, we can obtain $\mathtt{tr}\widetilde{\mathtt{tr}}\left(|0(\beta)\rangle\langle 0(\beta)|\right) = \mathtt{Tr}\left(|0(\beta)\rangle\langle 0(\beta)|\right) = \langle 0(\beta)|0(\beta)\rangle = 1$. Now, we use Eq. (3.14) to calculate the average number of photons in the chaotic light field. Since

$$a|0(\beta)\rangle = \mathrm{sech}\theta\left[a, \mathrm{e}^{\tilde{a}^\dagger a^\dagger \tanh\theta}\right]\left|0\tilde{0}\right\rangle = \tilde{a}^\dagger \tanh\theta\,|0(\beta)\rangle\,, \tag{3.16}$$

thus we have

$$\begin{aligned}\langle 0(\beta)|\,a^\dagger a\,|0(\beta)\rangle &= \tanh^2\theta\,\langle 0(\beta)|\,\tilde{a}\tilde{a}^\dagger\,|0(\beta)\rangle \\ &= \tanh^2\theta\left(\langle 0(\beta)|\,\tilde{a}^\dagger\tilde{a}\,|0(\beta)\rangle + 1\right).\end{aligned} \tag{3.17}$$

Considering that the state $|0(\beta)\rangle$ is symmetric about the operators $\tilde{a}^\dagger$ and $a^\dagger$, so we have

$$\langle 0(\beta)|\,a^\dagger a\,|0(\beta)\rangle = \langle 0(\beta)|\,\tilde{a}^\dagger\tilde{a}\,|0(\beta)\rangle \equiv x, \tag{3.18}$$

thus Eq. (3.17) becomes

$$x = \tanh^2\theta\,(x+1) = \mathrm{e}^{-\beta\hbar\omega}\,(x+1)\,, \tag{3.19}$$

its formal solution is

$$\begin{aligned}\langle 0(\beta)|\,a^\dagger a\,|0(\beta)\rangle &= x = \sinh^2\theta \\ &= \frac{1}{1-\mathrm{e}^{-\beta\hbar\omega}}\mathrm{e}^{-\beta\hbar\omega} \\ &= \left(\mathrm{e}^{\beta\hbar\omega}-1\right)^{-1},\end{aligned} \tag{3.20}$$

which is exactly the Bose-Einstein distribution.

Using the state $|0(\beta)\rangle$, we can easily calculate the entropy of the chaotic light field. According to the definition of quantum entropy, that is $S = -k_B \mathtt{tr}\left(\rho_c \ln\rho_c\right)$ [4, 5], we obtain

$$\begin{aligned}S &= -k_B\,\langle 0(\beta)|\ln\rho_c\,|0(\beta)\rangle \\ &= -k_B\,\langle 0(\beta)|\ln\left[(1-\mathrm{e}^{-\beta\hbar\omega})\mathrm{e}^{-\beta\hbar\omega a^\dagger a}\right]|0(\beta)\rangle \\ &= -k_B\ln\left(1-\mathrm{e}^{-\beta\hbar\omega}\right) + k_B\beta\hbar\omega\,\langle 0(\beta)|\,a^\dagger a\,|0(\beta)\rangle\,.\end{aligned} \tag{3.21}$$

Substituting Eq. (3.20) into Eq. (3.21) leads to the result

$$S = k_B\left[\ln\left(1-\mathrm{e}^{-\beta\hbar\omega}\right) + \frac{\beta\hbar\omega\mathrm{e}^{-\beta\hbar\omega}}{\mathrm{e}^{-\beta\hbar\omega}-1}\right]. \tag{3.22}$$

Clearly, the method of converting the mixed states into the pure states not only provides a physical approach, but also brings convenience to the other related calculations.

Next, we derive the expected value theorem of the binomial thermal vacuum state $|0\,(\beta)\rangle$. According to the negative binomial theorem

$$(1+x)^{-(s+1)} = \sum_{n=0}^{\infty} \frac{(n+s)!}{n!s!}(-x)^n \tag{3.23}$$

and Eq. (3.20), we have

$$\begin{aligned}\left(\langle 0(\beta)|\, a^{\dagger} a \,|0(\beta)\rangle\right)^p &= \mathrm{e}^{-p\frac{\hbar\omega}{k_B T}}\left(1-\mathrm{e}^{-\frac{\hbar\omega}{k_B T}}\right)^{-p} \\ &= \left(1-\mathrm{e}^{-\frac{\hbar\omega}{k_B T}}\right)\sum_{n=0}^{\infty}\frac{(m+p)!}{m!p!}\mathrm{e}^{-\frac{\hbar\omega(m+p)}{k_B T}}.\end{aligned} \tag{3.24}$$

Noting

$$a^p\,|m\rangle = \sqrt{\frac{m!}{(m-p)!}}\,|m-p\rangle\,, \tag{3.25}$$

thus

$$\langle m|\, a^{\dagger p} a^p \,|m\rangle = \frac{m!}{(m-p)!}. \tag{3.26}$$

Therefore, the average value of the operator $a^{\dagger p} a^p$ in the state $|0\,(\beta)\rangle$ can be expressed as

$$\begin{aligned}\langle 0(\beta)|\, a^{\dagger p} a^p \,|0(\beta)\rangle &= \left(1-\mathrm{e}^{-\frac{\hbar\omega}{\kappa T}}\right)\sum_{n=0}^{\infty}\mathrm{e}^{-m\frac{\hbar\omega}{\kappa T}}\,\langle m|\, a^{\dagger p} a^p \,|m\rangle \\ &= \gamma\sum_{n=0}^{\infty}(1-\gamma)^{m+p}\,\frac{(m+p)!}{m!}.\end{aligned} \tag{3.27}$$

Comparing Eqs. (3.24) and 3.27), we get

$$\langle 0(\beta)|\, a^{\dagger p} a^p \,|0(\beta)\rangle = p!\left(\langle 0(\beta)|\, a^{\dagger} a \,|0(\beta)\rangle\right)^p, \tag{3.28}$$

which is the expected value theorem of the states $|0\,(\beta)\rangle$.

On the other hand, using the form of $|0(\beta)\rangle$ in the expanded space, i.e., $|0(\beta)\rangle = \sum_{n=0}^{\infty}\sqrt{\gamma\,(1-\gamma)^m}\,|m,\tilde{m}\rangle$, we have

$$\begin{aligned}
\langle 0(\beta)| a^{\dagger p'} a^{p} |0(\beta)\rangle &= \sum_{m'=0}^{\infty} \sqrt{\gamma (1-\gamma)^{m'}} \langle m', \tilde{m}'| a^{\dagger p'} a^{p} \sum_{m=0}^{\infty} \sqrt{\gamma (1-\gamma)^{m}} |m, \tilde{m}\rangle \\
&= \sum_{m'=0}^{\infty} \sqrt{\gamma (1-\gamma)^{m'}} \langle m'| a^{\dagger p'} a^{p} \sum_{m=0}^{\infty} \sqrt{\gamma (1-\gamma)^{m}} |m\rangle \, \delta_{m,m'} \\
&= \sum_{m=0}^{\infty} \gamma (1-\gamma)^{m} \langle m| a^{\dagger p'} a^{p} |m\rangle \\
&= \sum_{m=0}^{\infty} \gamma (1-\gamma)^{m} \langle m-p'| \sqrt{\frac{m!m!}{(m-p')!\,(m-p)!}} |m-p\rangle \\
&= \sum_{m=0}^{\infty} \gamma (1-\gamma)^{m} \sqrt{\frac{m!m!}{(m-p')!\,(m-p)!}} \delta_{p',p},
\end{aligned} \tag{3.29}$$

it follows that

$$\begin{aligned}
\langle 0(\beta)| e^{fa^{\dagger}} e^{ga} |0(\beta)\rangle &= \sum_{p,p'=0}^{\infty} \frac{1}{p!p'!} \langle 0(\beta)| \left(fa^{\dagger}\right)^{p'} (ga)^{p} |0(\beta)\rangle \\
&= \sum_{p,p'=0}^{\infty} \frac{1}{p!p'!} f^{p'} g^{p} \sum_{m=0}^{\infty} \gamma (1-\gamma)^{m} \\
&\quad \times \sqrt{\frac{m!m!}{(m-p')!\,(m-p)!}} \delta_{p',p} \\
&= \sum_{p=0}^{\infty} \frac{1}{p!p!} (fg)^{p} \sum_{m=0}^{\infty} \frac{m! \gamma (1-\gamma)^{m}}{(m-p)!} \\
&= \sum_{p=0}^{\infty} \frac{1}{p!p!} (fg)^{p} \langle 0(\beta)| a^{\dagger p} a^{p} |0(\beta)\rangle \\
&= \sum_{p=0}^{\infty} \frac{(fg)^{p}}{p!} \left[\langle 0(\beta)| a^{\dagger} a |0(\beta)\rangle\right]^{p} \\
&= \exp\left[fg \langle 0(\beta)| a^{\dagger} a |0(\beta)\rangle\right].
\end{aligned} \tag{3.30}$$

Since

$$\langle 0(\beta)| a^{2} |0(\beta)\rangle = 0, \tag{3.31}$$

thus, using Eq. (3.30), we find that the thermal average value of the displaced operator $e^{fa+ga^{\dagger}}$ reads

$$\langle 0(\beta)| e^{fa+ga^{\dagger}} |0(\beta)\rangle = \exp\left(\frac{1}{2} \langle 0(\beta)| \left(fa+ga^{\dagger}\right)^{2} |0(\beta)\rangle\right), \tag{3.32}$$

which can be also simplified as

$$\left\langle e^{fa+ga^\dagger}\right\rangle = e^{\left\langle \frac{1}{2}\left(fa+ga^\dagger\right)^2\right\rangle}. \tag{3.33}$$

3.2 Thermal Vacuum State Corresponding to Negative Binomial Light Field

In mathematical statistics theory, the negative binomial distribution is

$$\binom{n+s}{n}\gamma^{s+1}(1-\gamma)^n,\ 0<\gamma<1,\ s\geq 0, \tag{3.34}$$

which corresponds to a negative binomial state of light field [6, 7], i.e.,

$$\rho_s = \sum_{n=0}^{\infty}\frac{(n+s)!}{n!s!}\gamma^{s+1}(1-\gamma)^n\left|n\right\rangle\left\langle n\right|, \tag{3.35}$$

which is generated during the process of s photons of a thermal photon beam being absorbed by an atom. On the other hand, the applications of the negative binomial states in many optical fields are very extensive, for example, after the number state $|m,0\rangle$ undergoing the action of the two-mode squeezing operator $\exp\left[\lambda\left(a_1^\dagger a_2^\dagger - a_1a_2\right)\right]$, it becomes

$$\begin{aligned} |\lambda\rangle_m &= \exp\left[\lambda\left(a_1^\dagger a_2^\dagger - a_1a_2\right)\right]|m,0\rangle \\ &= \operatorname{sech}^{m+1}\lambda\sum_{n=0}^{\infty}\sqrt{\frac{(n+m)!}{n!m!}}\tanh^n\lambda\left|n+m,n\right\rangle, \end{aligned} \tag{3.36}$$

which shows the negative binomial distribution, i.e.,

$$\operatorname{sech}^{2(m+1)}\lambda\frac{(n+m)!}{n!m!}\tanh^{2n}\lambda. \tag{3.37}$$

In the negative binomial state, the average photon number is

$$\begin{aligned} \mathtt{tr}\left(\rho_s a^\dagger a\right) &= \gamma^{n+1}\sum_{m=0}^{n}\binom{n+m}{m}(1-\gamma)^m m \\ &= \gamma^{n+1}(1-\gamma)\frac{\partial}{\partial(1-\gamma)}\gamma^{-(n+1)} \\ &= \frac{(n+1)(1-\gamma)}{\gamma}. \end{aligned} \tag{3.38}$$

Writing the operator ρ_s as its normal ordering product, that is

$$\rho_s = \sum_{m=0}^{\infty} \binom{m+s}{m} \gamma^{s+1} (1-\gamma)^m : \frac{\left(a^\dagger a\right)^m}{m!} e^{-a^\dagger a} :, \tag{3.39}$$

using the negative binomial theorem in Eq. (3.23) to obtain

$$\mathtt{tr}\rho_s = \gamma^{s+1} \sum_{n=0}^{\infty} \frac{(n+s)!}{n!s!} (1-\gamma)^n = 1. \tag{3.40}$$

Using the identity $a|m\rangle = \sqrt{m}|m-1\rangle$, we have

$$\begin{aligned}\rho_s &= \frac{\gamma^{s+1}}{(1-\gamma)^s\, s!} a^s \sum_{m=0}^{\infty} (1-\gamma)^m \, |m\rangle\langle m| a^{\dagger s} \\ &= \frac{\gamma}{s! n_c^s} a^s e^{a^\dagger a \ln(1-\gamma)} a^{\dagger s} \\ &= \frac{1}{s! n_c^s} a^s \rho_c a^{\dagger s} \quad \left(n_c = \frac{1-\gamma}{\gamma}\right),\end{aligned} \tag{3.41}$$

where

$$\rho_c = \gamma e^{a^\dagger a \ln(1-\gamma)} = \frac{\gamma}{1-\gamma} \vdots e^{-\frac{\gamma}{1-\gamma} a^\dagger a} \vdots \tag{3.42}$$

is the chaotic field. Obviously, Eq. (3.41) indicates that the negative binomial state light field is equivalent to the photon-subtracted chaotic light field. Using the operator identity

$$e^{-\lambda} \vdots e^{\left(1-e^{-\lambda}\right) a^\dagger a} \vdots = e^{\lambda a^\dagger a}, \tag{3.43}$$

we have

$$\begin{aligned}\sum_{s=0}^{\infty} \rho_s &= \sum_{m=0}^{\infty} \left[\sum_{s=0}^{\infty} \binom{s+m}{s} \gamma^{s+1} \right] (1-\gamma)^m : \frac{\left(a^\dagger a\right)^m}{m!} e^{-a^\dagger a} : \\ &= \gamma \sum_{m=0}^{\infty} (1-\gamma)^{-m-1} (1-\gamma)^m : \frac{\left(a^\dagger a\right)^m}{m!} e^{-a^\dagger a} : \\ &= \gamma (1-\gamma)^{-1} \sum_{m=0}^{\infty} : \frac{\left(a^\dagger a\right)^m}{m!} e^{-a^\dagger a} : \\ &= \frac{\gamma}{1-\gamma},\end{aligned} \tag{3.44}$$

or

$$\begin{aligned}\frac{1-\gamma}{\gamma}\sum_{s=0}^{\infty}\rho_s &= (1-\gamma)\sum_{s=0}^{\infty}\frac{1}{s!n_c^s}a^s{:}\frac{1}{1-\gamma}\mathrm{e}^{\left(1-\frac{1}{1-\gamma}\right)a^\dagger a}{:}a^{\dagger s}\\ &= \sum_{s=0}^{\infty}\frac{1}{s!n_c^s}{:}a^s\mathrm{e}^{\left(1-\frac{1}{1-\gamma}\right)a^\dagger a}a^{\dagger s}{:}\\ &= {:}\mathrm{e}^{\frac{\gamma}{1-\gamma}a^\dagger a}\mathrm{e}^{\left(1-\frac{1}{1-\gamma}\right)a^\dagger a}{:} = 1. \end{aligned} \tag{3.45}$$

Proof of $1=\frac{1-\gamma}{\gamma}\sum_{s=0}^{\infty}\rho_s$ as follows:

$$\begin{aligned}\frac{1-\gamma}{\gamma}\sum_{s=0}^{\infty}\rho_s &= \sum_{s=0}^{\infty}\frac{1}{s!n_c^{s-1}}a^s\rho_c a^{\dagger s}\\ &= \sum_{s=0}^{\infty}\frac{1}{s!n_c^s}a^s{:}\mathrm{e}^{\frac{-\gamma}{1-\gamma}a^\dagger a}{:}a^{\dagger s} = 1. \end{aligned} \tag{3.46}$$

Further, using the density operator $\rho_c=(1-\mathrm{e}^f)\mathrm{e}^{fa^\dagger a}$ and the completeness relation of coherent states, as well as the IWOP method, we obtain the normal ordering product of the state ρ_s as

$$\begin{aligned}\rho_s &= \frac{1}{s!n_c^s}a^s\rho_c a^{\dagger s} = \frac{1-\mathrm{e}^f}{s!n_c^s}a^s\mathrm{e}^{fa^\dagger a}a^{\dagger s}\\ &= \frac{1-\mathrm{e}^f}{s!n_c^s}\int\frac{d^2z}{\pi}a^s\mathrm{e}^{fa^\dagger a}\left|z\right\rangle\left\langle z\right| a^{\dagger s}\\ &= \frac{1-\mathrm{e}^f}{s!n_c^s}\int\frac{d^2z}{\pi}\mathrm{e}^{\frac{-|z|^2}{2}}a^s\mathrm{e}^{fa^\dagger a}\mathrm{e}^{za^\dagger}\mathrm{e}^{-fa^\dagger a}\left|0\right\rangle\left\langle z\right| z^{*s}\\ &= \frac{1-\mathrm{e}^f}{s!n_c^s}\int\frac{d^2z}{\pi}\mathrm{e}^{\frac{-|z|^2}{2}}a^s\mathrm{e}^{za^\dagger\mathrm{e}^f}\left|0\right\rangle\left\langle z\right| z^{*s}\\ &= \frac{1-\mathrm{e}^f}{s!n_c^s}\int\frac{d^2z}{\pi}\left(z\mathrm{e}^f\right)^s z^{*s}:\mathrm{e}^{-|z|^2+za^\dagger\mathrm{e}^f+z^*a-a^\dagger a}:\\ &= \frac{1-\mathrm{e}^f}{s!n_c^s}\mathrm{e}^{fs}:\sum_{l=0}^{\infty}\mathrm{e}^{(\mathrm{e}^f-1)a^\dagger a}\frac{(n!)^2\left(a^\dagger a\mathrm{e}^f\right)^{n-l}}{l!\,[(n-l)!]^2}:\\ &= \left(1-\mathrm{e}^f\right)^{s+1}:\mathrm{e}^{(\mathrm{e}^f-1)a^\dagger a}\mathrm{L}_s\left(-a^\dagger a\mathrm{e}^f\right):\\ &= \gamma^{s+1}:\mathrm{e}^{-\gamma a^\dagger a}\mathrm{L}_s\left[(\gamma-1)a^\dagger a\right]:, \end{aligned} \tag{3.47}$$

where we have used the normal ordering product $\left|0\right\rangle\left\langle 0\right| =: \mathrm{e}^{-a^\dagger a}:$ and the series expansions of Laguerre polynomials, i.e.,

$$\mathrm{L}_s(x) = \sum_{l=0}^{s} \frac{(-x)^l \, n!}{(l!)^2 (n-l)!}. \tag{3.48}$$

Using Eq. (3.47), it's not difficult to verify

$$\frac{1-\gamma}{\gamma} \sum_{s=0}^{\infty} \rho_s = (1-\gamma) \sum_{s=0}^{\infty} \gamma^s : \mathrm{e}^{-\gamma a^\dagger a} \mathrm{L}_s \left[(\gamma - 1) a^\dagger a \right] : \; = 1. \tag{3.49}$$

Next, we examine the thermal vacuum state corresponding to the negative binomial light field [1]. Using the IWOP method and the coherent states $|z\rangle$, the operator $a^s \mathrm{e}^{\lambda a^\dagger a} a^{\dagger s}$ can be rewritten as the integral form:

$$\begin{aligned} a^s \mathrm{e}^{\lambda a^\dagger a} a^{\dagger s} &= \int \frac{\mathrm{d}^2 z}{\pi} a^s : \exp(-|z|^2 + z^* a^\dagger \mathrm{e}^{\lambda/2} + z a \mathrm{e}^{\lambda/2} - a^\dagger a) : a^{\dagger s} \\ &= \int \frac{\mathrm{d}^2 z}{\pi} \mathrm{e}^{-|z|^2} a^s \left\| z^* \mathrm{e}^{\lambda/2} \right\rangle \left\langle z \mathrm{e}^{\lambda/2} \right\| a^{\dagger s}, \end{aligned} \tag{3.50}$$

where $\left\| z^* \mathrm{e}^{\lambda/2} \right\rangle$ is an unnormalized coherent state. Noting that $\left\langle \tilde{0} \, | \tilde{z} \right\rangle = \mathrm{e}^{-|z|^2/2}$, thus Eq. (3.50) becomes

$$\begin{aligned} a^s \mathrm{e}^{\lambda a^\dagger a} a^{\dagger s} &= \int \frac{\mathrm{d}^2 z}{\pi} z^s z^{*s} \mathrm{e}^{\lambda s} \mathrm{e}^{z^* a^\dagger \mathrm{e}^{\lambda/2}} |0\rangle \langle 0| \, \mathrm{e}^{z a \mathrm{e}^{\lambda/2}} \left\langle \tilde{z} | \tilde{0} \right\rangle \left\langle \tilde{0} | \tilde{z} \right\rangle \\ &= \int \frac{\mathrm{d}^2 z}{\pi} \langle \tilde{z} | \, z^s z^{*s} \mathrm{e}^{\lambda s} \mathrm{e}^{z^* a^\dagger \mathrm{e}^{\lambda/2}} \left| 0, \tilde{0} \right\rangle \left\langle 0, \tilde{0} \right| \mathrm{e}^{z a \mathrm{e}^{\lambda/2}} \, |\tilde{z}\rangle \\ &= \int \frac{\mathrm{d}^2 z}{\pi} \langle \tilde{z} | \, \tilde{a}^{\dagger s} \mathrm{e}^{\lambda s} \mathrm{e}^{\tilde{a}^\dagger a^\dagger \mathrm{e}^{\lambda/2}} \left| 0, \tilde{0} \right\rangle \left\langle 0, \tilde{0} \right| \mathrm{e}^{\tilde{a} a \mathrm{e}^{\lambda/2}} \tilde{a}^s \, |\tilde{z}\rangle \\ &= \mathrm{e}^{\lambda s} \, \widetilde{\mathrm{tr}} \left(\int \frac{\mathrm{d}^2 z}{\pi} \tilde{a}^{\dagger s} \mathrm{e}^{\tilde{a}^\dagger a^\dagger \mathrm{e}^{\lambda/2}} \left| 0, \tilde{0} \right\rangle \left\langle 0, \tilde{0} \right| \mathrm{e}^{\tilde{a} a \mathrm{e}^{\lambda/2}} \tilde{a}^s \, |\tilde{z}\rangle \langle \tilde{z}| \right) \\ &= (1-\gamma)^s \, \widetilde{\mathrm{tr}} \left(\tilde{a}^{\dagger s} \mathrm{e}^{\tilde{a}^\dagger a^\dagger \mathrm{e}^{\lambda/2}} \left| 0, \tilde{0} \right\rangle \left\langle 0, \tilde{0} \right| \mathrm{e}^{\tilde{a} a \mathrm{e}^{\lambda/2}} \tilde{a}^s \right). \end{aligned} \tag{3.51}$$

Substituting Eq. (3.51) into Eq. (3.41) leads to

$$\begin{aligned} \rho_s &= \frac{\gamma}{s! n_c^s} a^s \mathrm{e}^{\lambda a^\dagger a} a^{\dagger s} \\ &= \frac{\gamma^{s+1}}{s!} \widetilde{\mathrm{tr}} \left(\tilde{a}^{\dagger s} \mathrm{e}^{\tilde{a}^\dagger a^\dagger \mathrm{e}^{\lambda/2}} \left| 0, \tilde{0} \right\rangle \left\langle 0, \tilde{0} \right| \mathrm{e}^{\tilde{a} a \mathrm{e}^{\lambda/2}} \tilde{a}^s \right). \end{aligned} \tag{3.52}$$

Comparing with Eq. (3.5), we find that the thermal vacuum state corresponding to the negative binomial light field is expressed as

$$|\psi(\beta)\rangle_s = \sqrt{\frac{\gamma^{s+1}}{s!}}\tilde{a}^{\dagger s} e^{\tilde{a}^\dagger a^\dagger \sqrt{1-\gamma}} \left|0, \tilde{0}\right\rangle, \tag{3.53}$$

which is a fictious mode excitation on the thermal vacuum state belonging to a chaotic light field, which is a new perspective on the negative binomial state light field. Compared to the state $|\psi(\beta)\rangle_s$, the fictious mode excitation is equivalent to the annihilation of real modes, which is self consistent with $a^s e^{\lambda a^\dagger a} a^{\dagger s}$. Or, Eq. (3.53) can be rewritten as

$$\begin{aligned} |\psi(\beta)\rangle_s &= \sqrt{\gamma^{s+1}} \sum_{n=0}^{\infty} \frac{\left(\tilde{a}^\dagger a^\dagger \sqrt{1-\gamma}\right)^n}{n!} |0, \tilde{s}\rangle \\ &= \sum_{n=0}^{\infty} \sqrt{\binom{n+s}{n} \gamma^{s+1} (1-\gamma)^n} \, |n, \tilde{s}+\tilde{n}\rangle, \end{aligned} \tag{3.54}$$

where $|n, \tilde{s}+\tilde{n}\rangle = \tilde{a}^{\dagger s+n} a^{\dagger n} \left|0, \tilde{0}\right\rangle / \sqrt{n!\,(n+s)!}$. Comparing Eqs. (3.34) and (3.54), the factor $\sqrt{\binom{n+s}{n} \gamma^{s+1} (1-\gamma)^n}$ in Eq. (3.54) is just the half power of the coefficient of the negative binomial distribution, so $|\psi(\beta)\rangle_s$ is exactly a negative binomial pure state. So, we say that the thermal vacuum state corresponding to the negative binomial light field in Eq. (3.35) is just the pure state $|\psi(\beta)\rangle_s$ in Eq. (3.54).

Further, using $\mathrm{tr}\rho_s = 1$, we know

$$\begin{aligned} \mathrm{tr}\widetilde{\mathrm{tr}}\left(|\psi(\beta)\rangle_{ss}\langle\psi(\beta)|\right) &= \mathrm{Tr}\left(|\psi(\beta)\rangle_{ss}\langle\psi(\beta)|\right) \\ &= {}_s\langle\psi(\beta)|\psi(\beta)\rangle_s = 1, \end{aligned} \tag{3.55}$$

which can also be directly proven using Eq. (3.53), that is

$$\begin{aligned} {}_s\langle\psi(\beta)|\psi(\beta)\rangle_s &= \frac{\gamma^{s+1}}{s!}\left\langle\tilde{0}, 0\right| \tilde{a}^s e^{\tilde{a}a\sqrt{1-\gamma}} \int\int \frac{d^2z_1 d^2z_2}{\pi^2} |z_1 z_2\rangle\langle z_1 z_2| \\ &\quad \tilde{a}^{\dagger s} e^{\tilde{a}^\dagger a^\dagger \sqrt{1-\gamma}} \left|0, \tilde{0}\right\rangle \\ &= \frac{\gamma^{s+1}}{s!} \int\int \frac{d^2z_1 d^2z_2}{\pi^2} |z_2|^{2s} e^{z_1 z_2\sqrt{1-\gamma} + z_1^* z_2^* \sqrt{1-\gamma} - |z_2|^2 - |z_1|^2} \\ &= \frac{\gamma^{s+1}}{s!} \int \frac{d^2 z_2}{\pi} |z_2|^{2s} e^{-\gamma|z_2|^2} = 1, \end{aligned} \tag{3.56}$$

which shows that the pure state $|\psi(\beta)\rangle_s$ is normalized.

Besides, using the pure state $|\psi(\beta)\rangle_s$, it is easy to calculate the statistical functions of the negative binomial state ρ_s. For example, using Eq. (3.53), we obtain

$$a\left|\psi\left(\beta\right)\right\rangle_{s}=\sqrt{\frac{\gamma^{s+1}}{s!}}\sqrt{1-\gamma}\tilde{a}^{\dagger s+1}\mathrm{e}^{\tilde{a}^{\dagger}a^{\dagger}\sqrt{1-\gamma}}\left|0,\tilde{0}\right\rangle$$
$$=\sqrt{1-\gamma}\sqrt{\frac{s+1}{\gamma}}\left|\psi\left(\beta\right)\right\rangle_{s+1}, \tag{3.57}$$

it follows that the averager photon number of the state ρ_s is

$$_{s}\left\langle\psi\left(\beta\right)\right|a^{\dagger}a\left|\psi\left(\beta\right)\right\rangle_{s}=\left(1-\gamma\right)\frac{s+1}{\gamma}\,{}_{s+1}\left\langle\psi\left(\beta\right)\right|\psi\left(\beta\right)\rangle_{s+1}$$
$$=\left(1-\gamma\right)\frac{s+1}{\gamma}=\left(s+1\right)n_{c}. \tag{3.58}$$

Noting the identities

$$a^{2}\left|\psi\left(\beta\right)\right\rangle_{s}=\sqrt{\frac{\gamma^{s+1}}{s!}}\left(1-\gamma\right)\tilde{a}^{\dagger s+2}\mathrm{e}^{\tilde{a}^{\dagger}a^{\dagger}\sqrt{1-\gamma}}\left|0,\tilde{0}\right\rangle$$
$$=\frac{\left(1-\gamma\right)}{\gamma}\sqrt{\left(s+1\right)\left(s+2\right)}\left|\psi\left(\beta\right)\right\rangle_{s+2}, \tag{3.59}$$

and

$$_{s}\left\langle\psi\left(\beta\right)\right|a^{\dagger 2}a^{2}\left|\psi\left(\beta\right)\right\rangle_{s}=\frac{\left(1-\gamma\right)^{2}}{\gamma^{2}}\left(s+1\right)\left(s+2\right)$$
$$=\left(s+1\right)\left(s+2\right)n_{c}^{2}, \tag{3.60}$$

thus the photon number fluctuation of the state ρ_s reads

$$_{s}\left\langle\psi\left(\beta\right)\right|\left(a^{\dagger}a\right)^{2}\left|\psi\left(\beta\right)\right\rangle_{s}-\left[{}_{s}\left\langle\psi\left(\beta\right)\right|a^{\dagger}a\left|\psi\left(\beta\right)\right\rangle_{s}\right]^{2}=\left(s+1\right)\left(n_{c}+1\right)n_{c} \tag{3.61}$$

and its second-order coherence is

$$\frac{_{s}\left\langle\psi\left(\beta\right)\right|a^{\dagger 2}a^{2}\left|\psi\left(\beta\right)\right\rangle_{s}}{\left({}_{s}\left\langle\psi\left(\beta\right)\right|a^{\dagger}a\left|\psi\left(\beta\right)\right\rangle_{s}\right)^{2}}=\frac{s+2}{s+1}>1. \tag{3.62}$$

Earlier, we discussed the expected value theorem for the binomial thermal vacuum states [1]. So, at the end of this section, we naturally consider the expected value theorem of the negative binomial thermal vacuum states. Using Eq. (3.54), we have

$$a^{p}\left|\psi\left(\beta\right)\right\rangle_{s}=\left(\sqrt{\frac{1-\gamma}{\gamma}}\right)^{p}\sqrt{\left(s+1\right)\left(s+2\right)\cdots\left(s+p\right)}\left|\psi\left(\beta\right)\right\rangle_{s+p}, \tag{3.63}$$

so we have

$$ {}_s\langle\psi(\beta)|\, a^{\dagger p} a^p \,|\psi(\beta)\rangle_s = \left(\frac{1-\gamma}{\gamma}\right)^p (s+1)(s+2)\cdots(s+p) = \frac{(s+p)!}{s!} n_c^p. \tag{3.64} $$

Combining Eqs. (3.58) and (3.64) to show that

$$ {}_s\langle\psi(\beta)|\, a^{\dagger p} a^p \,|\psi(\beta)\rangle_s = \frac{(s+p)!}{(s+1)^p\, s!}\left({}_s\langle\psi(\beta)|\, a^{\dagger} a \,|\psi(\beta)\rangle_s\right)^p, \tag{3.65} $$

which is the expected value theorem of the negative binomial states $|\psi(\beta)\rangle_s$. Besides, using Eq. (3.54) to obtain

$$ \begin{aligned} {}_s\langle\psi(\beta)|\, a^{\dagger p} a^p \,|\psi(\beta)\rangle_s &= \sum_{n'=0}^{\infty}\sqrt{\gamma^{s+1}(1-\gamma)^{n'}\binom{n'+s}{s}}\left\langle n', \tilde{s}+\tilde{n}'\right|a^{\dagger p}a^p \\ &\times \sum_{n''=0}^{\infty}\sqrt{\gamma^{s+1}(1-\gamma)^{n''}\binom{n''+s}{s}}\left|n'', \tilde{s}+\tilde{n}''\right\rangle \\ &= \gamma^{s+1}\sum_{n',n''=0}^{\infty}\sqrt{(1-\gamma)^{n'+n''}\binom{n'+s}{s}\binom{n''+s}{s}} \\ &\times \left\langle n'\right|a^{\dagger p}a^p\left|n''\right\rangle \delta_{n'n''} \\ &= \gamma^{s+1}\sum_{n=0}^{\infty}(1-\gamma)^n\binom{n+s}{s}\frac{n!}{(n-p)!}, \end{aligned} \tag{3.66} $$

thus comparing Eqs. (3.64) and (3.66), we get

$$ \gamma^{s+1}\sum_{n=0}^{\infty}(1-\gamma)^n\binom{n+s}{s}\frac{n!}{(n-p)!} = \left(\frac{1-\gamma}{\gamma}\right)^p (s+1)(s+2)\cdots(s+p). \tag{3.67} $$

3.3 Thermal Vacuum State Corresponding to Squeezed Chaotic Light Field

The density operator for the squeezed chaotic light field is

$$ \rho_s = \left(1-e^{\lambda}\right) S(r)\, e^{\lambda a^{\dagger}a} S^{-1}(r) = S(r)\,\rho_c S^{-1}(r), \; \lambda = -\frac{\hbar\omega}{k_B T}, \tag{3.68} $$

where $S(r)$ is the single-mode squeezed operator

$$ S(r) = \exp\left[\frac{r}{2}\left(a^{\dagger 2}-a^2\right)\right] = \exp\left[-\frac{i}{2}(QP+PQ)\,r\right] \tag{3.69} $$

yielding the corresponding squeezed transformations

$$SPS^{-1} = \mathrm{e}^{r} P, \quad SQS^{-1} = \mathrm{e}^{-r} Q \tag{3.70}$$

with

$$Q = \frac{a^{\dagger} + a}{\sqrt{2}}, \quad P = \frac{a - a^{\dagger}}{\sqrt{2}\mathrm{i}}. \tag{3.71}$$

To obtain the thermal vacuum state corresponding to the state ρ_s, we first use the IWOP method to derive the normal ordering product of the density operator ρ_s. Using the formula that converts any operator A into its Weyl ordering product, i.e.,

$$A = 2 \int \frac{\mathrm{d}^2\beta}{\pi} \vdots \langle -\beta| A |\beta\rangle \exp\left[2\left(\beta^* a - \beta a^{\dagger} + a^{\dagger} a\right)\right] \vdots, \tag{3.72}$$

where $\vdots\ \vdots$ denotes the Weyl ordering symbol, $|\beta\rangle = \exp\left(-|\beta|^2/2 + \beta a^{\dagger}\right)|0\rangle$ is the coherent state, we obtain the Weyl ordering of the operator $\mathrm{e}^{\lambda a^{\dagger} a}$ as

$$\begin{aligned}
\mathrm{e}^{\lambda a^{\dagger} a} &= 2 \int \frac{\mathrm{d}^2\beta}{\pi} \langle -\beta| : \exp\left[\left(\mathrm{e}^{\lambda} - 1\right) a^{\dagger} a\right] : |\beta\rangle \vdots \exp\left[2\left(\beta^* a - \beta a^{\dagger} + a^{\dagger} a\right)\right] \vdots \\
&= 2 \int \frac{\mathrm{d}^2\beta}{\pi} \vdots \exp\left[-\left(\mathrm{e}^{\lambda} + 1\right)|\beta|^2 + 2\left(\beta^* a - \beta a^{\dagger} + a^{\dagger} a\right)\right] \vdots \\
&= \frac{2}{\mathrm{e}^{\lambda} - 1} \vdots \exp\left(2 \frac{\mathrm{e}^{\lambda} - 1}{\mathrm{e}^{\lambda} + 1} a^{\dagger} a\right) \vdots \\
&= \frac{2}{\mathrm{e}^{\lambda} - 1} \vdots \exp\left[\frac{\mathrm{e}^{\lambda} - 1}{\mathrm{e}^{\lambda} + 1}\left(P^2 + Q^2\right)\right] \vdots.
\end{aligned} \tag{3.73}$$

Considering that Weyl ordering operator has invariance under the similarity transformation, so using Eqs. (3.70) and (3.73), we get

$$\begin{aligned}
\rho_s &= \left(1 - \mathrm{e}^{\lambda}\right) S(r) \mathrm{e}^{\lambda a^{\dagger} a} S^{-1}(r) \\
&= \frac{2\left(1 - \mathrm{e}^{\lambda}\right)}{\mathrm{e}^{\lambda} + 1} \vdots \exp\left\{\frac{\mathrm{e}^{\lambda} - 1}{\mathrm{e}^{\lambda} + 1}\left[\mathrm{e}^{2r} P^2 + \mathrm{e}^{-2r} Q^2\right]\right\} \vdots.
\end{aligned} \tag{3.74}$$

If the operators P, Q in Eq. (3.74) are replaced by the classical c numbers p, q, respectively, thus we obtain the classical Weyl correspondence of ρ_s as

$$\frac{2\left(1 - \mathrm{e}^{\lambda}\right)}{\mathrm{e}^{\lambda} + 1} \exp\left[\frac{\mathrm{e}^{\lambda} - 1}{\mathrm{e}^{\lambda} + 1}\left(\mathrm{e}^{2r} p^2 + \mathrm{e}^{-2r} q^2\right)\right] \equiv h(q, p), \tag{3.75}$$

which means that the classical correspondence of the Weyl ordering operator $\vdots h(Q,P)\vdots$ can be obtained by making the substitutions $Q\rightarrow q,\ P\rightarrow p$ in $\vdots h(Q,P)\vdots$. So, using the Weyl correspondence rule [8]

$$\rho_s=\int\int_{-\infty}^{\infty}\mathrm{d}p\mathrm{d}qh(q,p)\,\Delta(q,p), \tag{3.76}$$

where $\Delta(q,p)$ is the Wigner operator, and its normal ordering product is

$$\Delta(q,p)=\frac{1}{\pi}:\mathrm{e}^{-(q-Q)^2-(p-P)^2}:, \tag{3.77}$$

and substituting Eqs. (3.75) and (3.77) into Eq. (3.76), we obtain the normal ordering product of ρ_s as

$$\begin{aligned}\rho_s&=\frac{2\left(1-\mathrm{e}^{\lambda}\right)}{\pi\left(\mathrm{e}^{\lambda}+1\right)}\int\int_{-\infty}^{\infty}\mathrm{d}p\mathrm{d}q:\mathrm{e}^{-(q-Q)^2-(p-P)^2}\\&\times:\exp\left[\frac{\mathrm{e}^{\lambda}-1}{\mathrm{e}^{\lambda}+1}\left(\mathrm{e}^{2r}p^2+\mathrm{e}^{-2r}q^2\right)\right]\\&=\frac{1}{\tau_1\tau_2}:\exp\left(-\frac{Q^2}{2\tau_1^2}-\frac{P^2}{2\tau_2^2}\right):,\end{aligned} \tag{3.78}$$

where

$$2\tau_1^2=(2\bar{n}+1)\,\mathrm{e}^{2r}+1,\ 2\tau_2^2=(2\bar{n}+1)\,\mathrm{e}^{-2r}+1,\ \bar{n}=\left(\mathrm{e}^{-\lambda}-1\right)^{-1}. \tag{3.79}$$

Using Eq. (3.78), thus the state ρ_s can be rewritten as

$$\begin{aligned}\rho_s&=\frac{1}{\tau_1\tau_2}\mathrm{e}^{-\frac{a^{\dagger 2}}{4}\left(\frac{1}{\tau_1^2}-\frac{1}{\tau_2^2}\right)}:\exp\left[a^{\dagger}a\left(-\frac{1}{2\tau_1^2}-\frac{1}{2\tau_2^2}\right)\right]:\mathrm{e}^{-\frac{a^2}{4}\left(\frac{1}{\tau_1^2}-\frac{1}{\tau_2^2}\right)}\\&=\frac{1}{\tau_1\tau_2}\mathrm{e}^{-\frac{a^{\dagger 2}}{4}\left(\frac{1}{\tau_1^2}-\frac{1}{\tau_2^2}\right)}\exp\left[a^{\dagger}a\ln\left(1-\frac{1}{2\tau_1^2}-\frac{1}{2\tau_2^2}\right)\right]\mathrm{e}^{-\frac{a^2}{4}\left(\frac{1}{\tau_1^2}-\frac{1}{\tau_2^2}\right)}.\end{aligned} \tag{3.80}$$

Introducing the coherent states of real mode

$$|z\rangle=\exp\left(-\frac{1}{2}|z|^2+za^{\dagger}\right)|0\rangle=\exp\left(-\frac{1}{2}|z|^2\right)\|z\rangle,\ \|z\rangle=\exp\left(za^{\dagger}\right)|0\rangle, \tag{3.81}$$

thus Eq. (3.80) can be expressed as

$$\rho_s = \frac{1}{\tau_1\tau_2} e^{-\frac{a^{\dagger 2}}{4}\left(\frac{1}{\tau_1^2}-\frac{1}{\tau_2^2}\right)} \int \frac{d^2 z}{\pi} : \exp\left(-|z|^2 - a^\dagger a\right)$$
$$\times \exp\left[\left(z^* a^\dagger + z a\right)\sqrt{1 - \frac{1}{2\tau_1^2} - \frac{1}{2\tau_2^2}}\right] : e^{-\frac{a^2}{4}\left(\frac{1}{\tau_1^2}-\frac{1}{\tau_2^2}\right)}. \quad (3.82)$$

Noting that the normal ordering product of vacuum projection operator, i.e., $: e^{-a^\dagger a} : = |0\rangle\langle 0|$ and using Eq. (3.81), thus Eq. (3.82) becomes

$$\rho_s = \frac{1}{\tau_1\tau_2} \int \frac{d^2 z}{\pi} e^{-|z|^2} e^{-\frac{a^{\dagger 2}}{4}\left(\frac{1}{\tau_1^2}-\frac{1}{\tau_2^2}\right)}$$
$$\left\| z^* \sqrt{1 - \frac{1}{2\tau_1^2} - \frac{1}{2\tau_2^2}} \right\rangle \left\langle z \sqrt{1 - \frac{1}{2\tau_1^2} - \frac{1}{2\tau_2^2}} \right\| e^{-\frac{a^2}{4}\left(\frac{1}{\tau_1^2}-\frac{1}{\tau_2^2}\right)}. \quad (3.83)$$

Besides, using the identities

$$\int \frac{d^2 z}{\pi} |\tilde{z}\rangle\langle\tilde{z}| = 1, \ \left\langle \tilde{0} \middle| \tilde{z} \right\rangle = e^{-|z|^2/2}, \ \tilde{a}|\tilde{z}\rangle = z|\tilde{z}\rangle, \quad (3.84)$$

where $|\tilde{z}\rangle$ is the coherent state of fictitious mode, thus Eq. (3.84) can be written as the following form:

$$\rho_s = \frac{1}{\tau_1\tau_2} \int \frac{d^2 z}{\pi} \langle\tilde{z}| e^{\tilde{a}^\dagger a^\dagger \sqrt{1-\frac{1}{2\tau_1^2}-\frac{1}{2\tau_2^2}}} e^{-\frac{a^{\dagger 2}}{4}\left(\frac{1}{\tau_1^2}-\frac{1}{\tau_2^2}\right)} \left|0\tilde{0}\right\rangle\left\langle 0\tilde{0}\right|$$
$$\times e^{-\frac{a^2}{4}\left(\frac{1}{\tau_1^2}-\frac{1}{\tau_2^2}\right)} e^{\tilde{a} a \sqrt{1-\frac{1}{2\tau_1^2}-\frac{1}{2\tau_2^2}}} |\tilde{z}\rangle$$
$$= \frac{1}{\tau_1\tau_2} \widetilde{\mathrm{tr}} \left[e^{\tilde{a}^\dagger a^\dagger \sqrt{1-\frac{1}{2\tau_1^2}-\frac{1}{2\tau_2^2}}} e^{-\frac{a^{\dagger 2}}{4}\left(\frac{1}{\tau_1^2}-\frac{1}{\tau_2^2}\right)} \left|0\tilde{0}\right\rangle\left\langle 0\tilde{0}\right| \right.$$
$$\left. e^{-\frac{a^2}{4}\left(\frac{1}{\tau_1^2}-\frac{1}{\tau_2^2}\right)} e^{\tilde{a} a \sqrt{1-\frac{1}{2\tau_1^2}-\frac{1}{2\tau_2^2}}} \right]$$
$$\equiv \widetilde{\mathrm{tr}} \left[|\psi(\beta)\rangle_{ss} \langle\psi(\beta)| \right], \quad (3.85)$$

it follows that

$$|\psi(\beta)\rangle_s = \sqrt{\frac{1}{\tau_1\tau_2}} \exp\left[\sqrt{1 - \frac{1}{2\tau_1^2} - \frac{1}{2\tau_2^2}} \tilde{a}^\dagger a^\dagger - \frac{1}{4}\left(\frac{1}{\tau_1^2} - \frac{1}{\tau_2^2}\right) a^{\dagger 2} \right] \left|0\tilde{0}\right\rangle, \quad (3.86)$$

which is the thermal vacuum state belonging to the squeezed chaotic light field [2]. Exactly, the state $|\psi(\beta)\rangle_s$ in Eq. (3.86) is a single- and two-mode combinatorial squeezed state in the real-fictitious space. In particular, when $r = 0$, no squeezing case, it follows from Eq. (3.80) that we get $\tau_1^2 = \tau_2^2 = \bar{n} + 1$, thus the state $|\psi(\beta)\rangle_s$ reduces to the state $|0(\beta)\rangle$, i.e.,

$$\frac{1}{\sqrt{\bar{n}+1}}e^{\tilde{a}^{\dagger}a^{\dagger}\sqrt{\frac{\bar{n}}{\bar{n}+1}}}\left|0\tilde{0}\right\rangle = \sqrt{1-e^{\lambda}}\exp\left(e^{\frac{\lambda}{2}}\tilde{a}^{\dagger}a^{\dagger}\right)\left|0\tilde{0}\right\rangle = \left|0\left(\beta\right)\right\rangle, \tag{3.87}$$

which means that the squeezed chaotic state ρ_s becomes the chaotic state ρ_c, as expected.

The thermal vacuum state $\left|\psi\left(\beta\right)\right\rangle_s$ is normalized since

$$\begin{aligned}
{}_s\left\langle\psi\left(\beta\right)\right|\left.\psi\left(\beta\right)\right\rangle_s &= \frac{1}{\tau_1\tau_2}\left\langle 0\tilde{0}\right| e^{\tilde{a}a\sqrt{1-\frac{1}{2\tau_1^2}-\frac{1}{2\tau_2^2}-\frac{\tau_2^2-\tau_1^2}{4\tau_1^2\tau_2^2}a^2}} \\
&\quad\times\int\frac{d^2z_1d^2z_2}{\pi^2}\left|z_1\tilde{z}_2\right\rangle\left\langle z_1\tilde{z}_2\right| e^{\tilde{a}^{\dagger}a^{\dagger}\sqrt{1-\frac{1}{2\tau_1^2}-\frac{1}{2\tau_2^2}-\frac{\tau_2^2-\tau_1^2}{4\tau_1^2\tau_2^2}a^{\dagger 2}}}\left|0\tilde{0}\right\rangle \\
&= \frac{1}{\tau_1\tau_2}\int\frac{d^2z_1d^2z_2}{\pi^2}\exp\left[-\left|z_1\right|^2-\left|z_2\right|^2\right. \\
&\quad\left.+\sqrt{1-\frac{1}{2\tau_1^2}-\frac{1}{2\tau_2^2}}\left(z_1^*z_2^*+z_1z_2\right)-\frac{\tau_2^2-\tau_1^2}{4\tau_1^2\tau_2^2}\left(z_1^2+z_1^{*2}\right)\right] \\
&= \frac{1}{\tau_1\tau_2}\int\frac{d^2z_1}{\pi}\exp\left[-\frac{\tau_1^2+\tau_2^2}{2\tau_1^2\tau_2^2}\left|z_1\right|^2-\frac{\tau_2^2-\tau_1^2}{4\tau_1^2\tau_2^2}\left(z_1^2+z_1^{*2}\right)\right] \\
&= 1,
\end{aligned} \tag{3.88}$$

which is equivalent to

$${}_s\left\langle\psi\left(\beta\right)\right|\left.\psi\left(\beta\right)\right\rangle_s = \mathtt{tr}\left[\widetilde{\mathtt{tr}}\left|\psi\left(\beta\right)\right\rangle_{ss}\left\langle\psi\left(\beta\right)\right|\right] = \mathtt{Tr}\left(\left|\psi\left(\beta\right)\right\rangle_{ss}\left\langle\psi\left(\beta\right)\right|\right) = 1. \tag{3.89}$$

In fact, the introduction of $\left|\psi\left(\beta\right)\right\rangle_s$ is very convenient for calculating the statistical average. For example, using Eq. (3.83) and the completeness relation of coherent states, i.e., $\pi^{-1}\int d^2z\left|z\right\rangle\left\langle z\right| = 1$, we get the average value of the operator $aa^{\dagger}$ in the state ρ_s, that is

$$\begin{aligned}
{}_s\left\langle\psi\left(\beta\right)\left|aa^{\dagger}\right|\psi\left(\beta\right)\right\rangle_s &= \frac{1}{\tau_1\tau_2}\int\frac{d^2z_1}{\pi}\left|z_1\right|^2\exp\left[-\frac{\tau_1^2+\tau_2^2}{2\tau_1^2\tau_2^2}\left|z_1\right|^2-\frac{\tau_2^2-\tau_1^2}{4\tau_1^2\tau_2^2}\left(z_1^2+z_1^{*2}\right)\right] \\
&= \frac{1}{\tau_1\tau_2}\frac{\partial^2}{\partial f\partial g}\int\frac{d^2z_1}{\pi}\exp\left[-\frac{\tau_1^2+\tau_2^2}{2\tau_1^2\tau_2^2}\left|z_1\right|^2\right. \\
&\quad\left.\left.+fz_1+gz_1^*-\frac{\tau_2^2-\tau_1^2}{4\tau_1^2\tau_2^2}\left(z_1^2+z_1^{*2}\right)\right]\right|_{f,g=0} \\
&= \frac{\partial^2}{\partial f\partial g}\exp\left.\left[\frac{fg}{2}\left(\tau_1^2+\tau_2^2\right)+\frac{1}{4}\left(\tau_1^2-\tau_2^2\right)\left(f^2+g^2\right)\right]\right|_{f,g=0} \\
&= \frac{1}{2}\left(\tau_1^2+\tau_2^2\right) = \left(\bar{n}+\frac{1}{2}\right)\cosh 2r+\frac{1}{2}.
\end{aligned} \tag{3.90}$$

So, the average photon number of the state ρ_s reads

$$_s\left\langle \psi\left(\beta\right)\left|a^{\dagger}a\right|\psi\left(\beta\right)\right\rangle_s = \left(\bar{n}+\frac{1}{2}\right)\cosh 2r - \frac{1}{2}, \tag{3.91}$$

which shows that the average photon number increases with the increase of the squeezing r.

3.4 Thermal Vacuum State Corresponding to Photon-Subtracted Squeezed Chaotic Light Field

What is the thermal vacuum state corresponding to the photon-abstracted squeezed chaotic light field? The density operator of the photon-subtracted squeezed chaotic light field is [9]

$$\rho = C_{s,r}a^s\left\{1-\mathrm{e}^{\lambda}\right\}\left\{S\left(r\right)\mathrm{e}^{\lambda a^{\dagger}a}S^{-1}\left(r\right)\right\}a^{\dagger s}, \tag{3.92}$$

where $C_{s,r}$ is determined by the normalization condition $\mathtt{tr}\rho = 1$, $S\left(r\right)$ is the single-mode squeezed operator. In fact, the density operator in $\{\}$ of Eq. (3.92) is the squeezed chaotic light field, that is

$$\left(1-\mathrm{e}^{\lambda}\right)S\left(r\right)\mathrm{e}^{\lambda a^{\dagger}a}S^{-1}\left(r\right) = S\left(r\right)\rho_c S^{-1}\left(r\right) \equiv \rho_s, \tag{3.93}$$

where

$$\rho_c = \left(1-\mathrm{e}^{\lambda}\right)\mathrm{e}^{\lambda a^{\dagger}a},\ \lambda = -\frac{\hbar\omega}{k_B T}. \tag{3.94}$$

On the other hand, for $r=0$, the value of $C_{s,r=0}$ should make $\rho \to \rho_{su}$, where ρ_{su} is the photon-subtracted chaotic light field, that is

$$\rho_{su} = \frac{\gamma^{s+1}}{s!(1-\gamma)^s}a^s\rho_c a^{\dagger s} \tag{3.95}$$

yielding the normalization condition $\mathtt{tr}\rho_{su} = 1$.

Using Eq. (3.94), the completeness relation of number states and $a\left|n\right\rangle = \sqrt{n}\left|n-1\right\rangle$, thus the state ρ_{su} can be written as

$$\rho_{su} = \sum_{n=0}^{\infty}\frac{(n+s)!}{n!s!}\gamma^{s+1}(1-\gamma)^n\left|n\right\rangle\left\langle n\right|. \tag{3.96}$$

Since the coefficient $(n+s)!\gamma^{s+1}(1-\gamma)^n/(n!s!)$ shows a negative binomial distribution, ρ_{su} is a negative binomial state. In experiment, when s photons of the squeezed chaotic light are absorbed by the atomic system, the output quantum state is the photon-subtracted squeezed chaotic light field. To simplify the calculation of the ensemble mean value of the light field, we further look for the thermal vacuum state corresponding to it. Same as the method in the previous section, it is implemented using IWOP method.

Letting $f = 1/\left(2\tau_1^2\right)$ and $g = 1/\left(2\tau_2^2\right)$, thus using Eq. (3.78), we obtain the normal ordering product of the density operator ρ_s, that is

$$\rho_s = 2\sqrt{fg}\colon \exp\left[\frac{1}{2}(g-f)\left(a^2+a^{\dagger 2}\right)-(f+g)a^{\dagger}a\right]\colon. \tag{3.97}$$

Using the formula that converts any operator into its anti-normal ordering product in Eq. (2.76), we have

$$\begin{aligned}
\rho_s &= \int \frac{d^2\beta}{\pi} \vdots \left\langle -\beta\right| \rho_s \left|\beta\right\rangle \exp\left(|\beta|^2+\beta^* a-\beta a^{\dagger}+a^{\dagger}a\right)\vdots \\
&= 2\sqrt{fg}\int \frac{d^2\beta}{\pi} \vdots \exp\left[-|\beta|^2-\frac{f(\beta-\beta^*)^2}{2}+\frac{g(\beta+\beta^*)^2}{2}+\beta^* a-\beta a^{\dagger}+a^{\dagger}a\right]\vdots \\
&= 2\sqrt{fg}\int \frac{d^2\beta}{\pi} \vdots \exp\left[(f+g-1)|\beta|^2-\beta a^{\dagger}+\beta^* a+\frac{g-f}{2}\left(\beta^2+\beta^{*2}\right)+a^{\dagger}a\right]\vdots \\
&= 2\sqrt{\frac{fg}{D}} \vdots \exp\left[\frac{g-f}{2D}\left(a^2+a^{\dagger 2}\right)+\frac{4fg-g-f}{D}aa^{\dagger}\right]\vdots,
\end{aligned} \tag{3.98}$$

where

$$D = (f+g-1)^2-(g-f)^2 = (2f-1)(2g-1). \tag{3.99}$$

Therefore, the anti-normal ordering product of the density operator ρ in Eq. (3.92) is obtained as

$$\begin{aligned}
\rho &= C_{r,s}\left(1-e^{\lambda}\right)a^s S e^{\lambda a^{\dagger}a} S^{-1} a^{\dagger s} \\
&= 2C_{r,s}\sqrt{\frac{fg}{D}} \vdots a^s \exp\left[\frac{g-f}{2D}\left(a^2+a^{\dagger 2}\right)+\frac{4fg-g-f}{D}aa^{\dagger}\right]a^{\dagger s}\vdots \\
&= 2C_{r,s}\sqrt{\frac{fg}{D}} a^s e^{\frac{g-f}{2D}a^2} \vdots e^{\frac{4fg-g-f}{D}aa^{\dagger}} \vdots e^{\frac{g-f}{2D}a^{\dagger 2}} a^{\dagger s}.
\end{aligned} \tag{3.100}$$

Next, let's determine the normalization coefficient $C_{r,s}$. For convenience, letting

$$e^{ua} e^{\frac{g-f}{2D}a^2} \vdots e^{\frac{4fg-g-f}{D}aa^{\dagger}} \vdots e^{\frac{g-f}{2D}a^{\dagger 2}} e^{va^{\dagger}} \equiv K, \tag{3.101}$$

thus we have

$$\rho = 2C_{r,s}\sqrt{\frac{fg}{D}} \frac{\partial^s}{\partial u^s}\frac{\partial^s}{\partial v^s} K\bigg|_{u,v=0}. \tag{3.102}$$

To find the normalization coefficient, we need to convert K into its normal ordering form. Using the P-representation of the operator K in Eq. (3.101) and the normal ordering form of the completeness relation of coherent states, i.e.,

$$\int \frac{d^2 z}{\pi} |z\rangle \langle z| = \int \frac{d^2 z}{\pi} : \exp\left(-|z|^2 + z a^\dagger + z^* a - a^\dagger a\right) : = 1, \tag{3.103}$$

thus the operator K can be written as

$$\begin{aligned}
K &= \int \frac{d^2 z}{\pi} e^{uz} e^{\frac{g-f}{2D} z^2} e^{\frac{4fg-g-f}{D}|z|^2} e^{\frac{g-f}{2D} z^{*2}} e^{vz^*} \ |z\rangle \langle z| \\
&= \int \frac{d^2 z}{\pi} e^{uz} e^{\frac{g-f}{2D} z^2} e^{\frac{4fg-g-f}{D}|z|^2} e^{\frac{g-f}{2D} z^{*2}} e^{vz^*} : e^{-|z|^2 + z a^\dagger + z^* a - a^\dagger a} : \\
&= \int \frac{d^2 z}{\pi} : \exp\left[\frac{g+f-1}{D} |z|^2 + z\left(a^\dagger + u\right)\right. \\
&\quad \left. + z^* (a + v) + \left(z^2 + z^{*2}\right) \frac{g-f}{2D} - a^\dagger a\right] : \\
&= \sqrt{D} : \exp\left\{-(g+f-1)\left(a^\dagger + u\right)(a+v)\right. \\
&\quad \left. + \frac{g-f}{2}\left[\left(a^\dagger + u\right)^2 + (a+v)^2\right] - a^\dagger a\right\} : .
\end{aligned} \tag{3.104}$$

Further, using the normalization condition and Eq. (3.102), we obtain

$$\begin{aligned}
1 &= \mathrm{tr}\rho = 2C_{r,s}\sqrt{\frac{fg}{D}} \frac{\partial^{2s}}{\partial u^s \partial v^s} \mathrm{tr} K \Big|_{u,v=0} \\
&= 2C_{r,s}\sqrt{fg} \frac{\partial^{2s}}{\partial u^s \partial v^s} \mathrm{tr}\left[\int \frac{d^2 \alpha}{\pi} |\alpha\rangle \langle\alpha| : \exp\left\{-(g+f-1)\left(a^\dagger + u\right)\right.\right. \\
&\quad \left.\left. \times (a+v) + \frac{g-f}{2}\left[\left(a^\dagger + u\right)^2 + (a+v)^2\right] - a^\dagger a\right\} : \right]\Big|_{u,v=0} \\
&= 2C_{r,s}\sqrt{fg} \frac{\partial^{2s}}{\partial u^s \partial v^s} \int \frac{d^2 \alpha}{\pi} \exp\left[-(g+f)|\alpha|^2 + \alpha u\right. \\
&\quad \left. + \alpha^* v + \frac{g-f}{2}\left(\alpha^{*2} + \alpha^2\right) - uv\right]\Big|_{u,v=0} \\
&= C_{r,s} \frac{\partial^{2s}}{\partial u^s \partial v^s} \exp\left[\frac{g-f}{8gf} u^2 + \frac{g-f}{8gf} v^2 + \frac{uv}{4gf}(g+f-4gf)\right]\Big|_{u,v=0} \\
&= C_{r,s}\left(\frac{f-g}{8gf}\right)^s 2^s s! \sum_{n=0}^{[s/2]} \frac{s!}{2^{2n}(n!)^2 (s-2n)!}\left(\frac{g+f-4gf}{f-g}\right)^{s-2n} \\
&= C_{r,s} s! \left(\frac{g+f-4gf}{4gf}\right)^s \mathrm{P}_s(1) = C_{r,s} s! \left(\frac{g+f-4gf}{4gf}\right)^s ,
\end{aligned} \tag{3.105}$$

where $\mathrm{P}_s(1) = 1$ is a s-order Laguerre polynomial. Therefore, the normalized coefficient of the state ρ is

$$C_{r,s}^{-1} = s!\left(\frac{g+f-4gf}{4gf}\right)^s \tag{3.106}$$

Using the operator identity $\mathrm{e}^{\lambda a^\dagger a} =: \exp\left[\left(\mathrm{e}^\lambda - 1\right) a^\dagger a\right]:$ and Eq. (3.100), thus the state ρ can be expressed as

$$\begin{aligned}\rho = &\frac{2\sqrt{fgD}}{s!\,(1-g-f)}\left(\frac{4gf}{g+f-4gf}\right)^s a^s \mathrm{e}^{\frac{g-f}{2D}a^2}\\ &\times : \exp\left[\left(\frac{D}{1-g-f} - 1\right) a^\dagger a\right] : \mathrm{e}^{\frac{g-f}{2D}a^{\dagger 2}} a^{\dagger s}.\end{aligned} \tag{3.107}$$

The normal ordering operator in Eq. (3.107) can be expressed by the integral

$$\begin{aligned}&a^s \mathrm{e}^{\frac{g-f}{2D}a^2} : \exp\left[\left(\frac{D}{1-g-f} - 1\right) a^\dagger a\right] : \mathrm{e}^{\frac{g-f}{2D}a^{\dagger 2}} a^{\dagger s}\\ &= \int \frac{\mathrm{d}^2 z}{\pi} a^s \mathrm{e}^{Aa^2} : \exp\left(-|z|^2 + z^* a^\dagger \sqrt{\frac{D}{1-g-f}}\right.\\ &\left. + za\sqrt{\frac{D}{1-g-f}} - a^\dagger a\right) : \mathrm{e}^{Aa^{\dagger 2}} a^{\dagger s},\end{aligned} \tag{3.108}$$

where $A = \frac{g-f}{2D}$. Using the identity $: \mathrm{e}^{-a^\dagger a} : = |0\rangle\langle 0|$ and the unnormalized coherent state $\|z\rangle = \exp\left(za^\dagger\right)|0\rangle$, thus the right side of Eq. (3.108) becomes

$$\begin{aligned}&\int \frac{\mathrm{d}^2 z}{\pi} \mathrm{e}^{-|z|^2} a^s \mathrm{e}^{Aa^2} \left\| z^* \sqrt{\frac{D}{1-g-f}} \right\rangle \left\langle z^* \sqrt{\frac{D}{1-g-f}} \right\| \mathrm{e}^{Aa^{\dagger 2}} a^{\dagger s}\\ &= \int \frac{\mathrm{d}^2 z}{\pi} \mathrm{e}^{-|z|^2} \left(z^* \sqrt{\frac{D}{1-g-f}}\right)^s \mathrm{e}^{\frac{AD}{1-g-f} z^{*2}}\\ &\times \left\| z^* \sqrt{\frac{D}{1-g-f}} \right\rangle \left\langle z^* \sqrt{\frac{D}{1-g-f}} \right\| \mathrm{e}^{\frac{AD}{1-g-f} z^2} \left(z\sqrt{\frac{D}{1-g-f}}\right)^s.\end{aligned} \tag{3.109}$$

Using Eq. (3.84), thus Eq. (3.107) is changed into

$$\rho = \frac{2\sqrt{fgD}}{s!\,(1-g-f)}\left[\frac{4gfD}{(g+f-4gf)\,(1-g-f)}\right]^{s}$$
$$\times \int \frac{d^2 z}{\pi} \left\langle \tilde{z} \left| \tilde{0} \right.\right\rangle z^{*s} e^{\frac{AD}{1-g-f} z^{*2}} e^{z^* a^\dagger \sqrt{\frac{D}{1-g-f}}} |0\rangle \langle 0| e^{za\sqrt{\frac{D}{1-g-f}}} z^s \langle \tilde{0} | \tilde{z} \rangle e^{\frac{AD}{1-g-f} z^2}$$
$$= \frac{2\sqrt{fgD}}{s!\,(1-g-f)}\left[\frac{4gfD}{(g+f-4gf)\,(1-g-f)}\right]^{s}$$
$$\times \int \frac{d^2 z}{\pi} \langle \tilde{z} | \tilde{a}^{\dagger s} e^{\frac{AD}{1-g-f} \tilde{a}^{\dagger 2}} e^{\tilde{a}^\dagger a^\dagger \sqrt{\frac{D}{1-g-f}}} \left| 0\tilde{0} \right\rangle \left\langle 0\tilde{0} \right| e^{\tilde{a} a \sqrt{\frac{D}{1-g-f}}} \tilde{a}^s e^{\frac{AD}{1-g-f} \tilde{a}^2} |\tilde{z}\rangle$$
$$= \frac{2\sqrt{fgD}}{s!\,(1-g-f)}\left[\frac{4gfD}{(g+f-4gf)\,(1-g-f)}\right]^{s}$$
$$\times \widetilde{\mathrm{tr}} \left(\tilde{a}^{\dagger s} e^{\frac{AD}{1-g-f} \tilde{a}^{\dagger 2}} e^{\tilde{a}^\dagger a^\dagger \sqrt{\frac{D}{1-g-f}}} \left| 0\tilde{0} \right\rangle \left\langle 0\tilde{0} \right| e^{\tilde{a} a \sqrt{\frac{D}{1-g-f}}} \tilde{a}^s e^{\frac{AD}{1-g-f} \tilde{a}^2} \right)$$
$$\equiv \widetilde{\mathrm{tr}} \left(|\psi(\beta)\rangle_{rs\,rs} \langle \psi(\beta)| \right), \tag{3.110}$$

it follows that the thermal vacuum state corresponding to the photon-subtracted squeezed chaotic light is given by

$$|\psi(\beta)\rangle_{rs} = \left\{ \frac{2\sqrt{fgD}}{s!\,(1-g-f)}\left[\frac{4gfD}{(g+f-4gf)\,(1-g-f)}\right]^{s} \right\}^{1/2}$$
$$\times \tilde{a}^{\dagger s} e^{\frac{AD}{1-g-f} \tilde{a}^{\dagger 2}} e^{\tilde{a}^\dagger a^\dagger \sqrt{\frac{D}{1-g-f}}} \left| 0\tilde{0} \right\rangle. \tag{3.111}$$

In fact, Eq. (3.111) shows that the state $|\psi(\beta)\rangle_{rs}$ can be viewed as the photon-added single- and two-combinatorial squeezed state in the expanded space [3].

Especially, for $r = 0$, no squeezing case, using Eqs. (3.9), (3.79) and (3.99) to obtain $\tau_1^2 = \tau_2^2 = \bar{n} + 1$, $f = g = \frac{\gamma}{2}$, $D = (1-\gamma)^2$, thus the state $|\psi(\beta)\rangle_{rs}$ reduces to the thermal vacuum state belonging to a negative binomial light field, that is

$$\sqrt{\frac{\gamma^{s+1}}{s!}} \tilde{a}^{\dagger s} e^{\tilde{a}^\dagger a^\dagger \sqrt{1-\gamma}} \left| 0\tilde{0} \right\rangle = |\psi(\beta)\rangle, \tag{3.112}$$

while for $s = 0$, according to Eq. (3.111), the state $|\psi(\beta)\rangle_{rs}$ becomes the thermal vacuum state $\left(\frac{2\sqrt{fgD}}{1-g-f}\right)^{1/2} e^{\frac{AD}{1-g-f} \tilde{a}^{\dagger 2}} e^{\tilde{a}^\dagger a^\dagger \sqrt{\frac{D}{1-g-f}}} \left| 0\tilde{0} \right\rangle \equiv |\psi(\beta)\rangle_s$ corresponding to the squeezed chaotic light field, this is another expression of $|\psi(\beta)\rangle_s$ in Eq. (3.86).

As mentioned above, the thermal vacuum state is convenient for calculating the ensemble average. The average photon number of the system can be calculated by ${}_{rs}\langle \psi(\beta)| a^\dagger a |\psi(\beta)\rangle_{rs}$. Using the thermal vacuum state $|\psi(\beta)\rangle_{rs}$, we have

$$a\,|\psi(\beta)\rangle_{rs} = \sqrt{\frac{D}{1-g-f}\left\{\frac{2\sqrt{fgD}}{s!\,(1-g-f)}\left[\frac{4gfD}{(g+f-4gf)(1-g-f)}\right]^s\right\}^{1/2}}$$
$$\times\,\tilde{a}^{\dagger s+1}\exp\left(\frac{AD}{1-g-f}\tilde{a}^{\dagger 2}\right)e^{\tilde{a}^{\dagger}a^{\dagger}\sqrt{\frac{D}{1-g-f}}}\left|0\tilde{0}\right\rangle$$
$$=\sqrt{\frac{(s+1)(g+f-4gf)}{4gf}}\,|\psi(\beta)\rangle_{rs+1}\,, \tag{3.113}$$

so its average photon number is

$$_{rs}\langle\psi(\beta)|\,a^{\dagger}a\,|\psi(\beta)\rangle_{rs} = \frac{(s+1)(g+f-4gf)}{4gf}$$
$$=(s+1)\left[\left(\bar{n}+\frac{1}{2}\right)\cosh 2r-\frac{1}{2}\right]. \tag{3.114}$$

3.5 Thermal Vacuum State Corresponding to Gaussian Enhanced Chaotic Light Field

In this chapter, we discuss the Gaussian enhanced chaotic light field (chaotic light is modulated and enhanced by Gaussian beams), whose density operator is [10, 11]

$$\rho_{gc} = Ce^{Da^{\dagger 2}}\rho_c e^{Da^2}. \tag{3.115}$$

Here, $\rho_c = (1-f)\,f^{a^{\dagger}a}$ is the chaotic light field with $f < 1$, C is the normalization coefficient determined by $\mathrm{tr}\rho_{gc} = 1$, and $e^{Da^{\dagger 2}}$ represents the Gaussian enhanced operator. For convenience, we set $D = A\,(f-1)$, where A is an independent Gaussian intensity parameter (also known as a squeezed parameter) and D is a quantity related to the parameter f of the chaotic field, thus we obtain

$$\rho_{gc} = Ce^{A(f-1)a^{\dagger 2}}\rho_c e^{A(f-1)a^2}. \tag{3.116}$$

Using the normal ordering product

$$f^{a^{\dagger}a} = e^{a^{\dagger}a\ln f} =:e^{(f-1)a^{\dagger}a}: \tag{3.117}$$

and the completeness relation of coherent states $|z\rangle$, that is $\pi^{-1}\int d^2z\,|z\rangle\langle z| = 1$, we have

$$
\begin{aligned}
1 = \mathtt{tr}\rho_{gc} &= \mathtt{tr}\left(\int \frac{d^2z}{\pi}|z\rangle\langle z|\rho\right) \\
&= C\int \frac{d^2z}{\pi}\langle z|\, e^{A(f-1)a^{\dagger 2}}\rho_c e^{A(f-1)a^2}|z\rangle \\
&= C(1-f)\int \frac{d^2z}{\pi}\langle z|\, e^{A(f-1)a^{\dagger 2}} : e^{(f-1)a^\dagger a} : e^{A(f-1)a^2}|z\rangle \\
&= C(1-f)\int \frac{d^2z}{\pi} e^{A(f-1)\left(z^{*2}+z^2\right)+(f-1)|z|^2} \\
&= C(1-f)\frac{1}{\sqrt{(1-f)^2-4A^2(1-f)^2}} \\
&= C\frac{1}{\sqrt{1-4A^2}},
\end{aligned}
\tag{3.118}
$$

thus the normalization coefficient is $C=\sqrt{1-4A^2}$.

According to the theory in this chapter, we derive the thermal vacuum state $|\phi(\beta)\rangle$ s corresponding to the state ρ_{gc}. Letting

$$
f \equiv \frac{B^2}{1-4A^2},
\tag{3.119}
$$

thus Eq. (3.116) becomes

$$
\begin{aligned}
\rho_{gc} &= \sqrt{1-4A^2}\,(1-f)\, e^{A(f-1)a^{\dagger 2}} f^{a^\dagger a} e^{A(f-1)a^2} \\
&= \sqrt{1-4A^2}\,(1-f) : e^{\left(\frac{B^2A}{1-4A^2}-A\right)a^{\dagger 2}+\left(\frac{B^2}{1-4A^2}-1\right)a^\dagger a+\left(\frac{B^2A}{1-4A^2}-A\right)a^2} : .
\end{aligned}
\tag{3.120}
$$

Using the integral formula

$$
\int \frac{d^2z}{\pi} e^{-h|z|^2+\xi z+\eta z^*} = \frac{1}{h}\exp\left(\frac{\xi\eta}{h}\right), \quad Reh < 0
\tag{3.121}
$$

and the IWOP method, we can write the density operator ρ_{gc} as its integral form, that is

$$
\begin{aligned}
\rho_{gc} = (1-4A^2)(1-f)\int \frac{d^2z}{\pi} &: \exp[-|z|^2 + A\left(z^{*2}+z^*\right) \\
&+B\left(a^\dagger z^* + az\right) - A\left(a^{\dagger 2}+a^2\right) - a^\dagger a] : .
\end{aligned}
\tag{3.122}
$$

Further, using the operator identity $: e^{-a^\dagger a} : = |0\rangle\langle 0|$, Eq. (3.122) can be expressed as

$$\rho_{gc} = (1-4A^2)(1-f)\int \frac{d^2z}{\pi} : \exp(-|z|^2 + Az^{*2} \\ -Aa^{\dagger 2} + Ba^{\dagger}z^*)\,|0\rangle\langle 0| \exp\left(Az^2 - Aa^2 + Baz\right). \tag{3.123}$$

Noting that $e^{-|z|^2/2} = \left\langle\tilde{0}\right|\tilde{z}\rangle$, $|\tilde{z}\rangle = e^{-|z|^2/2 + z\tilde{a}^{\dagger}}\left|\tilde{0}\right\rangle$, $\tilde{a}|\tilde{z}\rangle = z|\tilde{z}\rangle$, where $|\tilde{z}\rangle$ is the coherent state of fictitious mode, we can write Eq. (3.123) as

$$\rho_{gc} = (1-4A^2)(1-f)\int \frac{d^2z}{\pi}\langle\tilde{z}|\, e^{A(z^{*2}-a^{\dagger 2})+Ba^{\dagger}z^*} \\ \left|0\tilde{0}\right\rangle\left\langle 0\tilde{0}\right| e^{A(z^2-a^2)+Baz}\,|\tilde{z}\rangle \\ = (1-4A^2)(1-f)\int \frac{d^2z}{\pi}\langle\tilde{z}|\, e^{A((\tilde{a}^{\dagger 2}-a^{\dagger 2})+Ba^{\dagger}\tilde{a}^{\dagger}} \\ \left|0\tilde{0}\right\rangle\left\langle 0\tilde{0}\right| e^{A((\tilde{a}^2-a^2)+Ba\tilde{a}}\,|\tilde{z}\rangle. \tag{3.124}$$

Comparing Eqs. (3.122) and (3.124) and using the completeness relation of coherent states of fictitious mode, that is $\pi^{-1}\int d^2z\,|\tilde{z}\rangle\langle\tilde{z}| = 1$, we obtain

$$\rho_{gc} = (1-4A^2)(1-f)\widetilde{\mathrm{tr}}\left(\int \frac{d^2z}{\pi}|\tilde{z}\rangle\langle\tilde{z}|\, e^{A((\tilde{a}^{\dagger 2}-a^{\dagger 2})+Ba^{\dagger}\tilde{a}^{\dagger}} \right. \\ \left. \left|0\tilde{0}\right\rangle\left\langle 0\tilde{0}\right| e^{A((\tilde{a}^2-a^2)+Ba\tilde{a}}\right) \\ = \widetilde{\mathrm{tr}}\left(|\phi(\beta)\rangle_{ss}\langle\phi(\beta)|\right). \tag{3.125}$$

In terms of Eq. (3.5), we obtain the thermal vacuum state $|\phi(\beta)\rangle_s$ corresponding to the density operator ρ_{gc} as

$$|\phi(\beta)\rangle_s = \sqrt{(1-4A^2)(1-f)}\exp\left[A(\tilde{a}^{\dagger 2} - a^{\dagger 2}) + Ba^{\dagger}\tilde{a}^{\dagger}\right]\left|0\tilde{0}\right\rangle. \tag{3.126}$$

When $A = 0$, ρ_{gc} reduces to the chaotic field, letting $f = e^{-\beta\omega}$, $B = \sqrt{f} = e^{-\beta\omega/2} \equiv \tanh\lambda$, $\beta = 1/\kappa T$, κ is the Boltzmann constant, thus we have

$$|\phi(\beta)\rangle_s \to \sqrt{(1-f)}\exp\left(\sqrt{f}a^{\dagger}\tilde{a}^{\dagger}\right)\left|0\tilde{0}\right\rangle = \mathrm{sech}\lambda\exp\left(a^{\dagger}\tilde{a}^{\dagger}\tanh\lambda\right)\left|0\tilde{0}\right\rangle, \tag{3.127}$$

which is the thermal vacuum state belonging to the chaotic field.

Now, we calculate the normalization of the thermal vacuum state $|\phi(\beta)\rangle_s$, that is

$$
\begin{aligned}
\mathtt{Tr}\left(|\phi(\beta)\rangle_{ss}\langle\phi(\beta)|\right) &= \mathtt{tr}\left(\widetilde{\mathtt{tr}}\,|\phi(\beta)\rangle_{ss}\langle\phi(\beta)|\right) = {}_s\langle\phi(\beta)|\,\phi(\beta)\rangle_s \\
&= (1-4A^2)(1-f)\left\langle 0\tilde{0}\right| \mathrm{e}^{A((\tilde{a}^2-a^2)+Ba\tilde{a}} \\
&\int\int\frac{\mathrm{d}^2z_1\mathrm{d}^2z_2}{\pi^2}\,|\tilde{z}_1\tilde{z}_2\rangle\,\langle|\tilde{z}_1\tilde{z}_2\rangle|\,\mathrm{e}^{A((\tilde{a}^{\dagger 2}-a^{\dagger 2})+Ba^\dagger\tilde{a}^\dagger}\left|0\tilde{0}\right\rangle \\
&= (1-4A^2)(1-f)\int\int\frac{\mathrm{d}^2z_1\mathrm{d}^2z_2}{\pi^2}\exp\left[-|z_1|^2-|z_2|^2\right. \\
&\left.+Bz_1z_2+Bz_1^*z_2^*+A(z_1^2-z_2^2)+A(z_1^{*2}-z_2^{*2})\right] \\
&= \sqrt{(1-4A^2)(1-f)}\int\frac{\mathrm{d}^2z_1}{\pi} \\
&\exp\left[\frac{B_2-1+4A^2}{1-4A^2}|z_1|^2-\frac{A(B^2-1+4A^2)}{1-4A^2}(z_1^2+z_1^{*2})\right] \\
&= \sqrt{(1-4A^2)(1-f)}\int\frac{\mathrm{d}^2z_1}{\pi}\exp\left[-(1-f)\,|z_1|^2\right. \\
&\left.+A\,(1-f)\,(z_1^2+z_1^{*2})\right] \\
&= 1.
\end{aligned}
\tag{3.128}
$$

Letting

$$
L = 1-4A^2-B^2 = \frac{4}{(1-4A^2)(1-f)}, \tag{3.129}
$$

thus the state $|\phi(\beta)\rangle_s$ can be simplified as

$$
|\phi(\beta)\rangle_s = \frac{2}{\sqrt{L}}\exp\left[A(\tilde{a}^{\dagger 2}-a^{\dagger 2})+Ba^\dagger\tilde{a}^\dagger\right]\left|0\tilde{0}\right\rangle. \tag{3.130}
$$

On the other hand, the thermal vacuum state $|\psi(\beta)\rangle_s$ is helpful for calculating the average photon number and the photon number fluctuation of the state ρ_{gc}. Using Eq. (3.130) and the completeness relation of coherent states, we get the average value ${}_s\langle\psi(\beta)|\,aa^\dagger\,|\psi(\beta)\rangle_s$ as

$$
\begin{aligned}
&{}_s\langle\psi(\beta)|\,aa^\dagger\,|\psi(\beta)\rangle_s \\
&= \frac{4}{L}\left\langle 0\tilde{0}\right|\exp\left[A(\tilde{a}^2-a^2)+Ba\tilde{a}\right]a \\
&\int\int\frac{\mathrm{d}^2z_1\mathrm{d}^2z_2}{\pi^2}\,|\tilde{z}_1\tilde{z}_2\rangle\,\langle|\tilde{z}_1\tilde{z}_2\rangle|\,\tilde{a}^\dagger\exp\left[A(\tilde{a}^{\dagger 2}-a^{\dagger 2})+Ba^\dagger\tilde{a}^\dagger\right]\left|0\tilde{0}\right\rangle \\
&= \frac{4}{L}\int\int\frac{\mathrm{d}^2z_1\mathrm{d}^2z_2}{\pi^2}|z_1|^2\exp[-|z_1|^2-|z_2|^2+Bz_1z_2+Bz_1^*z_2^* \\
&+A(z_2^2-z_1^2)+A(z_2^{*2}-z_1^{*2})] \\
&= \frac{4}{L\sqrt{1-4A^2}}\int\frac{\mathrm{d}^2z_1}{\pi}|z_1|^2\exp\left[(f-1)|z_1|^2-A(f-1)(z_1^2+z_1^{*2})\right]
\end{aligned}
$$

$$
\begin{aligned}
&= \frac{4}{L\sqrt{1-4A^2}} \frac{\partial^2}{\partial k \partial g} \int \frac{d^2 z_1}{\pi} \exp[(f-1)|z_1|^2 \\
&\quad + k z_1 + g z_1^* - A(f-1)(z_1^2 + z_1^{*2})]|_{k,g=0} \\
&= \frac{\partial^2}{\partial k \partial g} \exp\left[-\frac{A(k^2+g^2)-kg}{(f-1)(1-4A^2)}\right]\Bigg|_{k,g=0} \\
&= \frac{1}{(1-f)(1-4A^2)},
\end{aligned}
\tag{3.131}
$$

which leads to the average photon number of the state ρ_{gc}, that is

$$
{}_s\langle\psi(\beta)|\, a^\dagger a\, |\psi(\beta)\rangle_s = \frac{1}{(1-f)(1-4A^2)} - 1.
\tag{3.132}
$$

As the squeezed parameter A increases, the average photon number also increases. When $A = 0$, Eq. (3.132) becomes $f/(1-f)$, which is the average photon number in chaotic light field. And clearly, there exists the inequality

$$
\frac{1}{(1-f)(1-4A^2)} - 1 > \frac{f}{1-f},
\tag{3.133}
$$

which means that comparing with the chaotic field, the average photon number of the state ρ_{gc} has indeed increased.

Further, by virtue of the state $|\psi(\beta)\rangle_s$, we can calculate ${}_s\langle\psi(\beta)|\, a^2 a^{\dagger 2}\, |\psi(\beta)\rangle_s$ as

$$
\begin{aligned}
{}_s\langle\psi(\beta)|\, a^2 a^{\dagger 2}\, |\psi(\beta)\rangle_s &= \frac{4}{L} \int\int \frac{d^2 z_1 d^2 z_2}{\pi^2} \left\langle 0\tilde{0}\right| e^{\left[A(a_2^2 - a_1^2) + B a_1 a_2\right]} \\
&\quad \times a^2 |z_1 z_2\rangle \langle z_1 z_2| \tilde{a}^{\dagger 2} e^{\left[A(a_2^{\dagger 2} - a_1^{\dagger 2}) + B a_1^\dagger a_2^\dagger\right]} \left| 0\tilde{0}\right\rangle \\
&= \frac{4}{L} \int\int \frac{d^2 z_1 d^2 z_2}{\pi^2} |z_1|^4 e^{\left[A(z_2^2 - z_1^2) + B z_1 z_2\right]} \\
&\quad \times e^{-|z_1^2| - |z_2^2|} e^{\left[A(z_2^{*2} - z_1^{*2}) + B z_1^* z_2^*\right]} \\
&= \frac{1}{\sqrt{1-4A^2}} \frac{4}{L} \left(\frac{\partial^2}{\partial k \partial g}\right)^2 \int \frac{d^2 z_1}{\pi^2} \exp\left[(f-1)|z_1|^2\right. \\
&\quad \left. + k z_1 + g z_1^* - A(f-1)(z_1^2 + z_1^{*2})\right]\Big|_{k,g=0} \\
&= \left(\frac{\partial^2}{\partial k \partial g}\right)^2 \exp\left[-\frac{(k^2+g^2)A + kg}{(f-1)(1-4A^2)}\right]\Bigg|_{k,g=0} \\
&= \frac{4A^2+2}{(f-1)^2(1-4A^2)^2}.
\end{aligned}
\tag{3.134}
$$

Since

$$
\begin{aligned}
(a^\dagger a)^2 &= (aa^\dagger - 1)^2 = aa^\dagger aa^\dagger - 2aa^\dagger + 1 \\
&= a(aa^\dagger - 1)a^\dagger - 2aa^\dagger + 1 \\
&= a^2 a^{\dagger 2} - 3aa^\dagger + 1,
\end{aligned}
\tag{3.135}
$$

thus we have

$$
{}_s\langle\psi(\beta)|(a^\dagger a)^2|\psi(\beta)\rangle_s = {}_s\langle\psi(\beta)|\left(a^2 a^{\dagger 2} - 3aa^\dagger + 1\right)|\psi(\beta)\rangle_s. \tag{3.136}
$$

Combining Eqs. (3.132), (3.134) and (3.136), we obtain the photon number fluctuation of the state ρ_{gc} as

$$
\begin{aligned}
&{}_s\langle\psi(\beta)|(a^\dagger a)^2|\psi(\beta)\rangle_s - \left({}_s\langle\psi(\beta)|a^\dagger a|\psi(\beta)\rangle_s\right)^2 \\
&= \frac{4A^2+2}{(f-1)^2(1-4A^2)^2} - \frac{3}{(1-f)(1-4A^2)} \\
&\quad +1 - \left[\frac{1}{(1-f)(1-4A^2)} - 1\right]^2 \\
&= \frac{8A^2 + f(1-4A^2)}{(f-1)^2(1-4A^2)^2}.
\end{aligned}
\tag{3.137}
$$

For $A = 0$, Eq. (3.137) becomes $f/(f-1)^2$, which is just the photon number fluctuation of the chaotic field. Besides, we obtain the inequality

$$
\frac{f}{(f-1)^2} < \frac{8A^2 + f(1-4A^2)}{(f-1)^2(1-4A^2)^2}, \tag{3.138}
$$

it means that the photon number fluctuation of the state ρ_{gc} also increases.

3.6 Thermo-invariant Coherent State

Noting that the annihilation pair operator $a\tilde{a}$ and the number-difference operator $a^\dagger a - \tilde{a}^\dagger\tilde{a}$ are commutative, that is

$$
\left[a\tilde{a}, a^\dagger a - \tilde{a}^\dagger\tilde{a}\right] = 0, \tag{3.139}
$$

which means that they have common eigen-states, denoted as $|z, \mathfrak{N}\rangle$, and yielding the following eigen-equations

$$
a\tilde{a}\,|z, \mathfrak{N}\rangle = z\,|z, \mathfrak{N}\rangle, \tag{3.140}
$$

$$
\left(a^\dagger a - \tilde{a}^\dagger\tilde{a}\right)|z, \mathfrak{N}\rangle = \mathfrak{N}\,|z, \mathfrak{N}\rangle, \tag{3.141}
$$

where z is complex, $\mathfrak{N}$ is an integer. Compared with the eigen-equations $a|z\rangle = z|z\rangle$ satisfied by coherent states, it can be seen that the states $|z,\mathfrak{N}\rangle$ are a kind of generalized coherent states. If $\tilde{a}^\dagger\tilde{a}$ is regarded as the number operator of negative energy particles in the fictitious space, then the physical meaning represented by quantum state $|z,\mathfrak{N}\rangle$ is that it annihilates a real particle in the system and also annihilates a fictitious particle in the fictitious space, without changing the energy value of the thermal equilibrium state (proportional to $\mathfrak{N}$). Therefore, this state $|z,\mathfrak{N}\rangle$ is called a thermo-invariant coherent state [12].

Using $N = a^\dagger a$, we know

$$af(N) = f(N+1)\,a,\ a^\dagger f(N) = f(N-1)\,a^\dagger, \tag{3.142}$$

it follows that

$$\left[a\tilde{a}, \frac{1}{N}a^\dagger\tilde{a}^\dagger\right] = 1, \tag{3.143}$$

and

$$\left[a^\dagger a - \tilde{a}^\dagger\tilde{a}, \frac{1}{N}a^\dagger\tilde{a}^\dagger\right] = 0, \tag{3.144}$$

thus the concise form of the state $|z,\mathfrak{N}\rangle$ is

$$|z,\mathfrak{N}\rangle = C\exp\left(\frac{z}{N}a^\dagger\tilde{a}^\dagger\right)\left|\mathfrak{N},\tilde{0}\right\rangle, \tag{3.145}$$

where $\left|\mathfrak{N},\tilde{0}\right\rangle = \frac{1}{\sqrt{\mathfrak{N}!}}a^{\dagger\mathfrak{N}}\left|0,\tilde{0}\right\rangle$, C is the normalization constant. In fact, using Eq. (3.145) to obtain

$$\tilde{a}|z,\mathfrak{N}\rangle = C\frac{z}{N}a^\dagger\exp\left(\frac{z}{N}a^\dagger\tilde{a}^\dagger\right)\left|\mathfrak{N},\tilde{0}\right\rangle = \frac{z}{N}a^\dagger|z,\mathfrak{N}\rangle. \tag{3.146}$$

Applying the annihilation operator a to Eq. (3.146), and using Eq. (3.142), we have

$$a\tilde{a}|a,\mathfrak{N}\rangle = a\frac{z}{N}a^\dagger|z,\mathfrak{N}\rangle = z|z,\mathfrak{N}\rangle. \tag{3.147}$$

On the other hand, using Eq. (3.145) to show that

$$\left(a^\dagger a - \tilde{a}^\dagger\tilde{a}\right)|z,\mathfrak{N}\rangle = C\exp\left(\frac{z}{N}a^\dagger\tilde{a}^\dagger\right)\left(a^\dagger a - \tilde{a}^\dagger\tilde{a}\right)\left|\mathfrak{N},\tilde{0}\right\rangle = \mathfrak{N}|z,\mathfrak{N}\rangle. \tag{3.148}$$

So far, we have proved Eqs. (3.140) and (3.141). For convenience, the state $|z,\mathfrak{N}\rangle$ in two-mode Fock state is often written as

$$
\begin{aligned}
|z,\mathfrak{N}\rangle &= C\sum_{n=0}^{\infty}\frac{1}{n!}\left(\frac{z}{N}a^{\dagger}\tilde{a}^{\dagger}\right)^{n}\left|\mathfrak{N},\tilde{0}\right\rangle \\
&= C_{\mathfrak{N}}\sum_{n=0}^{\infty}\frac{z^{n}}{\sqrt{n!\,(\mathfrak{N}+n)!}}\left|n+\mathfrak{N},\tilde{n}\right\rangle,
\end{aligned}
\tag{3.149}
$$

where $C_{\mathfrak{N}} = C\sqrt{\mathfrak{N}!}$.

3.7 Normalization of Photon-Subtracted Squeezed States

3.7.1 Single-Mode Case

In the Chap. 1, we have pointed out that the state $\operatorname{sech}^{\frac{1}{2}}\lambda e^{\frac{1}{2}a^{\dagger 2}\tanh\lambda}|0\rangle$ is a squeezed state, denoted as $S(\lambda)|0\rangle$, that is

$$
S(\lambda)|0\rangle = \operatorname{sech}^{\frac{1}{2}}\lambda e^{\frac{1}{2}a^{\dagger 2}\tanh\lambda}|0\rangle, \tag{3.150}
$$

where $S(\lambda) = e^{\lambda(a^{\dagger 2}-a^{2})}$ is a squeezed operator, and $S(\lambda)$ can lead to the squeezed transformation, i.e.,

$$
S^{\dagger}(\lambda)aS(\lambda) = a\cosh\lambda + a^{\dagger}\sinh\lambda. \tag{3.151}
$$

So, the photon-subtracted single-mode squeezed state is defined as

$$
a^{m}S(\lambda)|0\rangle \equiv |\lambda\rangle_{m}. \tag{3.152}
$$

To realize the normalization of the state $|\lambda\rangle_m$, we first calculate the normal ordering product of the operator $(a\cosh\lambda + a^{\dagger}\sinh\lambda)^m$. Using the Baker-Hausdorff operator identity

$$
e^{A}e^{B} = e^{B}e^{A}e^{[A,B]}, \tag{3.153}
$$

which holds for $[[A,B],A] = [[A,B],B] = 0$, we obtain

$$
\begin{aligned}
\exp[t(\mu a + va^{\dagger})] &= :e^{tva^{\dagger}}e^{t\mu a}\exp\left[-\frac{1}{2}\left(tva^{\dagger}+t\mu a\right)\right]: \\
&= :e^{tva^{\dagger}}e^{t\mu a}\exp\left(\frac{t^{2}}{2}\mu v\right): \\
&= :e^{-it\sqrt{2\mu v}\frac{i}{\sqrt{2\mu v}}(va^{\dagger}+\mu a)}:\exp\left[-\frac{\left(-it\sqrt{\mu v}\right)^{2}}{2}\right].
\end{aligned}
\tag{3.154}
$$

Using the generating function for the Hermite polynomial and Eq. (3.154), we get

$$(\mu a+va^{\dagger})^{m}=\left(-\mathrm{i}\sqrt{\frac{\mu v}{2}}\right)^{m}:\mathrm{H}_{m}\left(\mathrm{i}\sqrt{\frac{\mu}{2}}a+\mathrm{i}\sqrt{\frac{v}{2\mu}}a^{\dagger}\right):, \tag{3.155}$$

where $\mathrm{H}_m(\cdot)$ are m-order Hermite polynomials, so the normal ordering product of the operator $(a\cosh\lambda+a^{\dagger}\sinh\lambda)^m$ reads

$$\begin{aligned}&(a\cosh\lambda+a^{\dagger}\sinh\lambda)^{m}\\&=(-\mathrm{i})^{m}\frac{\sinh^{m/2}2\lambda}{2^{m}}:\mathrm{H}_{m}\left(\sqrt{\frac{\cosh\lambda}{2^{m}}}a+\mathrm{i}\sqrt{\frac{\tanh\lambda}{2^{m}}}a^{\dagger}\right):,\end{aligned} \tag{3.156}$$

which leads to the new form of the photon-subtracted single-mode squeezed state, that is

$$\begin{aligned}|\lambda\rangle_{m}&=S(\lambda)S^{\dagger}(\lambda)a^{m}S(\lambda)\left|0\right\rangle\\&=S(\lambda)(a\cosh\lambda+a^{\dagger}\sinh\lambda)^{m}\left|0\right\rangle\\&=(-\mathrm{i})^{m}\frac{\sinh^{m/2}2\lambda}{2^{m}}S(\lambda)\mathrm{H}_{m}\left(\mathrm{i}\sqrt{\frac{\tanh\lambda}{2^{m}}}a^{\dagger}\right)\left|0\right\rangle,\end{aligned} \tag{3.157}$$

which shows that the state $|\lambda\rangle_m$ is just a squeezed Hermite polynomial excited state.

Using the differential forms of Hermite polynomials

$$\mathrm{H}_{m}(x)=\frac{\partial^{m}}{\partial t^{m}}\exp\left(2xt-t^{2}\right)|_{t=0}, \tag{3.158}$$

we have

$$\begin{aligned}&{}_{m}\langle\lambda|\lambda\rangle_{m}\\&=\frac{\sinh^{m/2}2\lambda}{2^{2m}}\left\langle 0\right|\mathrm{H}_{m}\left(-\mathrm{i}\sqrt{\frac{\tanh\lambda}{2^{m}}}a\right)\mathrm{H}_{m}\left(\mathrm{i}\sqrt{\frac{\tanh\lambda}{2^{m}}}a^{\dagger}\right)\left|0\right\rangle\\&=\frac{\sinh^{m/2}2\lambda}{2^{2m}}\frac{\partial^{m}}{\partial t^{m}\partial\tau^{m}}\,\mathrm{e}^{-t^{2}-\tau^{2}}\left\langle 0\right|\mathrm{e}^{-\mathrm{i}\sqrt{2\tanh a t}}\mathrm{e}^{-\mathrm{i}\sqrt{2\tanh a^{\dagger}\tau}}\left|0\right\rangle\Big|_{t,\tau=0}\\&=\frac{\sinh^{m/2}2\lambda}{2^{2m}}\frac{\partial^{m}}{\partial t^{m}\partial\tau^{m}}\exp\left(-t^{2}-\tau^{2}+2\tau\tanh\lambda\right)|_{t,\tau=0},\end{aligned} \tag{3.159}$$

where

$$\begin{aligned}&\frac{\partial^{2m}}{\partial t^m \partial \tau^m} \exp\left(-t^2 - \tau^2 + 2x\tau t\right)|_{t,\tau=0}\\&= \sum_{n,l,k=0}^{\infty} \frac{(-1)^{n+l}}{n!l!k!} (2x)^k \frac{\partial^{2m}}{\partial t^m \partial \tau^m} \tau^{2n+k} t^{2l+k}\bigg|_{t,\tau=0}\\&= 2^m x^m m! \sum_{l=0}^{[m/2]} \frac{m!}{2^{2l} (l!)^2 (m-2l)!} \left(\frac{1}{x^2}\right)^l. \end{aligned} \tag{3.160}$$

So, Eq. (3.159) becomes

$$_m\langle\lambda|\lambda\rangle_m = \sinh^{2m} 2\lambda \sum_{l=0}^{[m/2]} \frac{(m!)^2}{2^{2l} (l!)^2 (m-2l)!} (\tanh\lambda)^{-2l}. \tag{3.161}$$

Using the new forms of Legendre polynomials, i.e.,

$$x^m \sum_{l=0}^{[m/2]} \frac{m!}{2^{2l} (l!)^2 (m-2l)!} \left(1 - \frac{1}{x^2}\right)^l = \mathrm{P}_m(x), \tag{3.162}$$

thus Eq. (3.161) is rewritten in the concise form:

$$\begin{aligned}_m\langle\lambda|\lambda\rangle_m &= \sinh^{2m}\lambda \sum_{l=0}^{[m/2]} \frac{(m!)^2}{2^{2l} (l!)^2 (m-2l)!} \left(1 - \frac{1}{\sinh^2\lambda}\right)^l\\&= m!(-\mathrm{i}\sinh\lambda)^m \mathrm{P}_m(\mathrm{i}\sinh\lambda). \end{aligned} \tag{3.163}$$

So, the normalized coefficient is

$$N_{\lambda,m} = [m!(-\mathrm{i}\sinh\lambda)^m \mathrm{P}_m(\mathrm{i}\sinh\lambda)]^{-\frac{1}{2}}, \tag{3.164}$$

and the normalized photon-subtracted single-mode squeezed state is

$$||\lambda\rangle_m = N_{\lambda,m} |\lambda\rangle_m. \tag{3.165}$$

Similarly, for the photon-added single-mode squeezed state, its normalization is expressed as

$$\langle 0| S^\dagger(\lambda) a^k a^{\dagger k} S(\lambda) |0\rangle = k! \cosh^k \lambda \mathrm{P}_k(\cosh\lambda). \tag{3.166}$$

Using the operator identities

$$a^{\dagger k} a^k = \vdots \mathrm{H}_{k,k}(a, a^\dagger) \vdots \tag{3.167}$$

and

$$\vdots \mathrm{H}_{k,k}(a, a^{\dagger}) \vdots = \sum_{l=0}^{k} \frac{(k!)^2}{(l!)^2 [(k-l)!]^2} (-1)^l a^{m-l} a^{\dagger m-l}, \tag{3.168}$$

we have the inner product [14]

$$\begin{aligned}
{}_m \langle \lambda | \lambda \rangle_m &= \langle 0 | S^{\dagger}(\lambda) a^{\dagger m} S(\lambda) | 0 \rangle \\
&= \langle 0 | S^{\dagger}(\lambda) \vdots \mathrm{H}_{m,m}(a, a^{\dagger}) \vdots S(\lambda) | 0 \rangle \\
&= \sum_{l=0}^{m} \frac{(m!)^2 (-1)^l}{(l!)^2 [(m-l)!]^2} \langle 0 | S^{\dagger}(\lambda) a^{m-l} a^{\dagger m-l} S(\lambda) | 0 \rangle \\
&= \sum_{l=0}^{m} \binom{m}{l} (-1)^l \cosh^{m-l} \lambda \mathrm{P}_{m-l}(\cosh \lambda).
\end{aligned} \tag{3.169}$$

Combining Eqs. (3.163) and (3.169), we finally see

$$(-\mathrm{i} \sinh \lambda)^m \mathrm{P}_m(\mathrm{i} \sinh \lambda) = \sum_{l=0}^{m} \binom{m}{l} (-1)^l \cosh^{m-l} \lambda \mathrm{P}_{m-l}(\cosh \lambda), \tag{3.170}$$

which leads to a binomial theorem involving Legendre polynomials, that is

$$\sum_{l=0}^{m} \binom{m}{l} (-1)^l x^{m-l} \mathrm{P}_{m-l}(x) = \left(-\mathrm{i}\sqrt{x-1}\right)^m \mathrm{P}_m \left(-\mathrm{i}\sqrt{x-1}\right). \tag{3.171}$$

3.7.2 Two-Mode Case

Operating the two-mode squeezed operator $S_2(\lambda) = \exp[\lambda(a^{\dagger} b^{\dagger} - ab)]$ on vacuum $|00\rangle$, we have [15, 16]

$$S_2(\lambda) |00\rangle = \mathrm{sech}\lambda \exp(a^{\dagger} b^{\dagger} \tanh \lambda) |00\rangle , \tag{3.172}$$

where λ is a real number, a, b are two-mode annihilation operators, $\left[a, a^{\dagger}\right] = \left[b, b^{\dagger}\right] = 1$. Applying the operators a^m and b^n to the state $S_2(\lambda) |00\rangle$ successively, we get the unnormalized state

$$|\lambda, m, n\rangle = a^m b^n S_2(\lambda) |00\rangle . \tag{3.173}$$

Noting the transformation relations of the two-mode squeezed operator $S_2(\lambda)$, i.e.,

$$\begin{aligned} S_2^{\dagger}(\lambda) a S_2(\lambda) &= a\cosh\lambda + b^{\dagger}\sinh\lambda, \\ S_2^{\dagger}(\lambda) b S_2(\lambda) &= b\cosh\lambda + a^{\dagger}\sinh\lambda, \end{aligned} \tag{3.174}$$

we have

$$\begin{aligned} |\lambda, m, n\rangle &= S_2(\lambda) S_2^{\dagger}(\lambda) a^m b^n S_2(\lambda) |00\rangle \\ &= S_2(\lambda)(a\cosh\lambda + b^{\dagger}\sinh\lambda)^m (b\cosh\lambda + a^{\dagger}\sinh\lambda)^n |00\rangle \\ &= S_2(\lambda)\sinh^{n+m}\lambda \sum_{l=0}^{m} \frac{m!\coth^l\lambda}{l!(m-l)!} b^{\dagger m-l} a^l a^{\dagger n} |00\rangle . \end{aligned} \tag{3.175}$$

Owing to $a^{\dagger n}|0\rangle = \sqrt{n!}\,|n\rangle$, $a^l|n\rangle = \sqrt{n!/(n-l)!}\,|n-l\rangle = \sqrt{n!}/(n-l)!a^{\dagger n-l}|0\rangle$, thus we have

$$a^l a^{\dagger n}|00\rangle = \frac{n!}{(n-l)!} a^{\dagger n-l}|00\rangle . \tag{3.176}$$

Therefore, we obtain

$$\begin{aligned} |\lambda, m, n\rangle =& S_2(\lambda)\sinh^{n+m}\lambda \sum_{l=0}^{\min(m,n)} \frac{m!n!\coth^l\lambda}{l!(m-l)!(n-l)!} a^{\dagger n-l} b^{\dagger n-l} |00\rangle \\ =& \frac{S_2(\lambda)\sinh^{(n+m)/2} 2\lambda}{\left(\mathrm{i}\sqrt{2}\right)^{n+m}} \sum_{l=0}^{\min(m,n)} \frac{(-1)\, m!n!}{l!(n-l)!(m-l)!} \\ & \times \left(\mathrm{i}\sqrt{\tanh\lambda} b^{\dagger}\right)^{m-l} \left(\mathrm{i}\sqrt{\tanh\lambda} a^{\dagger}\right)^{n-l} |00\rangle \\ =& \frac{\sinh^{(n+m)/2} 2\lambda}{\left(\mathrm{i}\sqrt{2}\right)^{n+m}} S_2(\lambda) \mathrm{H}_{m,n}\left(\mathrm{i}\sqrt{\tanh\lambda} b^{\dagger}, \mathrm{i}\sqrt{\tanh\lambda} a^{\dagger}\right) |00\rangle , \end{aligned} \tag{3.177}$$

which is equivalent to a two-mode squeezed Hermite polynomial excited vacuum state.

To derive the normalization coefficient $N_{\lambda,m,n}$ of the state $|\lambda, m, n\rangle$, we first calculate the inner product $\langle\lambda, m+s, n+t|\lambda, m, n\rangle$. For this, using Eq. (3.177) and the series expansions of two-variable Hermite polynomials, the state $|\lambda, m, n\rangle$ can be written as

$$|\lambda, m, n\rangle = S_2(\lambda) \sum_{l=0}^{\min(m,n)} \frac{m!n!\sinh^{n+m}\lambda\coth^l\lambda}{l!\sqrt{(m-l)!(n-l)!}} |n-l, m-l\rangle , \tag{3.178}$$

thus we have

$$\langle \lambda, m+s, n+t|\lambda, m, n\rangle = m!(n+s)!\delta_{s,t} \sinh^{2n+2m+2s} \lambda \times \sum_{l=0}^{\min(m,n)} \frac{(m+s)!n! \coth^{2l+s} \lambda}{l!(m-l)!(n-l)!(l+s)!}, \tag{3.179}$$

where $\delta_{s,t}$ is the Delta function. Without losing generality, letting $m < n$ and comparing Eq. (3.179) with the standard Jacobi polynomials

$$\mathrm{P}_m^{(\alpha,\beta)}(x) = \left(\frac{x-1}{2}\right)^m \sum_{k=0}^{m} \binom{m+\alpha}{k} \binom{m+\beta}{m-k} \left(\frac{x+1}{x-1}\right)^k, \tag{3.180}$$

thus Eq. (3.179) can further be written as [17]

$$\langle \lambda, m+s, n+t|\lambda, m, n\rangle = m!(n+s)!\delta_{s,t} \sinh^{2n+s} \lambda \cosh^s \lambda \mathrm{P}_m^{(n-m,s)} (\coth 2\lambda), \tag{3.181}$$

which shows that the inner product $\langle \lambda, m+s, n+t|\lambda, m, n\rangle$ is related to the Jacobi polynomial $\mathrm{P}_m^{(n-m,s)}(\cdot)$. Especially, when $s = t = 0$, the normalization coefficient of the state $|\lambda, m, n\rangle$ becomes

$$N_{\lambda,m,n} = \langle \lambda, m, n|\lambda, m, n\rangle = m!n! \sinh^{2n} \lambda \mathrm{P}_m^{(n-m,0)} (\coth 2\lambda). \tag{3.182}$$

When $m = n$, Eq. (3.182) is related to the Legendre polynomials with the squeezing parameter λ because $\mathrm{P}_n^{(0,0)}(x) = \mathrm{P}_n(x)$, $\mathrm{P}_0(x) = 1$; While for $n \neq 0, m = 0$, $\mathrm{P}_0^{(n,0)}(x) = 1$, thus we have $N_{\lambda,0,n} = n! \sinh^{2n} \lambda$. Therefore, the normalized photon-subtracted two-mode squeezed state reads

$$|\lambda, m, n\rangle \equiv [m!n! \sinh^{2n} \lambda \mathrm{P}_m^{(n-m,0)} (\coth 2\lambda)]^{-1/2} a^m b^n S_2(\lambda)|00\rangle. \tag{3.183}$$

Especially, we discuss the equal-photon-subtracted two-mode squeezed state. Noting that

$$\begin{aligned} a^n b^n e^{a^\dagger b^\dagger \tanh \lambda} &= \sum_{m=0}^{\infty} \frac{\tanh^m \lambda}{m!} a^n a^{\dagger m} b^n b^{\dagger m} \\ &= (-1)^n \sum_{m=0}^{\infty} \frac{(-\tanh \lambda)^m}{m!} : \mathrm{H}_{m,n}(\mathrm{i}a^\dagger, \mathrm{i}a) \mathrm{H}_{m,n}(\mathrm{i}b^\dagger, \mathrm{i}b) :, \end{aligned} \tag{3.184}$$

so we have to prove the following identity:

$$\sum_{m=0}^{\infty} \frac{s^m}{m!} \mathrm{H}_{m,n}(x, y)\mathrm{H}_{m,n}(x', y') = (-s)^n \mathrm{H}_{m,n}\left[\mathrm{i}\left(\frac{y}{s} - x'\right), \mathrm{i}\left(y' - sx\right)\right] \mathrm{e}^{sx'x}, \tag{3.185}$$

which is the summation formula of an index in the product of two different two-variable Hermite polynomials.

Proof Writing the generating function for the two-mode Hermite polynomial as

$$\sum_{m=0}^{\infty} \frac{s^m t^n}{m!n!} \mathrm{H}_{m,n}(x, y) = \exp\left[t(y - s) + sx\right] = \mathrm{e}^{sx} \sum_{m=0}^{\infty} \frac{t^n (y - s)^n}{n!}. \tag{3.186}$$

Comparing the coefficient $t^n/n!$ of both sides of Eq. (3.186), we can obtain the summation formula of an index in a two-variable Hermite polynomial, that is

$$\sum_{m=0}^{\infty} \frac{s^m}{m!} \mathrm{H}_{m,n}(x, y) = (y - s)^n \mathrm{e}^{sx}. \tag{3.187}$$

Next, we use the operator Hermite polynomial method to calculate

$$\sum_{m=0}^{\infty} \frac{s^m}{m!} \mathrm{H}_{m,n}(x, y) \vdots \mathrm{H}_{m,n}(a^\dagger, a) \vdots = ?. \tag{3.188}$$

Using the operator identity $\vdots \mathrm{H}_{m,n}(a^\dagger, a) \vdots =: a^{\dagger m} a^n :$, Eq. (3.188) can be expressed as

$$\begin{aligned} &\sum_{m=0}^{\infty} \frac{s^m}{m!} \mathrm{H}_{m,n}(x, y) \vdots \mathrm{H}_{m,n}(a^\dagger, a) \vdots \\ &= \sum_{m=0}^{\infty} \frac{: s^m a^{\dagger m} a^n :}{m!} \mathrm{H}_{m,n}(x, y) \\ &= (y - sa^\dagger)^n \mathrm{e}^{sa^\dagger x} a^n : \; = (y - sa^\dagger)^n (a - sx)^n \mathrm{e}^{sa^\dagger x}. \end{aligned} \tag{3.189}$$

On the other hand, using the formula that transforms the operator $\rho(a, a^\dagger)$ into its anti-normal ordering product, that is

$$\rho(a, a^\dagger) = \int \frac{\mathrm{d}^2}{\pi} \vdots \langle -\beta | \rho(a, a^\dagger) | \beta \rangle \exp\left(|\beta|^2 + \beta^* a - \beta a^\dagger + a^\dagger a\right) \vdots, \tag{3.190}$$

where $|\beta\rangle = \exp\left(-|\beta|^2/2 + \beta a^\dagger\right)|0\rangle$ is a coherent state, and $\langle -\beta | \beta \rangle = \mathrm{e}^{-2|\beta|^2}$, we can convert the operator $(y - sa^\dagger)^n (a - sx)^n$ into its anti-normal ordering product, that is

$$
\begin{aligned}
&(y - sa^\dagger)^n (a - sx)^n \\
&= \int \frac{d^2\beta}{\pi} \vdots \langle -\beta | (y - sa^\dagger)^n (a - sx)^n | \beta \rangle \exp\left(|\beta|^2 + \beta^* a - \beta a^\dagger + a^\dagger a\right) \vdots \\
&= s^n \int \frac{d^2\beta}{\pi} \left(\frac{y}{s} + \beta^*\right)^n (\beta - sx)^n \exp\left(|\beta|^2 + \beta^* a - \beta a^\dagger + a^\dagger a\right) \vdots \\
&= s^n e^{xy} \int \frac{d^2\beta}{\pi} |\beta|^{2n} \vdots \exp\left[-|\beta|^2 + \beta\left(\frac{y}{s} - a^\dagger\right)\right. \\
&\quad \left. + (a - sx)\beta^* - sxa^\dagger - \frac{y}{s} a + a^\dagger a\right] \vdots \\
&= (-s)^n \vdots H_{n,n}\left[i\left(\frac{y}{s} - a^\dagger\right), i(a - sx)\right] \vdots
\end{aligned}
\tag{3.191}
$$

where we have used the integral formula

$$
\int \frac{d^2 z}{\pi} z^m z^{*n} \exp\left(-|z|^2 + \xi z + \eta z^*\right) = (-i)^{m+n} e^{\eta\xi} H_{m,n}(i\eta, i\xi). \tag{3.192}
$$

Substituting Eq. (3.191) into Eq. (3.189) leads to

$$
\sum_{m=0}^{\infty} \frac{s^m}{m!} H_{m,n}(x, y) \vdots H_{m,n}(a^\dagger, a) \vdots = (-s)^n \vdots H_{n,n}\left[i\left(\frac{y}{s} - a^\dagger\right), i(a - sx)\right] \vdots e^{sa^\dagger x}. \tag{3.193}
$$

Since both sides of Eq. (3.193) are within the symbol $\vdots\ \vdots$, and the operators are commutative with each other in the anti-normal ordering product, that is, they are equivalent to classical c numbers. Thus, making the substitutions $a^\dagger \to x\prime$ and $a \to y\prime$ in Eq. (3.193) leads to the identity in Eq. (3.185). Noting the relations between the Hermite polynomials $H_{m,n}(x, y)$ and the Laguerre polynomials $L_m(xy)$, that is

$$
H_{m,m}(x, y) = (-1)^m m! L_m(xy), \tag{3.194}
$$

where the summation expression of $L_m(x)$ is

$$
L_m(x) = \sum_{m=0}^{\infty} (-1)^l \binom{m}{m-l} \frac{x^l}{l!}, \tag{3.195}
$$

thus Eq. (3.185) can be also expressed as

$$
\sum_{m=0}^{\infty} \frac{s^m}{m!} H_{m,n}(x, y) H_{m,n}(x', y') = s^n n! L_n\left[-\left(\frac{y}{s} - x'\right), (y' - sx)\right] e^{sx\prime x}. \tag{3.196}
$$

Similarly, it can also prove that

$$\sum_{m=0}^{\infty}\frac{t^n}{n!}\mathrm{H}_{m,n}(x,y)\mathrm{H}_{m,n}(x',y') = (-t)^m\mathrm{H}_{m,m}\left[\mathrm{i}\left(yt-x'\right),\mathrm{i}(y'-\frac{x}{t})\right]\mathrm{e}^{ty/y}$$
$$= t^m m!\mathrm{L}_m\left[-\left(yt-x'\right)(y'-\frac{x}{t})\right]\mathrm{e}^{ty/y}, \qquad (3.197)$$

so the operator $a^n b^n \mathrm{e}^{a^\dagger b^\dagger \tanh\lambda}$ can be expressed as

$$a^n b^n \mathrm{e}^{a^\dagger b^\dagger \tanh\lambda} = (-1)^n\sum_{m=0}^{\infty}\frac{(-\tanh\lambda)^m}{m!} : \mathrm{H}_{m,n}(\mathrm{i}a^\dagger, \mathrm{i}a)\mathrm{H}_{m,n}(\mathrm{i}b^\dagger, \mathrm{i}b):$$
$$= n!\tanh^n\lambda \mathrm{e}^{a^\dagger b^\dagger \tanh\lambda}$$
$$\times : \mathrm{L}_n\left(-a^\dagger a - b^\dagger b - ab\coth\lambda - a^\dagger b^\dagger\tanh\lambda\right):, \qquad (3.198)$$

and the equal-photon-subtracted two-mode squeezed vacuum state reads

$$|\lambda\rangle_n \equiv a^n b^n S_2(\lambda)|00\rangle = \mathrm{sech}\lambda a^n b^n \mathrm{e}^{a^\dagger b^\dagger \tanh\lambda}|00\rangle$$
$$= n!\mathrm{sech}\lambda\tanh^n\lambda \mathrm{e}^{a^\dagger b^\dagger \tanh\lambda}\mathrm{L}_n\left(-a^\dagger b^\dagger\tanh\lambda\right)|00\rangle$$
$$= n!\tanh^n\lambda\mathrm{L}_n\left(-a^\dagger b^\dagger\tanh\lambda\right)S_2(\lambda)|00\rangle. \qquad (3.199)$$

Therefore, the state $|\lambda\rangle_n$ is equivalent to a Laguerre polynomial excited two-mode squeezed vacuum state. In the previous section, we have calculated the normalization, that is

$$\langle 00|\,\mathrm{e}^{ab\tanh\lambda}a^{\dagger n}b^{\dagger n}a^n b^n\mathrm{e}^{a^\dagger b^\dagger\tanh\lambda}|00\rangle = (n!)^2\sinh^{2n}\lambda\mathrm{P}_n(\cosh 2\lambda), \qquad (3.200)$$

where $\mathrm{P}_n(x)$ is a Legendre polynomial, i.e.,

$$\mathrm{P}_n(x) = \sum_{m=0}^{[n/2]}\frac{(-1)^l(2n-2l)!x^{n-2l}}{2^n l!(n-l)!(n-2l)!}. \qquad (3.201)$$

Using Eqs. (3.199) and (3.200) and the completeness relation of two-mode coherent states, i.e., $\pi^{-2}\int\int \mathrm{d}^2\alpha\mathrm{d}^2\beta\,|\alpha,\beta\rangle\,\langle\alpha,\beta| = 1$, we obtain

$$\langle 00|\,\mathrm{e}^{ab\tanh\lambda}a^{\dagger n}b^{\dagger n}a^n b^n\mathrm{e}^{a^\dagger b^\dagger\tanh\lambda}|00\rangle$$
$$= (n!)^2\tanh^{2n}\lambda\,\langle 00|\,\mathrm{L}_n(-ab\tanh\lambda)\mathrm{e}^{ab\tanh\lambda}$$
$$\times\int\int\frac{\mathrm{d}^2\alpha\mathrm{d}^2\beta}{\pi^2}\,|\alpha,\beta\rangle\,\langle\alpha,\beta|\,\mathrm{e}^{a^\dagger b^\dagger\tanh\lambda}\mathrm{L}_n\left(-a^\dagger b^\dagger\tanh\lambda\right)|00\rangle, \quad (3.202)$$

which leads to a new integral formula

$$\int\int \frac{d^2\alpha d^2\beta}{\pi^2} |L_n(-\alpha\beta \tanh\lambda)|^2 e^{-|\alpha|^2|\beta|^2+(\alpha\beta+\alpha^*\beta^*)\tanh\lambda} = P_n(\cosh 2\lambda)\cosh^{2n}\lambda. \tag{3.203}$$

Using the generating function for the Laguerre polynomial in Eq. (2.34), we further perform the summation $\sum_{m=0}^{\infty}\frac{t^m}{m!}$ in Eq. (3.196), thus we have

$$\begin{aligned}
&\sum_{m,n=0}^{\infty}\frac{s^m t^n}{m!n!}H_{m,n}(x,y)H_{m,n}(x\prime,y\prime) \\
&=\sum_{n=0}^{\infty} t^n s^n L_n\left[-\left(\frac{y}{s}-x'\right)(y'-sx)\right]e^{sx\prime x} \\
&=\frac{1}{1-ts}\exp\left[\frac{sxx'+tyy'-ts(xy+x'y')}{1-ts}\right].
\end{aligned} \tag{3.204}$$

This formula is very useful. For example, using it, we can get

$$\sum_{m,n=0}^{\infty}\frac{s^m t^n}{m!n!}H_{m,n}(x,y)\vdots H_{m,n}(a^\dagger,a)\vdots = \frac{1}{1-ts}\vdots\exp\left[\frac{sxa^\dagger+tya-ts(xy+a^\dagger a)}{1-ts}\right]\vdots. \tag{3.205}$$

On the other hand, noting the operator identity $\vdots H_{m,n}(a^\dagger,a)\vdots =: a^{\dagger m}a^n:$, thus Eq. (3.205) becomes

$$\begin{aligned}
\sum_{m,n=0}^{\infty}\frac{s^m t^n}{m!n!}H_{m,n}(x,y)\vdots H_{m,n}(a^\dagger,a)\vdots &= \sum_{m,n=0}^{\infty}\frac{s^m a^{\dagger m}t^n a^n}{m!n!}H_{m,n}(x,y) \\
&= \exp(-sta^\dagger a+sa^\dagger x+tay):.
\end{aligned} \tag{3.206}$$

Therefore, we get an operator identity connecting the normal and anti-normal products of operators, that is

$$:\exp(-sta^\dagger a+sa^\dagger x+tay): = \frac{1}{1-ts}\vdots\exp\left[\frac{sxa^\dagger+tya-ts(xy+a^\dagger a)}{1-ts}\right]\vdots. \tag{3.207}$$

Especially, if $x=y=0$, thus we have

$$:\exp(-sta^\dagger a): = \frac{1}{1-ts}\vdots\exp\left(\frac{-tsa^\dagger a}{1-ts}\right)\vdots. \tag{3.208}$$

Using Eqs. (2.29) and (3.204) to obtain

$$\begin{aligned} e^{sab} e^{ta^\dagger b^\dagger} &= \sum_{m=0}^{\infty} \frac{s^m}{m!} a^m b^m \sum_{n=0}^{\infty} \frac{t^n}{n!} a^{\dagger n} b^{\dagger n} \\ &= \sum_{m,n=0}^{\infty} \frac{(-1)^{m+n} s^m t^n}{m! n!} : \mathrm{H}_{m,n}(ia^\dagger, ia) \mathrm{H}_{m,n}(ib^\dagger, ib) : \\ &= \frac{1}{1-ts} : \exp\left[\frac{ts(a^\dagger a + b^\dagger b) + sa^\dagger b^\dagger + tab}{1-ts}\right] :, \end{aligned} \tag{3.209}$$

so the normal ordering product of the operator $e^{sab} e^{ta^\dagger b^\dagger}$ is

$$e^{sab} e^{ta^\dagger b^\dagger} = \frac{1}{1-ts} e^{\frac{s}{1-ts} a^\dagger b^\dagger} e^{-(a^\dagger a + b^\dagger b)\ln(1-ts)} e^{\frac{t}{1-ts} ab}, \tag{3.210}$$

while the normal ordering product of the operator $a^n b^n e^{a^\dagger b^\dagger \tanh\lambda}$ can be seen in Eq. (3.198).

References

1. Wan ZL, Fan HY (2016) Expectation value theorem for thermo vacuum states of optical chaotic field and negative-binomial field. Int J Theor Phys 55:3164–3172
2. Wan ZL, Fan HY, Wang Z (2015) Thermal vacuum state corresponding to squeezed chaotic light and its application. Chin Phys B 24(12):120301
3. Wan ZL, Fan HY, Li HM, Wang Z (2017) Thermo vacuum state for describing the density operator of photon-subtracted squeezed chaotic light. Int J Theor Phys 56:3188–201
4. Gour G, Wilde MM (2021) Entropy of a quantum channel. Phys Rev Res 3(2):023096
5. Chu YJ, Huang F, Li MX, Zheng ZJ (2022) An entropy function of a quantum channel. Quantum Inf Process 22(1):27–30
6. Fu HC, Sasaki R (1997) Negative binomial states of quantized radiation fields. J Phys Soc Jpn 66(7):1989–1994
7. Chen F, Fan HY (2014) Evolution law of a negative binomial state in an amplitude dissipative channel. Chin Phys B 23(3):030304
8. Fan HY (2008) Newton-Leibniz integration for ket-bra operators in quantum mechanics (IV)-integrations within Weyl ordered product of operators and their applications. Ann Phys 323(2):500–526
9. Wang S, Xu XX, Yuan HC, Hu LY, Fan HY (2011) Coherent operation of photon subtraction and addition for squeezed thermal states: analysis of nonclassicality and decoherence. J Opt Soc Am B 28(9):2149–2158
10. Wu WF, Fan HY (2017) Thermo vacuum state for Gaussian-enhanced chaotic light. Mod Phys Lett B 31(13):1750151
11. Wang TT, Fan HY (2018) Dissipation of Gaussian-enhenced chaotic state. Int J Theor Phys 57(12):3643–3650
12. Xu XF, Fan HY (2006) On the thermo-invariant coherent state in thermo field dynamics. Mod Phys Lett A 21(11):911–918
13. Meng XG, Wang Z, Fan HY, Wang JS (2012) Nonclassicality and decoherence of photon-subtracted squeezed vacuum states. J Opt Soc Am B 29(11):3141–3149
14. Fan HY, Hu LY, Xu XX (2009) Legendre polynomials as the normalization of photon-subtracted squeezed states. Mod Phys Lett A 24(20):1597–1603

15. Hiroshima T (2001) Decoherence and entanglement in two-mode squeezed vacuum states. Phys Rev A 63(2):022305
16. Lee CT (1990) Nonclassical photon statistics of two-mode squeezed states. Phys Rev A 42(3):1608–1610
17. Xu XX, Hu LY, Fan HY (2009) On the normalized two-mode photon-subtracted squeezed vacuum state. Mod Phys Lett A 24(32):2623–2630

Chapter 4
Weyl Ordering, *s*-Ordering, Quantum Phase Operator and Radon Transform of Wigner Operator

4.1 Weyl Ordering Formula

According to Eq. (1.92), the Weyl correspondence of the coherent state projector $|z\rangle\langle z|$ is [1, 2]

$$2\pi \mathrm{tr}\left[|z\rangle\langle z|\,\Delta\left(\alpha,\alpha^*\right)\right]=2\langle z|:\mathrm{e}^{-2(a-\alpha)\left(a^{\dagger}-\alpha^*\right)}:|z\rangle=2\mathrm{e}^{-2(z^*-\alpha^*)(z-\alpha)},\quad(4.1)$$

where $\Delta(\alpha,\alpha^*)$ is the Wigner operator in the coherent state representation, thus using Eq. (1.90), we have

$$|z\rangle\langle z|=4\int \mathrm{d}^2\alpha \mathrm{e}^{-2(z^*-\alpha^*)(z-\alpha)}\Delta\left(\alpha,\alpha^*\right).\quad(4.2)$$

Noting that the Weyl ordering product of $|z\rangle\langle z|$ reads

$$|z\rangle\langle z|=2\,\vdots\,\mathrm{e}^{-2\left(z^*-a^{\dagger}\right)(z-a)}\,\vdots,\quad(4.3)$$

thus the Weyl ordering product of the Wigner operator $\Delta(\alpha,\alpha^*)$ is

$$\Delta\left(\alpha,\alpha^*\right)=\frac{1}{2}\,\vdots\,\delta\left(\alpha^*-a^{\dagger}\right)\delta(\alpha-a)\,\vdots.\quad(4.4)$$

Another method is using Eq. (1.83) and the IWOP method to obtain the Weyl ordering form of the Wigner operator $\Delta(x,p)$, that is

$$\begin{aligned}\Delta(x,p)&=\frac{1}{\pi}:\mathrm{e}^{-(x-X)^2-(p-P)^2}\\&=\int\int\frac{\mathrm{d}u\mathrm{d}v}{4\pi^2}\mathrm{e}^{-\frac{u^2}{4}+\mathrm{i}u(x-X)-\frac{u^2}{4}+\mathrm{i}v(p-P)}:\end{aligned}$$

H. Fan and L. Hu, *Dissipation and Diffusion of Quantum Light Field*,
https://doi.org/10.1007/978-981-95-1528-8_4

$$
\begin{aligned}
&= \int\int \frac{\mathrm{d}u\mathrm{d}v}{4\pi^2} \mathrm{e}^{\mathrm{i}u(x-X)+\mathrm{i}v(p-P)} \\
&= \int\int \frac{\mathrm{d}u\mathrm{d}v}{4\pi^2} \vdots\, \mathrm{e}^{\mathrm{i}u(x-X)+\mathrm{i}v(p-P)} \,\vdots \\
&= \vdots\, \delta(x-X)\,\delta(p-P) \,\vdots \\
&= \vdots\, \delta(p-P)\,\delta(x-X) \,\vdots\,. \qquad (4.5)
\end{aligned}
$$

The Weyl ordering form of the Wigner operator $\Delta(x, p)$ is exactly the Delta δ function, which was first derived by Fan. It not only highlights the beauty and conciseness of the theory, but also has a wide range of applications. For example, the Weyl correspondence (1.90) can be rewritten as

$$
\begin{aligned}
H(a^\dagger, a) &= 2\int \mathrm{d}^2\alpha \Delta(\alpha, \alpha^*)\, h(\alpha, \alpha^*) \\
&= \int \mathrm{d}^2\alpha \,\vdots\, \delta(a^\dagger - \alpha^*)\,\delta(\alpha - a) \,\vdots\, h(\alpha, \alpha^*) \\
&= \vdots\, h(a, a^\dagger) \,\vdots\,, \qquad (4.6)
\end{aligned}
$$

which shows that, in order to obtain the Weyl ordering form of $H(a^\dagger, a)$, we only need to make the substitutions $a \rightarrow$, $\alpha^* \rightarrow a^\dagger$ in its classical Weyl corresponding function $h(\alpha, \alpha^*)$.

Using the P-representation of the operator ρ, that is $\rho = \pi^{-1}\int \mathrm{d}^2z \mathrm{P}(z, z^*)\,|z\rangle\,\langle z|$, where

$$
\mathrm{P}(z, z^*) = \frac{\mathrm{e}^{|z|^2}}{\pi}\int \frac{\mathrm{d}^2\beta}{\pi}\langle -\beta|\,\rho\,|\beta\rangle\, \mathrm{e}^{|\beta|^2}\mathrm{e}^{-\beta z^* + \beta^* z}, \qquad (4.7)
$$

$|\beta\rangle$ is a coherent state. Substituting Eqs. (4.3) and (4.7) into Eq. (4.6) to show that

$$
\begin{aligned}
\rho &= 2\int \frac{\mathrm{d}^2z}{\pi}\mathrm{P}(z)\frac{\mathrm{e}^{|z|^2}}{\pi^2}\int \frac{\mathrm{d}^2\beta}{\pi}\langle -\beta|\,\rho\,|\beta\rangle\, \mathrm{e}^{|\beta|^2 - \beta z^* + \beta^* z} \\
&\quad \times \vdots\, \exp\left[-2(z^* - \alpha^*)(z - \alpha)\right] \,\vdots \\
&= 2\int \frac{\mathrm{d}^2\beta}{\pi}\langle -\beta|\,\rho\,|\beta\rangle \,\vdots\, \exp\left[-2\left(\beta^*\alpha - \beta a^\dagger + a^\dagger a\right)\right] \,\vdots\,, \qquad (4.8)
\end{aligned}
$$

which is just the formula that converts the operator ρ into its Weyl ordering product. In particular, for $\rho = 1$, i.e., a unit operator, noting that $\langle -\beta|\,\beta\rangle = \mathrm{e}^{-2|\beta|^2}$, Eq. (4.8) reduces to

$$1 = 2\int \frac{d^2\beta}{\pi} \vdots \exp\left[-2\left(\beta^* + a^\dagger\right)(\beta - a)\right] \vdots. \tag{4.9}$$

In addition, the general formula of Weyl ordering can also be obtained by the following method. Using the normal ordering product of $\Delta(\alpha, \alpha^*)$, we have

$$\begin{aligned} \langle -\beta|\, \Delta\left(\alpha, \alpha^*\right)|\beta\rangle &= \frac{1}{\pi}\langle -\beta| : e^{-2\left(\alpha^* - a^\dagger\right)(\alpha - a)} : |\beta\rangle \\ &= \frac{1}{\pi} e^{-2(\alpha^* + \beta^*)(\alpha - \beta)}\langle -\beta|\, \beta\rangle \\ &= \frac{1}{\pi} e^{-2\alpha^*\alpha - 2\beta^*\alpha + 2\beta\alpha^*}, \end{aligned} \tag{4.10}$$

so we have

$$\begin{aligned} \langle -\beta|\, H(a^\dagger, a)\,|\beta\rangle &= 2\int d^2\alpha \,\langle -\beta|\, \Delta\left(\alpha, \alpha^*\right)|\beta\rangle\, h\left(\alpha, \alpha^*\right) \\ &= 2\int \frac{d^2\alpha}{\pi} h\left(\alpha, \alpha^*\right) e^{-2\alpha^*\alpha} e^{2\beta\alpha^* - 2\beta^*\alpha}, \end{aligned} \tag{4.11}$$

where $e^{2\beta\alpha^* - 2\beta^*\alpha}$ is a Fourier integral transform kernel, so its inverse transform is given by

$$h\left(\alpha, \alpha^*\right) = 2e^{2\alpha^*\alpha}\int \frac{d^2\beta}{\pi}\langle -\beta|\, H(a^\dagger, a)\,|\beta\rangle\, e^{2\beta^*\alpha - 2\beta\alpha^*}, \tag{4.12}$$

which is also the formula for calculating the classical Weyl correspondence for the known operator $H(a^\dagger, a)$. Substituting Eqs. (4.12) into (1.90), we have

$$\begin{aligned} H(a^\dagger, a) &= 4\int d^2\alpha\Delta\left(\alpha, \alpha^*\right) e^{2\alpha^*\alpha}\int \frac{d^2\beta}{\pi}\langle -\beta|\, H(a^\dagger, a)\,|\beta\rangle\, e^{2\beta^*\alpha - 2\beta\alpha^*} \\ &= 2\int d^2\alpha \vdots \delta\left(\alpha^* - a^\dagger\right)\delta(\alpha - a) \vdots \\ &\quad \times e^{2\alpha^*\alpha}\int \frac{d^2\beta}{\pi}\langle -\beta|\, H(a^\dagger, a)\,|\beta\rangle\, e^{2\beta^*\alpha - 2\beta\alpha^*} \\ &= 2\int \frac{d^2\beta}{\pi} \vdots \langle -\beta|\, H(a^\dagger, a)\,|\beta\rangle\, e^{2\left(\beta^* a - \beta a^\dagger + a^\dagger a\right)} \vdots, \end{aligned} \tag{4.13}$$

which is consistent with Eq. (4.8) [3].

The classical Weyl correspondence of the operator $e^{\lambda a^\dagger + \mu a}$ is $e^{\lambda\alpha^* + \mu\alpha}$, which can be verified by Eq. (1.90). Using Eq. (4.4), we obtain

$$\begin{aligned} e^{\lambda a^\dagger + \mu a} &= 2\int d^2\alpha \frac{1}{2} \vdots \delta\left(\alpha^* - a^\dagger\right)\delta(\alpha - a) \vdots e^{\lambda\alpha^* + \mu\alpha} \\ &= \vdots e^{\lambda a^\dagger + \mu a} \vdots. \end{aligned} \tag{4.14}$$

Similarly, we have

$$\mathrm{e}^{\sigma Q+\lambda P}=\vdots\,\mathrm{e}^{\sigma Q+\lambda P}\,\vdots. \tag{4.15}$$

That is to say, the Weyl ordering products of the operators $\mathrm{e}^{\lambda a^{\dagger}+\mu a}$ and $\mathrm{e}^{\sigma Q+\lambda P}$ are themselves. In terms of Eq. (4.15), we easily obtain

$$\vdots\,\mathrm{e}^{\sigma Q+\lambda P}\,\vdots=\mathrm{e}^{-\mathrm{i}\frac{1}{2}\lambda\sigma}\mathrm{e}^{\sigma Q}\mathrm{e}^{\lambda P}=\mathfrak{Q}\left[\mathrm{e}^{-\mathrm{i}\frac{1}{2}\lambda\sigma+\lambda P+\sigma Q}\right], \tag{4.16}$$

where $\mathfrak{Q}$ represents the Q-P ordering operator, that is, the ordering form where all coordinate operators are placed to the left of the momentum operators, and the operators within the $\mathfrak{Q}$-ordering are commutative.

If we want to transform the Weyl ordering operator $\vdots\, P^n Q^m \,\vdots$ into its $\mathfrak{Q}$-ordering product, we can expand the right side of Eq. (4.16) as

$$\mathfrak{Q}\left[\mathrm{e}^{-\mathrm{i}\frac{1}{2}\lambda\sigma+\lambda P+\sigma Q}\right]=\sum_{n=0}^{\infty}\frac{\left(\mathrm{i}\frac{\lambda}{\sqrt{2}}\right)^n\left(\frac{\sigma}{\sqrt{2}}\right)^m}{n!m!}\mathfrak{Q}\left[\mathrm{H}_{n,m}\left(-\mathrm{i}\sqrt{2}P,\sqrt{2}Q\right)\right]. \tag{4.17}$$

Comparing Eq. (4.17) and

$$\vdots\,\mathrm{e}^{\sigma Q+\lambda P}\,\vdots=\sum_{n=0}^{\infty}\frac{\lambda^n\sigma^m}{n!m!}\,\vdots\, P^n Q^m\,\vdots \tag{4.18}$$

leads to

$$\vdots\, P^n Q^m\,\vdots=\mathrm{i}^n\left(\frac{1}{\sqrt{2}}\right)^{n+m}\mathfrak{Q}\left[\mathrm{H}_{n,m}\left(-\mathrm{i}\sqrt{2}P,\sqrt{2}Q\right)\right], \tag{4.19}$$

which is the $\mathfrak{Q}$-ordering product of the Weyl ordering operator $\vdots\, P^n Q^m\,\vdots$. According to Eqs. (4.16) and (4.19), it's not difficult to obtain

$$\begin{aligned}(fP+gQ)^n&=\vdots\,(fP+gQ)^n\,\vdots\\&=\sum_{l=0}^{n}\binom{n}{l}\vdots\,(fP)^{n-l}(gQ)^l\,\vdots\\&=\left(\frac{1}{\sqrt{2}}\right)^n\sum_{l=0}^{n}\binom{n}{l}(\mathrm{i}f)^{n-l}g^l\mathfrak{Q}\left[\mathrm{H}_{n-l,l}\left(-\mathrm{i}\sqrt{2}P,\sqrt{2}Q\right)\right].\end{aligned} \tag{4.20}$$

Using the binomial theorem involving two-variable Hermite polynomials, i.e., $\sum_{l=0}^{n}\binom{n}{l}z^{l}\mathrm{H}_{n-l,l}(x,y)=\sqrt{z^{n}}\mathrm{H}_{n}\left(x/\sqrt{z}+\sqrt{z}y\right)$, we obtain

$$(fP+gQ)^{n}=\left(\frac{\mathrm{i}f}{\sqrt{2}}\right)^{n}\sqrt{\left(\frac{g}{\mathrm{i}f}\right)^{n}}\,\mathfrak{Q}\left[\mathrm{H}_{n}\left(\frac{-\mathrm{i}\sqrt{2}P}{\sqrt{\frac{g}{\mathrm{i}f}}+\sqrt{\frac{g}{\mathrm{i}f}}\sqrt{2}Q}\right)\right]. \tag{4.21}$$

Letting $g=1$ and $f=\mathrm{i}$, thus Eq. (4.21) becomes

$$(\mathrm{i}P+Q)^{n}=\left(\frac{-\mathrm{i}}{\sqrt{2}}\right)^{n}\mathfrak{Q}\left[\mathrm{H}_{n}\left(-\sqrt{2}P-\mathrm{i}\sqrt{2}Q\right)\right]. \tag{4.22}$$

Noting that $a=(-\mathrm{i}P+Q)/\sqrt{2}$, so we have

$$a^{n}=\left(\frac{-\mathrm{i}}{2}\right)^{n}\mathfrak{Q}\left[\mathrm{H}_{n}\left(-\sqrt{2}P-\mathrm{i}\sqrt{2}Q\right)\right], \tag{4.23}$$

which shows that the $\mathfrak{Q}$-ordering product of the annihilation operator a^n is just a single-variable Hermite polynomial. Thus, we get

$$\langle q|a^{n}|p\rangle=\left(\frac{\mathrm{i}}{2}\right)^{n}\mathrm{H}_{n}\left(\sqrt{2}p+\mathrm{i}\sqrt{2}q\right)\langle q|p\rangle. \tag{4.24}$$

4.2 s-Ordering Formula

In the previous chapters, we obtained the Weyl ordering formula in Eq. (4.8) and the anti-normal ordering formula in Eq. (2.76). Now, we introduce a parameter s to unify the two formulas as [4, 5]

$$\begin{aligned}\rho(a,a^{\dagger})=&\frac{2}{1-s}\int\frac{\mathrm{d}^{2}\beta}{\pi}\langle-\beta|\,\rho(a,a^{\dagger})\,|\beta\rangle\\&\times Ⓢ\exp\left[\frac{2}{s-1}\left(s|\beta|^{2}+\beta a^{\dagger}-\beta^{*}a-a^{\dagger}a\right)\right]Ⓢ,\end{aligned} \tag{4.25}$$

where the symbol Ⓢ refers to s-ordering. For $s=0$, Ⓢ $\rightarrow$ $\vdots\;\vdots$, Eq. (4.25) becomes Eq. (4.8); for $s=-1$, Ⓢ $\rightarrow$ $\vdots\;\vdots$, Eq. (4.25) reduces to Eq. (2.76). For example, the s-ordering of the vacuum projection operator is

$$|0\rangle\langle 0| = \circledS \exp\left(\frac{-2}{s+1}a^{\dagger}a\right)\circledS. \tag{4.26}$$

As another example, the s-ordering of the parity operator $(-1)^N$ is

$$\begin{aligned}(-1)^N &= \frac{2}{1-s}\int\frac{\mathrm{d}^2\beta}{\pi}\langle -\beta| : \mathrm{e}^{-2a^{\dagger}a} : |\beta\rangle \\ &\quad\times \circledS \exp\left[\frac{2}{s-1}\left(s|\beta|^2 + \beta a^{\dagger} - \beta^* a - a^{\dagger}a\right)\right]\circledS \\ &= \frac{2}{1-s}\int\frac{\mathrm{d}^2\beta}{\pi}\circledS \exp\left[\frac{2}{s-1}\left(s|\beta|^2 + \beta a^{\dagger} - \beta^* a - a^{\dagger}a\right)\right]\circledS \\ &= \frac{1}{s}\circledS \exp\left[\frac{2}{s(s-1)}a^{\dagger}a - \frac{2}{s-1}a^{\dagger}a\right]\circledS \\ &= \frac{1}{s}\circledS \exp\left(-\frac{2}{s}a^{\dagger}a\right)\circledS. \end{aligned} \tag{4.27}$$

For $s=-1$, $\circledS \to \vdots\ \vdots$, $(-1)^N = -\vdots \mathrm{e}^{2aa^{\dagger}} \vdots$; while for $=1$, $\circledS \to : :$, $(-1)^N =: \mathrm{e}^{-2a^{\dagger}a} :$, where we have used the identity

$$\delta(x) = \lim_{\epsilon\to 0}\frac{1}{\sqrt{\pi\epsilon}}\mathrm{e}^{-x^2/\epsilon}. \tag{4.28}$$

4.3 Quantum Phase Operator

4.3.1 Classical Correspondence of Phase Operator

In classical harmonic oscillator theory, phase is one of the three elements (amplitude, phase, frequency) of vibration. The quantum phase operator was first defined by Dirac, $a = \sqrt{N}\mathrm{e}^{\mathrm{i}\phi}$ $(N = a^{\dagger}a)$, which can be seen as the quantum correspondence of a complex polar decomposition. Later, in order to avoid $N^{-1/2}|0\rangle$, Susskind and Glogower will modify the phase operator to

$$\mathrm{e}^{\mathrm{i}\phi} = (N+1)^{-\frac{1}{2}}a, \quad \mathrm{e}^{-\mathrm{i}\phi} = a^{\dagger}(N+1)^{\frac{1}{2}}. \tag{4.29}$$

It is easy to show that

$$\mathrm{e}^{\mathrm{i}\phi}\mathrm{e}^{-\mathrm{i}\phi} = 1, \quad \mathrm{e}^{-\mathrm{i}\phi}\mathrm{e}^{\mathrm{i}\phi} = 1 - |0\rangle\langle 0|, \tag{4.30}$$

so the operators $\mathrm{e}^{\mathrm{i}\phi}$ and $\mathrm{e}^{-\mathrm{i}\phi}$ are not commutative. Besides, there exist the following commutation relations

$$[N, e^{i\phi}] = -e^{i\phi}, \quad [N, e^{-i\phi}] = e^{-i\phi},$$
$$[N, \cos\phi] = -i\sin\phi, \ [N, \sin\phi] = i\cos\phi, \tag{4.31}$$

so we obtain

$$\Delta N \Delta \cos\phi \geqslant \frac{1}{2} | \langle \sin\phi \rangle |, \ \Delta N \Delta \sin\phi \geqslant \frac{1}{2} | \langle \cos\phi \rangle |, \tag{4.32}$$

which shows that the coherent states are quantum states that minimize the number-phase uncertainty relation.

So, what is the classical Weyl correspondence of the phase operator $e^{i\phi}$? What is the specific form of $h(\alpha, \alpha^*)$ in

$$e^{i\phi} = 2 \int d^2\alpha h\left(\alpha, \alpha^*\right) \Delta\left(\alpha, \alpha^*\right) ? \tag{4.33}$$

Can it display a classical phase factor? Our approach is: as long as we find the Weyl ordering form of $e^{i\phi}$, i.e., $e^{i\phi} = \vdots h\left(a^\dagger, a\right) \vdots$, then we obtain the result by taking $a^\dagger \to \alpha^*, a \to \alpha$ in $h\left(a^\dagger, a\right)$ because $\Delta(\alpha, \alpha^*) = \frac{1}{2} \vdots \delta\left(a^\dagger - \alpha^*\right) \delta(a - \alpha) \vdots$.

Expanding $e^{i\phi}$ in the Fock space, that is

$$e^{i\phi} = \sum_{m,n}^{\infty} |m\rangle \langle n| \langle m| e^{i\phi} |n\rangle, \tag{4.34}$$

where $|n\rangle = \frac{a^{+n}}{\sqrt{n!}} |0\rangle$, and

$$\langle m| e^{i\phi} |n\rangle = \langle m| (N+1)^{-1/2} a |n\rangle = \delta_{m,n-1}. \tag{4.35}$$

Using Eq. (4.8), the Weyl ordering of the operator $|m\rangle \langle n|$ can be obtained as

$$\begin{aligned} |m\rangle \langle n| &= 2 \int \frac{d^2\beta}{\pi} \vdots \langle -\beta| m\rangle \langle n| \beta\rangle \exp\left[2\left(\beta^* a - a^\dagger \beta + a^\dagger a\right)\right] \vdots \\ &= 2 \int \frac{d^2\beta}{\pi} \vdots \frac{\beta^n (-\beta^*)^m}{\sqrt{n!m!}} \exp\left[2\left(\beta^* a - a^\dagger \beta + a^\dagger a\right)\right] \vdots . \end{aligned} \tag{4.36}$$

By virtue of the integral formula of two-variable Hermite polynomials $H_{m,n}(\xi, \eta)$, i.e.,

$$H_{m,n}(\xi, \eta) = (-1)^n e^{\xi\eta} \int \frac{d^2 z}{\pi} z^n z^{*m} e^{-|z|^2 + \xi z - \eta z^*}, \tag{4.37}$$

where the series expansions of $H_{m,n}(\xi, \eta)$ are

$$\mathrm{H}_{m,n}(\xi,\eta)=\sum_{l=0}^{\min(m,n)} l!\binom{m}{l}\binom{n}{l}(-1)^l \xi^{m-l}\eta^{n-l}, \tag{4.38}$$

thus the Weyl ordering of the operator $|m\rangle\langle n|$ reads

$$|m\rangle\langle n|=\frac{2}{\sqrt{n!m!}} \vdots \mathrm{H}_{m,n}\left(2a^\dagger,2a\right)\mathrm{e}^{-2a^\dagger a} \vdots. \tag{4.39}$$

Substituting Eq. (4.39) into Eq. (4.34) leads to the Weyl ordering of $\mathrm{e}^{\mathrm{i}\phi}$, that is

$$\begin{aligned}\mathrm{e}^{\mathrm{i}\phi}&=\sum_{m,n}^{\infty}\delta_{m,n-1}\frac{2}{\sqrt{n!m!}} \vdots \mathrm{H}_{m,n}\left(2a^\dagger,2a\right)\mathrm{e}^{-2a^\dagger a} \vdots\\&=\sum_{n=0}^{\infty}\frac{2}{n!\sqrt{n+1}} \vdots \mathrm{H}_{n+1,n}\left(2a^\dagger,2a\right)\exp\left(-2a^\dagger a\right) \vdots.\end{aligned} \tag{4.40}$$

Therefore, the Weyl correspondence of $\mathrm{e}^{\mathrm{i}\phi}$ is

$$\begin{aligned}h\left(\alpha,\alpha^*\right)&=\sum_{n=0}^{\infty}\frac{4}{n!\sqrt{n+1}}\mathrm{H}_{n+1,n}\left(2\alpha^*,2\alpha\right)\exp\left(-2|\alpha|^2\right)\\&=\sum_{n=0}^{\infty}\frac{8}{n!\sqrt{n+1}}\sum_{l=0}^{n}l!\binom{n+1}{l}\binom{n}{l}\\&\quad\times(-1)^l\left(4|\alpha|^2\right)^{n-l+1/2}|\alpha|\,\mathrm{e}^{\mathrm{i}\varphi}.\end{aligned} \tag{4.41}$$

Clearly, in addition to the amplitude $|\alpha|^2$, the classical phase $\mathrm{e}^{\mathrm{i}\varphi}$ indeed appears here.
Similarly, we obtain

$$\begin{aligned}\mathrm{e}^{-\mathrm{i}\phi}&=\sum_{m,n}^{\infty}|m\rangle\langle m|\mathrm{e}^{-\mathrm{i}\phi}|n\rangle\langle n|\\&=\sum_{m,n}^{\infty}\delta_{m,n+1}\frac{2}{\sqrt{n!m!}} \vdots \mathrm{H}_{m,n}\left(2a^\dagger,2a\right)\exp\left(-2a^\dagger a\right) \vdots\\&=\sum_{n=0}^{\infty}\frac{2}{n!\sqrt{n+1}} \vdots \mathrm{H}_{n+1,n}\left(2a^\dagger,2a\right)\exp\left(-2a^\dagger a\right) \vdots,\end{aligned} \tag{4.42}$$

its classical Weyl correspondence is

$$
\begin{aligned}
\mathrm{e}^{-\mathrm{i}\phi} \rightarrow & \sum_{n=0}^{\infty} \frac{4}{n!\sqrt{n+1}} \mathrm{H}_{n+1,n}\left(2\alpha^{*}, 2\alpha\right) \exp\left(-2|\alpha|^{2}\right) \\
= & \sum_{n=0}^{\infty} \frac{8}{n!\sqrt{n+1}} \sum_{l=0}^{n} l! \binom{n+1}{l} \binom{n}{l} \\
& \times (-1)^{l} \left(4|\alpha|^{2}\right)^{n-l+1/2} |\alpha| \mathrm{e}^{-\mathrm{i}\varphi},
\end{aligned} \tag{4.43}
$$

where the classical phase $\mathrm{e}^{-\mathrm{i}\varphi}$ corresponds to $\mathrm{e}^{-\mathrm{i}\phi}$.

4.3.2 Eigen-States of Phase Operator

In Chap. 1, we have pointed out that the dynamic Hamiltonians for generating the coherent states of harmonic oscillators are $H_0 = \omega a^{\dagger}a + \mathrm{i}fa - \mathrm{i}f^{*}a^{\dagger}$, where $\mathrm{i}fa - \mathrm{i}f^{*}a^{\dagger}$ represent a kind of linear external sources. In nonlinear optics, when the external source is related to the field strength of light, the corresponding Hamiltonian is

$$
H = \omega a^{\dagger}a + \mathrm{i}\left(f\sqrt{N+1}a - f^{*}a^{\dagger}\sqrt{N+1}\right), \quad f = |f|\mathrm{e}^{\mathrm{i}\omega t}. \tag{4.44}
$$

Performing the unitary transformation $\exp\left(\mathrm{i}\omega t a^{\dagger}a\right)$ on H, we can obtain the Hamiltonian H_I in the interaction picture, that is

$$
H_I = \mathrm{i}|f|\left(\sqrt{N+1}a - a^{\dagger}\sqrt{N+1}\right), \tag{4.45}
$$

which leads to the time-evolution operator

$$
U_I = \mathrm{e}^{t|f|\left(\sqrt{N+1}a - a^{\dagger}\sqrt{N+1}\right)}. \tag{4.46}
$$

If the initial state is vacuum, thus the evolved state at time t reads

$$
U_I|0\rangle = \mathrm{e}^{t|f|\left(\sqrt{N+1}a - a^{\dagger}\sqrt{N+1}\right)}|0\rangle. \tag{4.47}
$$

Now, we discuss the more general nonlinear operator

$$
V \equiv \mathrm{e}^{\lambda a^{\dagger}\sqrt{N+1} - \lambda^{*}\sqrt{N+1}a}, \tag{4.48}
$$

where $\lambda = \mathrm{e}^{\mathrm{i}\varphi}|\lambda|$. Noting that

$$\left[a^\dagger\sqrt{N+1}, \sqrt{N+1}a\right] = -2\left(N+\frac{1}{2}\right),$$
$$\left[\sqrt{N+1}a, N+\frac{1}{2}\right] = \sqrt{N+1}a,$$
$$\left[a^\dagger\sqrt{N+1}, N+\frac{1}{2}\right] = -a^\dagger\sqrt{N+1} \tag{4.49}$$

form SU (1, 1) Lie algebras, so the operator V can be decomposed into

$$V = \exp\left(e^{i\varphi}a^\dagger\sqrt{N+1}\tanh|\lambda|\right)\operatorname{sech}^{2N+1}|\lambda|\exp\left(-e^{-i\varphi}\sqrt{N+1}a\tanh|\lambda|\right). \tag{4.50}$$

Acting the operator V on vacuum leads to

$$V|0\rangle = \operatorname{sech}|\lambda|\exp\left(e^{i\varphi}a^\dagger\sqrt{N+1}\tanh|\lambda|\right)|0\rangle \equiv |\lambda\rangle_\varphi, \tag{4.51}$$

which is normalized, i.e., ${}_\varphi\langle\lambda|\lambda\rangle_\varphi = 1$.

Using Eq. (4.29), we can obtain the following commutative relation

$$\left[e^{i\phi}, a^\dagger\sqrt{N+1}\right] = \left[\sqrt{N+1}^{-1/2}a, a^\dagger\sqrt{N+1}\right] = 1, \tag{4.52}$$

and we can prove that $|\lambda\rangle_\varphi$ is an eigen-state of the phase operator $e^{i\phi}$, that is

$$e^{i\phi}|\lambda\rangle_\varphi = \operatorname{sech}|\lambda|\left[e^{i\phi}, \exp\left(e^{i\varphi}a^\dagger\sqrt{N+1}\tanh|\lambda|\right)\right]|0\rangle = e^{i\varphi}\tanh|\lambda|\,|\lambda\rangle_\varphi. \tag{4.53}$$

When $|\lambda| \to \infty$, i.e., $\tanh|\lambda| \to 1$, and

$$\frac{\left(a^\dagger\sqrt{N+1}\right)^n}{n!}|0\rangle = |n\rangle, \tag{4.54}$$

so we have

$$|\lambda\rangle_\varphi \to \exp\left(e^{i\varphi}a^\dagger\sqrt{N+1}\right)|0\rangle = \sum_{n=0}^{\infty} e^{in\varphi}|n\rangle \equiv \left|e^{i\varphi}\right\rangle, \tag{4.55}$$

where $\left|e^{i\varphi}\right\rangle$ is called the phase state and yields the following eigen-equation

$$e^{i\phi}\left|e^{i\varphi}\right\rangle = e^{i\varphi}\left|e^{i\varphi}\right\rangle \tag{4.56}$$

and the completeness relation

$$\int_0^{2\pi}\frac{d\varphi}{2\pi}\left|e^{i\varphi}\right\rangle\left\langle e^{i\varphi}\right| = 1. \tag{4.57}$$

Besides, we examine the fluctuations of the photon number operator and the phase operator in the phase state $|\lambda > \varphi$. For convenience, letting

$$R = \sqrt{N+1}a,\ \ R^{\dagger} = a^{\dagger}\sqrt{N+1}. \tag{4.58}$$

Using Eq. (4.49), we get

$$V^{-1}RV = R\cosh^2|\lambda| + R^{\dagger}\sinh^2|\lambda|\,\mathrm{e}^{\mathrm{i}2\varphi} + \left(N+\frac{1}{2}\right)\mathrm{e}^{\mathrm{i}\varphi}\sinh 2|\lambda|, \tag{4.59}$$

$$V^{-1}(2N+1)V = (2N+1)\cosh 2|\lambda| + \left(\mathrm{e}^{\mathrm{i}\varphi}R^{\dagger} + \mathrm{e}^{-\mathrm{i}\varphi}R\right)\sinh 2|\lambda|, \tag{4.60}$$

which leads to

$$\begin{aligned} {}_{\varphi}\langle\lambda|\,(2N+1)\,|\lambda\rangle_{\varphi} &= \langle 0|\,V^{-1}\,(2N+1)\,V\,|0\rangle \\ &= \langle 0|\left[(2N+1)\cosh 2|\lambda| + \left(\mathrm{e}^{\mathrm{i}\varphi}R^{\dagger} + \mathrm{e}^{-\mathrm{i}\varphi}R\right)\sinh 2|\lambda|\right]|0\rangle \\ &= \cosh 2|\lambda|. \end{aligned} \tag{4.61}$$

Therefore, the average photon number in the state $|\lambda\rangle_{\varphi}$ is

$${}_{\varphi}\langle\lambda|\,N\,|\lambda\rangle_{\varphi} = \frac{\cosh 2|\lambda| - 1}{2} = \sinh^2|\lambda|. \tag{4.62}$$

Similarly, we have

$$\begin{aligned} {}_{\varphi}\left\langle\lambda|(2N+1)^2|\lambda\right\rangle_{\varphi} &= \langle 0|\,[(2N+1)\cosh 2\,|\lambda| + (\mathrm{e}^{\mathrm{i}\varphi}R^{\dagger} + \mathrm{e}^{-\mathrm{i}\varphi}R)\sinh 2\,|\lambda|]^2\,|0\rangle \\ &= \cosh^2 2\,|\lambda| + \langle 0|(\mathrm{e}^{\mathrm{i}\varphi}R^{\dagger} + \mathrm{e}^{-\mathrm{i}\varphi}R)^2|0\rangle\sinh^2 2|\lambda| \\ &= \cosh^2 2\,|\lambda| + \langle 0|RR^{\dagger}|0\rangle\sinh^2 2\,|\lambda| = \cosh 4|\lambda|. \end{aligned} \tag{4.63}$$

Noting that

$$\frac{1}{4}\left[{}_{\varphi}\left\langle\lambda|(2N+1)^2|\lambda\right\rangle_{\varphi} - 4_{\varphi}\left\langle\lambda|N|\lambda\right\rangle_{\varphi} - 1\right] = {}_{\varphi}\left\langle\lambda|N^2|\lambda\right\rangle_{\varphi}, \tag{4.64}$$

thus we have

$${}_{\varphi}\left\langle\lambda|N^2|\lambda\right\rangle_{\varphi} = \frac{1}{4}\cosh 4\,|\lambda| - \frac{\cosh 2\,|\lambda|}{2} + \frac{1}{4} = \cosh\left(2\,|\lambda|\right)\sinh^2|\lambda|\,. \tag{4.65}$$

According to Eqs. (4.62) and (4.65), the photon number fluctuation in the state $|\lambda\rangle_{\varphi}$ is

$$(\Delta N)^2 = {}_{\varphi}\left\langle\lambda|N^2|\lambda\right\rangle_{\varphi} - ({}_{\varphi}\langle\lambda|\,N\,|\lambda\rangle_{\varphi})^2 = \frac{1}{4}\sinh^2 2\,|\lambda|\,. \tag{4.66}$$

which shows that, the larger $|\lambda|$, the more pronounced the photon number fluctuation in the phase state $|\lambda\rangle_\varphi$.

On the other hand, using Eq. (4.53), we get

$$ {}_\varphi\langle\lambda|\mathrm{e}^{\mathrm{i}\phi}\,|\lambda\rangle_\varphi = \mathrm{e}^{\mathrm{i}\varphi}\tanh|\lambda|\,, \quad {}_\varphi\langle\lambda|\mathrm{e}^{-\mathrm{i}\phi}\,|\lambda\rangle_\varphi = \mathrm{e}^{-\mathrm{i}\varphi}\tanh|\lambda|\,, \tag{4.67} $$

which leads to

$$ {}_\varphi\langle\lambda|\cos\phi\,|\lambda\rangle_\varphi = {}_\varphi\langle\lambda|\,\frac{\mathrm{e}^{\mathrm{i}\phi}+\mathrm{e}^{-\mathrm{i}\phi}}{2}\,|\lambda\rangle_\varphi = \cos\varphi\tanh|\lambda|\,. \tag{4.68} $$

Similarly, we have

$$ {}_\varphi\langle\lambda|\mathrm{e}^{2\mathrm{i}\phi}\,|\lambda\rangle_\varphi = \mathrm{e}^{2\mathrm{i}\varphi}\tanh^2|\lambda|\,, \quad {}_\varphi\langle\lambda|\mathrm{e}^{-2\mathrm{i}\phi}\,|\lambda\rangle_\varphi = \mathrm{e}^{-2\mathrm{i}\varphi}\tanh^2|\lambda|\,, \tag{4.69} $$

it follows that

$$ \begin{aligned} {}_\varphi\langle\lambda|\cos 2\phi\,|\lambda\rangle_\varphi &= {}_\varphi\langle\lambda|\frac{\mathrm{e}^{2\mathrm{i}\phi}+\mathrm{e}^{-2\mathrm{i}\phi}}{2}\,|\lambda\rangle_\varphi \\ &= \cos 2\varphi\tanh^2|\lambda| \\ &= 2{}_\varphi\langle\lambda|\cos^2\phi\,|\lambda\rangle_\varphi - 1, \end{aligned} \tag{4.70} $$

which leads to

$$ {}_\varphi\langle\lambda|\cos^2\phi\,|\lambda\rangle_\varphi = \frac{\cos 2\varphi\tanh^2|\lambda|+1}{2}, \tag{4.71} $$

so the the fluctuation of the phase operator $\cos\phi$ in the phase state $|\lambda\rangle_\varphi$ is given by

$$ \begin{aligned} (\Delta\cos\phi)^2 &\equiv {}_\varphi\langle\lambda|\cos^2\phi\,|\lambda\rangle_\varphi - ({}_\varphi\langle\lambda|\cos\phi\,|\lambda\rangle_\varphi)^2 \\ &= \frac{\cos 2\varphi\tanh^2|\lambda|+1}{2} - (\cos\varphi\tanh|\lambda|)^2 \\ &= \frac{1-\tanh^2|\lambda|}{2} = \frac{\mathrm{sech}^2|\lambda|}{2}, \end{aligned} \tag{4.72} $$

which shows that the larger $|\lambda|$, the smaller the fluctuation of the phase operator $\cos\phi$.

From Eqs. (4.66) and (4.72), it can be seen that for the state $|\lambda\rangle_\varphi$, the number-phase uncertainty relation is

$$ \Delta N\Delta\cos\phi = \frac{1}{2}\sinh 2\,|\lambda|\,\frac{\mathrm{sech}\,|\lambda|}{\sqrt{2}} = \frac{1}{\sqrt{2}}\sinh|\lambda|\,. \tag{4.73} $$

Similarly, we get

$$
\begin{aligned}
{}_{\varphi}\langle\lambda|\cos 2\phi\,|\lambda\rangle_{\varphi} &= 1-2{}_{\varphi}\langle\lambda|\sin^2\phi\,|\lambda\rangle_{\varphi}\,,\\
{}_{\varphi}\langle\lambda|\sin^2\phi\,|\lambda\rangle_{\varphi} &= \frac{1-\cos 2\varphi\tanh^2|\lambda|}{2},\\
{}_{\varphi}\langle\lambda|\sin\phi\,|\lambda\rangle_{\varphi} &= \sin\varphi\tanh|\lambda|\,,
\end{aligned}
\tag{4.74}
$$

so the fluctuation of the phase operator $\sin\phi$ is

$$
(\Delta\sin\phi)^2 = {}_{\varphi}\langle\lambda|\sin^2\phi\,|\lambda\rangle_{\varphi} - ({}_{\varphi}\langle\lambda|\sin\phi\,|\lambda\rangle_{\varphi})^2 = \frac{\operatorname{sech}^2|\lambda|}{2}
\tag{4.75}
$$

and the number-phase uncertainty relation is

$$
\Delta N\Delta\sin\phi = \frac{1}{2}\sinh 2\,|\lambda|\,\frac{1}{\sqrt{2}}\operatorname{sech}^2|\lambda| = \frac{1}{\sqrt{2}}\sinh|\lambda|\,.
\tag{4.76}
$$

According to Heisenberg's uncertainty principle, and using Eqs. (4.66), (4.68) and (4.74), it can be inferred that

$$
\begin{aligned}
\Delta N\Delta\cos\phi &= \frac{1}{\sqrt{2}}\sinh|\lambda| > \frac{\tanh|\lambda|}{2}|\sin\varphi|,\\
\Delta N\Delta\sin\phi &= \frac{1}{\sqrt{2}}\sinh|\lambda| > \frac{\tanh|\lambda|}{2}|\cos\varphi|,
\end{aligned}
\tag{4.77}
$$

that is, the above results of the number-phase operator satisfy the uncertainty principle. Therefore, the phase state $|\lambda\rangle_{\varphi}$ is actually a number-phase squeezed state, but the fluctuation $\Delta N\Delta\cos\phi$ cannot reach its minimum value in the state $|\lambda\rangle_{\varphi}$, which is different from the usual coherent state that is the minimum uncertainty state of the number-phase operator.

4.3.3 Two-Mode Phase Operator

Since the existing single-mode phase operator $e^{i\phi}$ is not unitary, we seek whether there is a unitary phase operator in the two-mode Fock space. In terms of the entangled state

$$
|\xi\rangle = \exp\left(-\frac{1}{2}|\xi|^2 + \xi a^\dagger + \xi^* b^\dagger - a^\dagger b^\dagger\right)|00\rangle
\tag{4.78}
$$

and its completeness relation

$$
1 = \int\frac{d^2\xi}{\pi} : e^{-[\xi^*-(a^\dagger+b)][\xi-(a+b^\dagger)]} : = \int\frac{d^2\xi}{\pi}|\xi\rangle\langle\xi|\,,
\tag{4.79}
$$

as well as the eigen-equations satisfied by the state $|\xi\rangle$, i.e.,

$$\left(a^\dagger + b\right)|\xi\rangle = \xi^* |\xi\rangle, \quad \left(a + b^\dagger\right)|\xi\rangle = \xi |\xi\rangle, \quad \xi = |\xi| e^{i\theta}, \tag{4.80}$$

we get

$$\sqrt{\frac{a+b^\dagger}{a^\dagger + b}}\,|\xi\rangle = e^{i\theta}|\xi\rangle, \quad \sqrt{\frac{a^\dagger + b}{a + b^\dagger}}\,|\xi\rangle = e^{i\theta}|\xi\rangle. \tag{4.81}$$

Therefore, $\sqrt{\frac{a+b^\dagger}{a^\dagger+b}} \equiv e^{i\Theta}$ is a unitary phase operator [6, 7], i.e.,

$$e^{i\Theta} = \int \frac{d^2\xi}{\pi} e^{i\theta}|\xi\rangle\langle\xi|, \quad e^{-i\Theta} = \int \frac{d^2\xi}{\pi} e^{-i\theta}|\xi\rangle\langle\xi|, \tag{4.82}$$

and the phase angle Θ is a Hermite operator, that is

$$\Theta = -\frac{1}{2}\ln\frac{a+b^\dagger}{a^\dagger + b}, \quad \Theta|\xi\rangle = \theta|\xi\rangle. \tag{4.83}$$

The operator conjugate with Θ is

$$h = a^\dagger a - b^\dagger b, \tag{4.84}$$

because

$$\begin{aligned} h\left|\xi = |\xi| e^{i\theta}\right\rangle &= |\xi|\left(e^{i\theta}a^\dagger - e^{-i\theta}b^\dagger\right)\exp\left(-\frac{|\xi|^2}{2} + |\xi|e^{i\theta}a^\dagger \right. \\ &\quad \left. + |\xi|e^{-i\theta}b^\dagger - a^\dagger b^\dagger\right)|00\rangle \\ &= -i\frac{\partial}{\partial\theta}|\xi\rangle, \end{aligned} \tag{4.85}$$

so we have

$$[\Theta, h] = i. \tag{4.86}$$

4.4 Radon Transform of Wigner Operator

4.4.1 Single-Mode Case

In recent years, the tomography method based on Radon transform for reconstructing Wigner function has attracted much attention. Using the normal ordering product of the Wigner operator $\Delta(x, p)$, that is

$$\Delta(x, p) = \frac{1}{\pi} : e^{-(x-X)^2 - (p-P)^2} :, \tag{4.87}$$

and the IWOP method, we obtain

$$\begin{aligned}&\iint_{-\infty}^{\infty} \mathrm{d}x\mathrm{d}p\delta\left(u-\lambda x-\tau p\right)\Delta\left(x,p\right)\\&=\left[\pi\left(\lambda^2+\tau^2\right)\right]^{-1/2}:\exp\left[-\frac{1}{\lambda^2+\tau^2}\left(u-\lambda X-\tau P\right)^2\right]:\\&=|u\rangle_{\lambda,\tau\lambda,\tau}\langle u|\,,\end{aligned} \tag{4.88}$$

where

$$|u\rangle_{\lambda,\tau}=\left[\pi\left(\lambda^2+\tau^2\right)\right]^{-1/4}\mathrm{e}^{-\frac{u^2}{2\left(\lambda^2+\tau^2\right)}+\frac{\sqrt{2}ua^\dagger}{\lambda-\mathrm{i}\tau}-\frac{\lambda+\mathrm{i}\tau}{2(\lambda-\mathrm{i}\tau)}a^{\dagger2}}|0\rangle \tag{4.89}$$

is just the eigen-state of the operator $\lambda X+\tau P$ [8, 9], that is

$$\left(\lambda X+\tau P\right)|u\rangle_{\lambda,\tau}=u\,|u\rangle_{\lambda,\tau}\,, \tag{4.90}$$

$X=\left(a+a^\dagger\right)/\sqrt{2}$, $P=\left(a-a^\dagger\right)/(\mathrm{i}\sqrt{2})$, and $[a,a^\dagger]=1$. The states $|u\rangle_{\lambda,\tau}$ also possess the completeness relation, that is

$$\int_{-\infty}^{\infty}\mathrm{d}u\,|u\rangle_{\lambda,\tau\lambda,\tau}\langle u|=\frac{1}{\sqrt{\pi\left(\lambda^2+\tau^2\right)}}\int_{-\infty}^{\infty}\mathrm{d}u:\mathrm{e}^{-\frac{1}{\lambda^2+\tau^2}\left(u-\lambda\hat{X}-\tau\hat{P}\right)^2}:=1, \tag{4.91}$$

and the orthogonality, i.e.,

$$_{\lambda,\tau}\langle u'|\,u\rangle_{\lambda,\tau}=\delta\left(u-u'\right). \tag{4.92}$$

Therefore, the states $|u\rangle_{\lambda,\tau}$ constitute a new representation of quantum mechanics. It is called the coordinate-momentum intermediate representation in terms of the specific expression in Eq. (4.89). Besides, according to Eq. (4.88), we find that the classical Weyl correspondence of the operator $|u\rangle_{\lambda,\tau\lambda,\tau}\langle u|$ is $\delta\left(u-\lambda x-\tau p\right)$, which represents a ray in the x-p phase space.

4.4.2 Two-Mode Case

Since $[X_1+X_2, P_1-P_2]=0$, thus we obtain their common eigen-states, denoted as $|\beta_1,\beta_2\rangle$, by using the Radon transformation of two-mode Wigner operator. The state $|\beta_1,\beta_2\rangle$ obeys the following eigen-equations [10]

$$\begin{aligned}\left(X_1+X_2\right)|\beta_1,\beta_2\rangle&=\beta_1\,|\beta_1,\beta_2\rangle\,,\\\left(P_1-P_2\right)|\beta_1,\beta_2\rangle&=\beta_2\,|\beta_1,\beta_2\rangle\,.\end{aligned} \tag{4.93}$$

Noting that

$$\begin{aligned}
&(X_2+X_1)(P_1-P_2)\\
&=\vdots(X_2+X_1)(P_1-P_2)\vdots\\
&=\vdots(P_1-P_2)(X_2+X_1)\vdots,
\end{aligned} \tag{4.94}$$

so the Weyl classical correspondence of the operator $(X_2+X_1)(P_1-P_2)$ is $(x_2+x_1)(p_1-p_2)$. Therefore, referring to Eqs. (4.87), (4.88) and (4.93), we can construct the following Radon integral transformation:

$$\begin{aligned}
&\iiint\int_{-\infty}^{\infty}\mathrm{d}p_1\mathrm{d}x_1\mathrm{d}p_2\mathrm{d}x_2\delta\left[\beta_1-(x_1+x_2)\right]\delta\left[\beta_2-(p_1-p_2)\right]\\
&\times\Delta_1(x_1,p_1)\Delta_2(x_2,p_2)\\
&=\iiint\int_{-\infty}^{\infty}\mathrm{d}p_1\mathrm{d}x_1\mathrm{d}p_2\mathrm{d}x_2\delta\left[\beta_1-(x_1+x_2)\right]\delta\left[\beta_2-(p_1-p_2)\right]\\
&\times\frac{1}{\pi^2}:\mathrm{e}^{-(x_1-X_1)^2-(p_1-P_1)^2-(x_2-X_2)^2-(p_2-P_2)^2}:\\
&=\frac{1}{\pi^2}\int\int_{-\infty}^{\infty}\mathrm{d}p_2\mathrm{d}x_2:\mathrm{e}^{-(\beta_1-x_2-X_1)^2-(\beta_2+p_2-P_1)^2-(x_2-X_2)^2-(p_2-P_2)^2}:.
\end{aligned} \tag{4.95}$$

Using the convolution formula of integral

$$\int_{-\infty}^{\infty}\frac{\mathrm{d}x}{\sqrt{\pi}}\mathrm{e}^{-(x-y)^2-x^2}=\frac{\mathrm{e}^{-y^2/2}}{\sqrt{2}}, \tag{4.96}$$

we obtain

$$\begin{aligned}
\text{Equation (4.95)}&=\frac{1}{\pi^2}\int\int_{-\infty}^{\infty}\mathrm{d}p_2\mathrm{d}x_2:\mathrm{e}^{-(\beta_1-x_2-X_1-X_2)^2-(\beta_2+p_2+P_2-P_1)^2-x_2^2-p_2^2}:\\
&=\frac{1}{2\pi}:\mathrm{e}^{-\frac{1}{2}[\beta_1-(X_1+X_2)]^2-\frac{1}{2}[\beta_2-(P_1-P_2)]^2}:\ \left|\beta_1,\beta_2\right\rangle\left\langle\beta_1,\beta_2\right|.
\end{aligned} \tag{4.97}$$

Using the identities $X_j=\left(a_j+a_j^\dagger\right)/\sqrt{2}$, $P_j=\left(a_j-a_j^\dagger\right)/(\mathrm{i}\sqrt{2})$, $(j=1,2)$, and $:\exp(-a_1^\dagger a_1-a_2^\dagger a_2):\ =|00\rangle\langle 00|$, it follows from Eq. (4.97) that

$$\left|\beta_1,\beta_2\right\rangle=\frac{1}{\sqrt{2\pi}}\mathrm{e}^{-\frac{\beta_1^2+\beta_2^2}{4}+\frac{\beta_1+\mathrm{i}\beta_2}{\sqrt{2}}a_1^\dagger+\frac{\beta_1-\mathrm{i}\beta_2}{\sqrt{2}}a_2^\dagger-a_1^\dagger a_2^\dagger}\left|00\right\rangle. \tag{4.98}$$

Letting $\xi=(\beta_1+\mathrm{i}\beta_2)/\sqrt{2}$, thus Eq. (4.98) becomes

$$\left|\xi\right\rangle\equiv\sqrt{2\pi}\left|\beta_1,\beta_2\right\rangle=\mathrm{e}^{-\frac{|\xi|^2}{2}+\xi a_1^\dagger+\xi^* a_2^\dagger-a_1^\dagger a_2^\dagger}\left|00\right\rangle, \tag{4.99}$$

substituting it into Eq. (4.97) and completing the integral, we obtain

$$\int\int_{-\infty}^{\infty} \mathrm{d}\beta_1 \mathrm{d}\beta_2 \left|\beta_1, \beta_2\right\rangle \left\langle\beta_1, \beta_2\right| = 1, \tag{4.100}$$

or

$$\int \frac{\mathrm{d}^2\xi}{\pi} \left|\xi\right\rangle \left\langle\xi\right| = 1, \tag{4.101}$$

that is, the states $|\xi\rangle$ have the completeness.

On the other hand, owing to $[X_1 + P_2, P_1 + X_2] = 0$, they also possess the common eigen-states, denoted as $|\kappa_1, \kappa_2\rangle$, and yield the following eigen-equation:

$$\begin{aligned} (X_1 + P_2) \left|\kappa_1, \kappa_2\right\rangle &= \kappa_1 \left|\kappa_1, \kappa_2\right\rangle, \\ (P_1 + X_2) \left|\kappa_1, \kappa_2\right\rangle &= \kappa_2 \left|\kappa_1, \kappa_2\right\rangle. \end{aligned} \tag{4.102}$$

Refering to Eq. (4.102), we construct the following Radon integral transformation:

$$\begin{aligned} \left|\kappa_1, \kappa_2\right\rangle \left\langle\kappa_1, \kappa_2\right| &= \frac{1}{\pi^2} \int\int\int\int_{-\infty}^{\infty} \mathrm{d}p_1 \mathrm{d}x_1 \mathrm{d}p_2 \mathrm{d}x_2 \delta\left[\kappa_1 - (x_1 + p_2)\right] \delta\left[\kappa_2 - (p_1 + x_2)\right] \\ &\quad \times : \mathrm{e}^{-(x_1 - X_1)^2 - (p_1 - P_1)^2 - (x_2 - X_2)^2 - (p_2 - P_2)^2} : \\ &= \frac{1}{\pi^2} \int\int_{-\infty}^{\infty} \mathrm{d}p_2 \mathrm{d}x_2 : \mathrm{e}^{-(\kappa_1 - p_2 - X_1)^2 - (\kappa_2 - x_2 - P_1)^2 - (x_2 - X_2)^2 - (p_2 - P_2)^2} : \\ &= \frac{1}{2\pi} : \mathrm{e}^{-\frac{1}{2}[\kappa_1 - (P_2 + X_1)]^2 - \frac{1}{2}[\kappa_2 - (P_1 + X_2)]^2} : . \end{aligned} \tag{4.103}$$

Similar to obtaining Eq. (4.98), we have

$$\left|\kappa_1, \kappa_2\right\rangle = \frac{1}{\sqrt{2\pi}} \mathrm{e}^{-\frac{\kappa_2^2 + \kappa_1^2}{4} + \frac{\kappa_1 + \mathrm{i}\kappa_2}{\sqrt{2}} a_1^\dagger + \frac{\mathrm{i}(\kappa_1 - \mathrm{i}\kappa_2)}{\sqrt{2}} a_2^\dagger - \mathrm{i} a_1^\dagger a_2^\dagger} \left|00\right\rangle. \tag{4.104}$$

Assuming $\gamma = (\kappa_1 + \mathrm{i}\kappa_2)/\sqrt{2} = \gamma_1 + \mathrm{i}\gamma_2$, thus we have

$$\sqrt{2\pi} \left|\kappa_1, \kappa_2\right\rangle = \mathrm{e}^{-\frac{|\gamma|^2}{2} + \gamma a_1^\dagger + \mathrm{i}\gamma^* a_2^\dagger - \mathrm{i} a_1^\dagger a_2^\dagger} \left|00\right\rangle \equiv \left|\gamma\right\rangle. \tag{4.105}$$

Similarly, integrating Eq. (4.103), we have

$$\int\int_{-\infty}^{\infty} \mathrm{d}\kappa_1 \mathrm{d}\kappa_2 \left|\kappa_1, \kappa_2\right\rangle \left\langle\kappa_1, \kappa_2\right| = 1 \tag{4.106}$$

or

$$\int \frac{\mathrm{d}^2\gamma}{\pi} \left|\gamma\right\rangle \left\langle\gamma\right| = 1. \tag{4.107}$$

The above discussion shows that it is convenient to construct the entangled state representation of quantum mechanics by using the Radon transform of Wigner operator and the Weyl correspondence of Wigner operator. This provides a new way for constructing the quantum mechanics representations. Interested readers can extend the above discussion to the multi-mode case.

4.5 Thermal Wigner Function

This section uses the thermal entangled state representation $|\eta\rangle$ to introduce the thermal Wigner operator. The entangled states $|\eta\rangle$ with fictitious mode read

$$|\eta\rangle = \exp\left(\frac{1}{2}|\eta|^2 + \eta a^\dagger - \eta^* \tilde{a}^\dagger + a^\dagger \tilde{a}^\dagger\right)|00\rangle, \tag{4.108}$$

where $\tilde{a}^\dagger$ is the creation operator of fictitious mode, $[\tilde{a}, \tilde{a}^\dagger] = 1$, $\tilde{a}\left|\tilde{0}\right\rangle = 0$, and the state $|\eta\rangle$ yields the following eigen-equations

$$\langle\eta|(a^\dagger - \tilde{a}) = \eta^* \langle\eta|, \quad \langle\eta|(a - \tilde{a}^\dagger) = \eta^* \langle\eta| \tag{4.109}$$

and the completeness relation

$$\begin{aligned}\int \frac{d^2\eta}{\pi}|\eta\rangle\langle\eta| &= \int \frac{d^2\eta}{\pi} : \exp\left[-|\eta|^2 + \eta(a^\dagger - \tilde{a})\right.\\ &\quad \left.+\eta^*(a - \tilde{a}^\dagger) - a^\dagger a - \tilde{a}^\dagger \tilde{a}\right] :\\ &= 1.\end{aligned} \tag{4.110}$$

Using the thermal entangled state representation, the thermal Wigner operator can be expressed as [11, 12]

$$\Delta_T(\sigma, \gamma) = \int \frac{d^2\eta}{\pi^3}|\sigma - \eta\rangle\langle\sigma + \eta| e^{\eta\gamma^* - \gamma\eta^*}, \tag{4.111}$$

where σ, η are complex, the subscript T represents thermal. To confirm that $\Delta_T(\sigma, \gamma)$ is indeed a thermal Wigner operator, we use the normal ordering product of the projection operator $\left|0, \tilde{0}\right\rangle\left\langle 0, \tilde{0}\right|$, that is

$$\left|0, \tilde{0}\right\rangle\left\langle 0, \tilde{0}\right| =: \exp(-a^\dagger a - \tilde{a}^\dagger \tilde{a}):, \tag{4.112}$$

and the IWOP method to carry out the integral in Eq. (4.111), thus we obtain the normal ordering product of $\Delta_T(\sigma, \gamma)$ as

$$\Delta_T(\sigma,\gamma) = \int \frac{d^2\eta}{\pi^3} \exp\left[-|\sigma|^2 - |\eta|^2 + (\sigma-\eta)a^\dagger - (\sigma^*-\eta^*)\tilde{a}^\dagger + a^\dagger\tilde{a}^\dagger\right]$$
$$\left|0,\tilde{0}\right\rangle\left\langle 0,\tilde{0}\right| \exp\left[(\sigma^*+\eta^*)a - (\sigma+\eta)\,\tilde{a} + a\tilde{a} + \eta\gamma^* - \gamma\eta^*\right]$$
$$= :\ \exp\left[-|\sigma|^2 + \sigma(a^\dagger - \tilde{a}) + \sigma(a - \tilde{a}^\dagger)\right] \int \frac{d^2\eta}{\pi^3}$$
$$:\ \exp[-|\eta|^2 - \eta(a^\dagger + \tilde{a} - \gamma^*)$$
$$+\eta^*(\tilde{a}^\dagger + a - \gamma) + a^\dagger\tilde{a}^\dagger + a\tilde{a} - a^\dagger a - a^\dagger\tilde{a}]:$$
$$= \frac{1}{\pi^2}:\ \exp[-|\sigma|^2 - |\gamma|^2 + \sigma(a^\dagger - \tilde{a}) + \sigma(a - \tilde{a}^\dagger) + \gamma(a^\dagger + \tilde{a})$$
$$+\gamma^*(\tilde{a}^\dagger + a) - 2a^\dagger a - 2\tilde{a}^\dagger\tilde{a}]:. \tag{4.113}$$

Letting

$$\gamma = \alpha + \tau^*, \sigma = \alpha - \tau^*, \tag{4.114}$$

thus Eq. (4.113) becomes

$$\Delta_T(\sigma,\gamma) = \pi^{-2}:\ \exp[-2\left(a^\dagger - \alpha^*\right)(a-\alpha) - 2\left(\tilde{a}^\dagger - \tau^*\right)(\tilde{a}-\tau):, \tag{4.115}$$

which is the normal ordering product of the thermal Wigner operator $\Delta_T(\sigma,\gamma)$ of real and fictitious modes. It is not difficult to see that it is formally the same as the two-mode Wigner operator. Therefore, Eq. (4.115) can be expressed as the product of two single-mode Wigner operators, i.e.

$$\Delta_T(\sigma,\gamma) = \Delta(\alpha)\otimes\tilde{\Delta}(\tau),\quad \alpha = \frac{\gamma+\sigma}{2},\quad \tau^* = \frac{\gamma-\sigma}{2}. \tag{4.116}$$

For the entangled state $|\eta\rangle$, its Wigner function reads

$$\langle\eta|\,\Delta_T(\sigma,\gamma)\,|\eta\rangle = \langle\eta| \int \frac{d^2\eta'}{\pi^3} \left|\sigma-\eta'\right\rangle\left\langle\sigma+\eta'\right| e^{\eta'\gamma^*-\gamma\eta'^*}\,|\eta\rangle$$
$$= \frac{1}{\pi}\delta^{(2)}(2\eta - 2\sigma). \tag{4.117}$$

Of course, we can also directly introduce the direct product form $\Delta(\alpha)\otimes\tilde{\Delta}(\tau)$ as the thermal Wigner operator, but this makes it difficult to see the entangled form of $\Delta(\alpha)\otimes\tilde{\Delta}(\tau)$ in the entangled state representation. It should be noted that using the entangled state representation of the thermal Wigner operator in Eq. (4.111) and the orthogonal relation of $|\eta\rangle$, it is not difficult to prove

$$4\pi^2 \mathtt{tr}[\Delta_T(\sigma,\gamma)\Delta_T(\sigma',\gamma'] = \delta(\sigma-\sigma')\delta(\sigma^*-\sigma'^*)\delta(\gamma-\gamma')\delta(\gamma^*-\gamma'^*) \tag{4.118}$$

and

$$\begin{aligned} 2\pi \mathtt{tr}_{\tilde{a}}\Delta_T(\sigma,\gamma) &= 2\int \frac{d^2\tau}{\pi^2} : \exp[-2\left(a^\dagger-\alpha^*\right)(a-\alpha) - 2\left(\tilde{a}^\dagger-\tau^*\right)(\tilde{a}-\tau) : \\ &= \Delta(\alpha). \end{aligned} \tag{4.119}$$

Thus, the Wigner function for any quantum state ρ can be obtained by virtue of the thermal Wigner function, that is

$$W_\rho(\alpha) = \mathtt{tr}\,(\rho\Delta(\alpha)) = 2\pi \mathtt{tr}_a \mathtt{tr}_{\tilde{a}}[\Delta_T(\sigma,\gamma)\rho], \tag{4.120}$$

where $\mathtt{tr} \equiv \mathtt{tr}_a \mathtt{tr}_{\tilde{a}}$. Using the method of thermal field dynamics, the density operator ρ can be rewritten as the corresponding thermal pure state, i.e., $\rho = \mathtt{tr}_{\tilde{a}}\,(|\psi\rangle\,\langle\psi|)$, thus

$$W_\rho(\alpha) = 2\pi \mathtt{tr}_{\tilde{a}}\,\langle\psi|\,\Delta_T(\sigma,\gamma)\,|\psi\rangle = 2\pi \mathtt{tr}_{\tilde{a}} W_T(\sigma,\gamma), \tag{4.121}$$

where $W_T(\sigma,\gamma)$ is the thermal Wigner function.

Next, using the thermal Wigner operator $\Delta_T(\sigma,\gamma)$, we can easily calculate the Wigner functions for some thermal states. Firstly, using Eq. (4.111), we can obtain

$$\begin{aligned} \left\langle 0\tilde{0}\right|\Delta_T(\sigma,\gamma)\left|0\tilde{0}\right\rangle &= \left\langle 0\tilde{0}\right| \int \frac{d^2\eta'}{\pi^3} \left|\sigma-\eta\right\rangle \left\langle\sigma+\eta'\right| e^{\eta'\gamma^*-\eta^*\gamma} \left|0\tilde{0}\right\rangle \\ &= \int \frac{d^2\eta'}{\pi^3} \exp\left(-|\sigma|^2 - |\eta|^2 + \eta\gamma^* - \eta^*\gamma\right) \\ &= \frac{1}{\pi^2}\exp\left(-|\sigma|^2-|\gamma|^2\right), \end{aligned} \tag{4.122}$$

thus the thermal Wigner function for the thermal vacuum state $|0(\beta)\rangle$ in Eq. (3.8) is obtained as

$$\begin{aligned} \langle 0(\beta)|\,\Delta_T(\sigma,\gamma)\,|0(\beta)\rangle &= \left\langle 0\tilde{0}\right| S^\dagger(\mu)\Delta_T(\sigma,\gamma)S(\mu)\left|0\tilde{0}\right\rangle \\ &= \left\langle 0\tilde{0}\right|\Delta_T\left(\mu\sigma,\frac{\gamma}{\mu}\right)\left|0\tilde{0}\right\rangle \\ &= \frac{1}{\pi^2}\exp\left(-\mu^2|\sigma|^2 - \frac{1}{\mu^2}|\gamma|^2\right), \end{aligned} \tag{4.123}$$

which leads to the Wigner function for the state $|0(\beta)\rangle$ (letting $\gamma+\sigma = 2\alpha$, $\gamma-\sigma = 2\tau^*$), that is

$$\begin{aligned}\langle 0(\beta)|\,\Delta(\alpha)\,|0(\beta)\rangle &= 2\pi \mathtt{tr}_{\tilde{a}} W_T(\sigma,\gamma)\\ &= 2\int \frac{\mathrm{d}^2\tau}{\pi^2} :\, \exp\left(-\mu^2|\alpha-\tau^*|^2 - \frac{1}{\mu^2}|\alpha+\tau^*|^2\right)\\ &= \frac{2\mu^2}{\pi(\mu^4+1)}\exp\left(-\frac{4\mu^2}{\mu^4+1}|\alpha|^2\right).\end{aligned} \tag{4.124}$$

Letting

$$\mu^2 = \frac{1+\tanh\theta}{1-\tanh\theta},\quad \tanh\theta = \exp\left(\frac{\hbar\omega}{2k_BT}\right), \tag{4.125}$$

thus Eq. (4.124) becomes

$$\begin{aligned}\langle 0(\beta)|\,\Delta(\alpha)\,|0(\beta)\rangle &= \frac{1}{2\pi}\mathrm{sech}2\theta \exp\left(-2|\alpha|^2\mathrm{sech}2\theta\right)\\ &= \frac{1-\mathrm{e}^{-\omega\beta}}{\pi(1+\mathrm{e}^{-\omega\beta})}\exp\left[\frac{-2(1-\mathrm{e}^{-\omega\beta})}{1+\mathrm{e}^{-\omega\beta}}|\alpha|^2\right],\end{aligned} \tag{4.126}$$

which is the Wigner function for the state $|0(\beta)\rangle$ with the Bose statistical distribution. To verify the correctness of the above method, we use the original definition of the Wigner operator to calculate the Wigner function for the state $|0(\beta)\rangle$. The thermal state is defined as

$$\rho = (1-\mathrm{e}^{-\omega\beta})\sum_{n=0}^{\infty}\mathrm{e}^{-n\omega\beta}\,|n\rangle\,\langle n|\,, \tag{4.127}$$

where $|n\rangle = \frac{a^{\dagger n}}{\sqrt{n!}}\,|0\rangle$. Noting that the coordinate state representation of the Wigner operator is

$$\Delta(p,q) = \frac{1}{2\pi}\int_{-\infty}^{\infty}\mathrm{d}\mu \left|q-\frac{u}{2}\right\rangle\left\langle q+\frac{u}{2}\right|\mathrm{e}^{\mathrm{i}pu}, \tag{4.128}$$

thus the Wigner function for the thermal state ρ is calculated as

$$\begin{aligned}W(p,q) &= \frac{1}{2\pi}\left(1-\mathrm{e}^{-\omega\beta}\right)\sum_{n=0}^{\infty}\mathrm{e}^{-n\omega\beta}\int_{-\infty}^{\infty}\mathrm{d}\mu\left\langle q+\frac{u}{2}\right|n\Big\rangle\Big\langle n\left|q-\frac{u}{2}\right\rangle \mathrm{e}^{\mathrm{i}pu}\\ &= \frac{1-\mathrm{e}^{-\omega\beta}}{\pi\left(1+\mathrm{e}^{-\omega\beta}\right)}\exp\left[-\frac{(1-\mathrm{e}^{-\omega\beta})}{1+\mathrm{e}^{-\omega\beta}}(q^2+p^2)\right].\end{aligned} \tag{4.129}$$

Since $\alpha = (q+\mathrm{i}p)/\sqrt{2}$, thus we find that Eq. (4.129) is the same as Eq. (4.126).

References

1. Fan HY (2008) Newton-Leibniz integration for ket-bra operators in quantum mechanics (IV)-integrations within Weyl ordered product of operators and their applications. Ann Phys 323(2):500–526
2. Fan HY, Ruan TN (1983) Coherent state formulation of the Weyl correspondence and the Wigner function. Commun Theor Phys 2(6):1563–1574
3. Fan HY, Yuan HC, Jiang NQ (2010) Deriving new operator identities by alternately using normally, antinormally, and Weyl ordered integration technique. Sci China Phys Mech Astron 53:1626–1630
4. Fan HY, Xu YJ, Yuan HC (2011) s-ordering operator expansions of quantum-mechanical fundamental representations and their applications. Sci China Phys Mech Astron 54:2150–2154
5. Fan HY (2010) Unifying the theory of integration within normal-, Weyl- and antinormal-ordering of operators and the s-ordered operator expansion formula of density operators. Chin Phys B 19(5):050303
6. Fan HY, Fan Y (2000) Entanglement of amplitude and optical phase involved in the EPR eigenstate. Mod Phys Lett B 14(27n28):967–973
7. Fan HY, Zou H (1999) Equivalence between SWH phase operator and Ban's phase operator revealed by virtue of $\langle|$ representation. Phys Lett A 255(1–2):7–10
8. Fan HY, Wang JS (2007) New approach for calculating tomograms of quantum states by virtue of the IWOP technique. Commun Theor Phys 47(3):431–436
9. Fan HY (2003) Operator ordering in quantum optics theory and the development of Dirac'ssymbolic method. J Opt B Quantum Semiclassical Opt 5(4):R147–R163
10. Fan HY, Chen HL (2001) Two-parameter Radon transformation of the Wigner operator and its inverse. Chin Phys Lett 18(7):850–853
11. Fan HY, Jiang NQ (2008) Thermo Wigner operator in thermo field dynamics: its introduction and application. Phys Scr 78(4):045402
12. Fan HY (2003) New application of thermo field dynamics in simplifying the calculation of Wigner functions. Mod Phys Lett A 18(11):733–742

Chapter 5
Light Fields Caused by Quantum Diffusion

5.1 Quantum Diffusion Equation and Its Kraus Operator-Sum Solution

The classical diffusion equation is [1]

$$\frac{\partial \mathrm{P}(z,t)}{\partial t}=-\kappa\frac{\partial^2\mathrm{P}(z,t)}{\partial z\partial z^*}, \tag{5.1}$$

where κ is the diffusion rate, thus what is the corresponding quantum diffusion equation? Using the P-representation of the density operator ρ, i.e., $\rho=\pi^{-1}\int \mathrm{d}^2z\mathrm{P}(z)|z\rangle\langle z|$, thus the time-evolution of the density operator ρ satisfies the equation

$$\frac{\mathrm{d}\rho(t)}{\mathrm{d}t}=\int\frac{\mathrm{d}^2z}{\pi}\frac{\partial \mathrm{P}(z,t)}{\partial t}|z\rangle\langle z|. \tag{5.2}$$

Substituting Eq. (5.1) into Eq. (5.2) leads to

$$\frac{\mathrm{d}\rho(t)}{\mathrm{d}t}=-\kappa\int\frac{\mathrm{d}^2z}{\pi}\frac{\partial^2\mathrm{P}(z,t)}{\partial z\partial z^*}|z\rangle\langle z|. \tag{5.3}$$

Using the normal ordering product of the coherent state projector, i.e., $|z\rangle\langle z|=:\mathrm{e}^{-|z|^2+za^\dagger+z^*a-a^\dagger a}:$, we have

$$a^\dagger|z\rangle\langle z|=a^\dagger:\mathrm{e}^{-|z|^2+za^\dagger+z^*a-a^\dagger a}:=\left(z^*+\frac{\partial}{\partial z}\right)|z\rangle\langle z| \tag{5.4}$$

$$|z\rangle\langle z|a=\left(z+\frac{\partial}{\partial z^*}\right)|z\rangle\langle z|, \tag{5.5}$$

H. Fan and L. Hu, *Dissipation and Diffusion of Quantum Light Field*,
https://doi.org/10.1007/978-981-95-1528-8_5

it follows that

$$
\begin{aligned}
-\frac{\partial^2}{\partial z\partial z^*}|z\rangle\langle z| = & \; z\left(z^*+\frac{\partial}{\partial z}\right)|z\rangle\langle z| - \left(z^*+\frac{\partial}{\partial z}\right)\left(z+\frac{\partial}{\partial z^*}\right) \\
& |z\rangle\langle z| - |z|^2|z\rangle\langle z| + \left(z+\frac{\partial}{\partial z^*}\right)(z^*|z\rangle\langle z|) \\
= & \; za^\dagger|z\rangle\langle z| - \left(z^*+\frac{\partial}{\partial z}\right)|z\rangle\langle z|a \\
& -|z|^2|z\rangle\langle z| + \left(z+\frac{\partial}{\partial z^*}\right)|z\rangle\langle z|a^\dagger \\
= & \; a^\dagger a|z\rangle\langle z| - a^\dagger|z\rangle\langle z|a - a|z\rangle\langle z|a^\dagger + |z\rangle\langle z|aa^\dagger. \qquad (5.6)
\end{aligned}
$$

Using the fractional integration method, and noting that $\mathrm{P}(z,t)$ disappears at infinity, we get

$$
\int \frac{\mathrm{d}^2 z}{\pi}\frac{\partial^2 \mathrm{P}(z,t)}{\partial z\partial z^*}|z\rangle\langle z| = \int \frac{\mathrm{d}^2 z}{\pi}\mathrm{P}(z,t)\frac{\partial^2}{\partial z\partial z^*}|z\rangle\langle z|. \qquad (5.7)
$$

In fact, substituting Eq. (5.6) into Eq. (5.7) leads to

$$
\frac{\mathrm{d}\rho(t)}{\mathrm{d}t} = \kappa\int \frac{\mathrm{d}^2 z}{\pi}\mathrm{P}(z,t)(a^\dagger a|z\rangle\langle z| - a^\dagger|z\rangle\langle z|a - a|z\rangle\langle z|a^\dagger + |z\rangle\langle z|aa^\dagger), \qquad (5.8)
$$

which leads to the quantum diffusion equation

$$
\frac{\mathrm{d}\rho(t)}{\mathrm{d}t} = \kappa(a^\dagger a\rho(t) - a^\dagger\rho(t)a - a\rho(t)a^\dagger + \rho(t)aa^\dagger). \qquad (5.9)
$$

Next, we want to obtain the solution of quantum diffusion equation [2–4]. Using the entangled state with fictitious mode in Eq. (4.108), denoted as $|\eta=0\rangle \equiv |I\rangle$, we find [5, 6]

$$
a\,|I\rangle = \tilde{a}^\dagger\,|I\rangle\,, \qquad a^\dagger\,|I\rangle = \tilde{a}\,|I\rangle\,, (a^\dagger a)^n\,|I\rangle = (\tilde{a}^\dagger\tilde{a})^n\,|I\rangle\,. \qquad (5.10)
$$

Acting the state $|I\rangle$ on the both sides of Eq. (5.9), and noting Eq. (5.10), and denoted as $|\rho(t)\rangle = \rho(t)\,|I\rangle$, we get a like Schrödinger equation about the state $|\rho(t)\rangle$, i.e.,

$$
\begin{aligned}
\frac{\mathrm{d}}{\mathrm{d}t}\,|\rho(t)\rangle &= -\kappa(a^\dagger a\rho(t) - a^\dagger\rho(t)a - a\rho(t)a^\dagger + \rho(t)aa^\dagger)\,|I\rangle \\
&= -\kappa(a^\dagger - \tilde{a})(a - \tilde{a}^\dagger)\,|\rho(t)\rangle\,, \qquad (5.11)
\end{aligned}
$$

its formal solution reads

$$
|\rho(t)\rangle = \exp[-\kappa t(a^\dagger - \tilde{a})(a - \tilde{a}^\dagger]|\rho(0)\rangle. \qquad (5.12)
$$

Using Eqs. (4.109) and (5.12) to obtain the inner product $\langle\eta|\rho(t)\rangle$, that is

$$\langle\eta|\rho(t)\rangle = \langle\eta|\exp[-\kappa t(a^\dagger - \tilde{a})(a - \tilde{a}^\dagger)]|\rho(0)\rangle = \mathrm{e}^{-\kappa t|\eta|^2}\langle\eta|\rho(0)\rangle. \tag{5.13}$$

Further, using the completeness relation of the states $|\eta\rangle$ and the operator identity

$$:\exp[f(a^\dagger a + \tilde{a}^\dagger\tilde{a})]: = (f+1)^{a^\dagger a + \tilde{a}^\dagger\tilde{a}}, \tag{5.14}$$

we get

$$\begin{aligned}
|\rho(t)\rangle &= \int \frac{\mathrm{d}^2\eta}{\pi}\mathrm{e}^{-\kappa t|\eta|^2}|\eta\rangle\langle\eta|\rho(0)\rangle \\
&= \int \frac{\mathrm{d}^2\eta}{\pi} : \exp[-(1+\kappa t)|\eta|^2 + \eta(a^\dagger - \tilde{a}) \\
&+ \eta^*(a - \tilde{a}^\dagger) + a^\dagger\tilde{a}^\dagger + a\tilde{a} - a^\dagger a - \tilde{a}^\dagger\tilde{a}] : |\rho(0)\rangle \\
&= \frac{1}{1+\kappa t} : \exp\left[\frac{\kappa t}{1+\kappa t}(a^\dagger\tilde{a}^\dagger + a\tilde{a} - a^\dagger a - \tilde{a}^\dagger\tilde{a})\right] : |\rho(0)\rangle \\
&= \frac{1}{1+\kappa t}\mathrm{e}^{\frac{\kappa t}{1+\kappa t}a^\dagger\tilde{a}^\dagger}\left(\frac{1}{1+\kappa t}\right)^{a^\dagger a + \tilde{a}^\dagger\tilde{a}}\mathrm{e}^{\frac{\kappa t}{1+\kappa t}a\tilde{a}}|\rho(0)\rangle.
\end{aligned} \tag{5.15}$$

Using Eq. (5.10) again, we have

$$\mathrm{e}^{\frac{\kappa t}{1+\kappa t}a\tilde{a}}|\rho(0)\rangle = \sum_{n=0}^{\infty}\frac{1}{n!}\left(\frac{\kappa t}{1+\kappa t}a\right)^n \rho(0)a^{\dagger n}|I\rangle, \tag{5.16}$$

thus Eq. (5.15) can be changed into

$$\begin{aligned}
|\rho(t)\rangle &= \mathrm{e}^{\frac{\kappa t}{1+\kappa t}a^\dagger\tilde{a}^\dagger}\left(\frac{1}{1+\kappa t}\right)^{a^\dagger a+1}\sum_{n=0}^{\infty}\frac{1}{n!}\left(\frac{\kappa t}{1+\kappa t}a\right)^n\rho(0)a^{\dagger n}\left(\frac{1}{1+\kappa t}\right)^{a^\dagger a}|I\rangle \\
&= \sum_{m,n=0}^{\infty}\frac{1}{m!n!}\frac{(\kappa t)^{m+n}}{(\kappa t+1)^{m+n+1}}a^{\dagger m}\left(\frac{1}{1+\kappa t}\right)^{a^\dagger a}a^n\rho(0)a^{\dagger n}\left(\frac{1}{1+\kappa t}\right)^{a^\dagger a}a^m|I\rangle.
\end{aligned} \tag{5.17}$$

Therefore, the density operator $\rho(t)$ is equivalent to

$$\begin{aligned}
\rho(t) &= \sum_{m,n=0}^{\infty}\frac{1}{m!n!}\frac{(\kappa t)^{m+n}}{(\kappa t+1)^{m+n+1}}a^{\dagger m}\left(\frac{1}{1+\kappa t}\right)^{a^\dagger a}a^n\rho(0)a^{\dagger n}\left(\frac{1}{1+\kappa t}\right)^{a^\dagger a}a^m \\
&\equiv \sum_{m,n=0}^{\infty}M_{m,n}\rho(0)M_{m,n}^\dagger,
\end{aligned} \tag{5.18}$$

where $M_{m,n}$ is a Kraus operator

$$M_{m,n} = \sqrt{\frac{1}{m!n!}\frac{(\kappa t)^{m+n}}{(\kappa t+1)^{m+n+1}}} a^{\dagger m}\left(\frac{1}{1+\kappa t}\right)^{a^\dagger a} a^n, \tag{5.19}$$

and yields the identity $\sum_{m,n=0}^{\infty} M_{m,n}^\dagger M_{m,n} = 1$. Equation (5.18) is just the quantum solution of the diffusion equation.

The proof of $\sum_{m,n=0}^{\infty} M_{m,n}^\dagger M_{m,n} = 1$ is given below. Since

$$\vdots e^{xaa^\dagger} \vdots = \frac{1}{1-x} e^{a^\dagger a \ln \frac{1}{1-x}}, \tag{5.20}$$

we have

$$\begin{aligned}\sum_{m,n=0}^{\infty} \frac{1}{m!}\frac{(\kappa t)^m}{(\kappa t+1)^m} a^m a^{\dagger m} &= \vdots \exp\left(\frac{\kappa t}{\kappa t+1} aa^\dagger\right) \vdots \\ &= (\kappa t+1)\, e^{a^\dagger a \ln(\kappa t+1)},\end{aligned} \tag{5.21}$$

thus the summation $\sum_{m,n=0}^{\infty} M_{m,n}^\dagger M_{m,n}$ can be expressed as

$$\begin{aligned}\sum_{m,n=0}^{\infty} M_{m,n}^\dagger M_{m,n} &= \sum_{m,n=0}^{\infty} \frac{1}{m!n!}\frac{(\kappa t)^{m+n}}{(\kappa t+1)^{m+n+1}} a^{\dagger n} \\ &\quad \times \left(\frac{1}{1+\kappa t}\right)^{a^\dagger a} a^m a^{\dagger m} \left(\frac{1}{1+\kappa t}\right)^{a^\dagger a} a^n \\ &= \sum_{n=0}^{\infty} \frac{1}{n!}\frac{(\kappa t)^n}{(\kappa t+1)^n} a^{\dagger n} \left(\frac{1}{1+\kappa t}\right)^{2a^\dagger a} e^{a^\dagger a \ln(\kappa t+1)} a^n \\ &= \sum_{n=0}^{\infty} \frac{1}{n!}\frac{(\kappa t)^n}{(\kappa t+1)^n} a^{\dagger n} e^{a^\dagger a \ln(\kappa t+1)} e^{2a^\dagger a \ln \frac{1}{1+\kappa t}} a^n \\ &= \sum_{n=0}^{\infty} \frac{1}{n!}\frac{(\kappa t)^n}{(\kappa t+1)^n} a^{\dagger n} e^{a^\dagger a\left[\ln(\kappa t+1) + 2\ln \frac{1}{1+\kappa t}\right]} a^n \\ &= \sum_{n=0}^{\infty} \frac{1}{n!}\frac{(\kappa t)^n}{(\kappa t+1)^n} a^{\dagger n} e^{a^\dagger a \ln \frac{1}{1+\kappa t}} a^n \\ &= \sum_{n=0}^{\infty} \frac{1}{n!}\frac{(\kappa t)^n}{(\kappa t+1)^n} a^{\dagger n} :e^{a^\dagger a\left(\frac{1}{1+\kappa t}-1\right)}: a^n \\ &=: e^{a^\dagger a \frac{\kappa t}{1+\kappa t}} e^{a^\dagger a \ln \frac{-\kappa t}{1+\kappa t}} : = 1.\end{aligned} \tag{5.22}$$

Thus, we have completed the proof of $\sum_{m,n=0}^{\infty} M_{m,n}^\dagger M_{m,n} = 1$.

5.2 Diffusion Equation of Wigner Function

This section examines how the Wigner functions diffuse in the diffusion process [5, 7]. The normal ordering product of the Wigner operator $\Delta(\alpha, \alpha^*)$ in the coherent state $|\alpha\rangle$ representation is given by

$$\Delta\left(\alpha, \alpha^*\right) = \frac{1}{\pi} : \mathrm{e}^{-2\left(\alpha^* - a^\dagger\right)(\alpha - a)} : , \tag{5.23}$$

so we have

$$\begin{aligned} \frac{\partial}{\partial \alpha} \Delta\left(\alpha, \alpha^*\right) &= \frac{1}{\pi} \frac{\partial}{\partial \alpha} : \mathrm{e}^{-2\left(\alpha^* - a^\dagger\right)(\alpha - a)} : \\ &= 2\left(a^\dagger - \alpha^*\right) \Delta\left(\alpha, \alpha^*\right), \end{aligned} \tag{5.24}$$

$$\frac{\partial}{\partial \alpha^*} \Delta\left(\alpha, \alpha^*\right) = \Delta\left(\alpha, \alpha^*\right) 2(a - \alpha), \tag{5.25}$$

it follows that

$$a^\dagger \Delta\left(\alpha, \alpha^*\right) = \left(\frac{\partial}{2\partial \alpha} + \alpha^*\right) \Delta\left(\alpha, \alpha^*\right), \tag{5.26}$$

$$\Delta\left(\alpha, \alpha^*\right) a = \left(\frac{\partial}{2\partial \alpha^*} + \alpha\right) \Delta\left(\alpha, \alpha^*\right). \tag{5.27}$$

On the other hand, using the anti-normal ordering product of the Wigner operator $\Delta(\alpha, \alpha^*)$, i.e.,

$$\Delta\left(\alpha, \alpha^*\right) = -\frac{1}{\pi} \vdots \mathrm{e}^{2\left(\alpha^* - a^\dagger\right)(\alpha - a)} \vdots, \tag{5.28}$$

we obtain

$$\frac{\partial}{\partial \alpha^*} \Delta\left(\alpha, \alpha^*\right) = 2(\alpha - a) \Delta\left(\alpha, \alpha^*\right) = \Delta\left(\alpha, \alpha^*\right) 2(a - \alpha), \tag{5.29}$$

$$\frac{\partial}{\partial \alpha} \Delta\left(\alpha, \alpha^*\right) = 2\Delta\left(\alpha, \alpha^*\right)\left(\alpha^* - a^\dagger\right) = 2\left(a^\dagger - \alpha^*\right) \Delta\left(\alpha, \alpha^*\right), \tag{5.30}$$

which leads to the identities

$$\left(\alpha - \frac{\partial}{2\partial \alpha^*}\right) \Delta\left(\alpha, \alpha^*\right) = a \Delta\left(\alpha, \alpha^*\right), \tag{5.31}$$

$$\left(\alpha^* - \frac{\partial}{2\partial \alpha}\right) \Delta\left(\alpha, \alpha^*\right) = \Delta\left(\alpha, \alpha^*\right) a^\dagger. \tag{5.32}$$

Therefore, substituting Eqs. (5.26), (5.27), (5.31) and (5.32) into the definition of the Wigner function, i.e., $W(\alpha, \alpha^*) = \mathtt{tr}(\rho\Delta(\alpha, \alpha^*))$, inserting

$$\begin{aligned}\Delta(\alpha, \alpha^*) a^\dagger a &= \left(\alpha^* - \frac{\partial}{2\partial\alpha}\right)\Delta(\alpha, \alpha^*) a \\ &= \left(\alpha^* - \frac{\partial}{2\partial\alpha}\right)\left(\frac{\partial}{2\partial\alpha^*} + \alpha\right)\Delta(\alpha, \alpha^*) \end{aligned} \tag{5.33}$$

and

$$\begin{aligned}\mathtt{tr}\left(\Delta(\alpha, \alpha^*) a^\dagger \rho a\right) &= \mathtt{tr}\left(a\Delta(\alpha, \alpha^*) a^\dagger \rho\right) \\ &= \mathtt{tr}\left[\left(\alpha - \frac{\partial}{2\partial\alpha^*}\right)\Delta(\alpha, \alpha^*) a^\dagger \rho\right] \\ &= \mathtt{tr}\left[\left(\alpha - \frac{\partial}{2\partial\alpha^*}\right)\left(\alpha^* - \frac{\partial}{2\partial\alpha}\right)\Delta(\alpha, \alpha^*)\rho\right] \\ &= \left(\alpha - \frac{\partial}{2\partial\alpha^*}\right)\left(\alpha^* - \frac{\partial}{2\partial\alpha}\right) W(\alpha, \alpha^*), \end{aligned} \tag{5.34}$$

$$\begin{aligned}\mathtt{tr}\left(\Delta(\alpha, \alpha^*) a\rho a^\dagger\right) &= \mathtt{tr}\left(a^\dagger\Delta(\alpha, \alpha^*) a\rho\right) \\ &= \mathtt{tr}\left[\left(\frac{\partial}{2\partial\alpha} + \alpha^*\right)\Delta(\alpha, \alpha^*) a\rho\right], \\ &= \mathtt{tr}\left[\left(\frac{\partial}{2\partial\alpha} + \alpha^*\right)\left(\frac{\partial}{2\partial\alpha^*} + \alpha\right)\Delta(\alpha, \alpha^*)\rho\right] \\ &= \left(\frac{\partial}{2\partial\alpha} + \alpha^*\right)\left(\frac{\partial}{2\partial\alpha^*} + \alpha\right) W(\alpha, \alpha^*) \\ \mathtt{tr}\left(\Delta(\alpha, \alpha^*)\rho a a^\dagger\right) &= \mathtt{tr}\left(a a^\dagger\Delta(\alpha, \alpha^*)\rho\right) \\ &= \mathtt{tr}\left[\left(\frac{\partial}{2\partial\alpha} + \alpha^*\right)\left(\alpha - \frac{\partial}{2\partial\alpha^*}\right)\Delta(\alpha, \alpha^*)\rho\right] \\ &= \left(\frac{\partial}{2\partial\alpha} + \alpha^*\right)\left(\alpha - \frac{\partial}{2\partial\alpha^*}\right) W(\alpha, \alpha^*), \end{aligned} \tag{5.35}$$

as well as

$$\left[\alpha, \frac{\partial}{2\partial\alpha}\right] = -\frac{1}{2} \tag{5.36}$$

into Eq. (5.9), we finally obtain the diffusion equation of the Wigner function $W(\alpha, \alpha^*)$ as

$$
\begin{aligned}
&\frac{\partial W\left(\alpha, \alpha^{*}, t\right)}{\partial t} \\
&= \operatorname{tr}\left(\Delta\left(\alpha, \alpha^{*}\right) \frac{\partial}{\partial t} \rho(t)\right) \\
&= \kappa \operatorname{tr}\left[\Delta\left(\alpha, \alpha^{*}\right)\left(a^{\dagger} a \rho(t)-a^{\dagger} \rho(t) a-a \rho(t) a^{\dagger}+\rho(t) a a^{\dagger}\right)\right] \\
&= \kappa\left[\left(\alpha^{*}-\frac{\partial}{2 \partial \alpha}\right)\left(\frac{\partial}{2 \partial \alpha^{*}}+\alpha\right)-\left(\alpha-\frac{\partial}{2 \partial \alpha^{*}}\right)\left(\alpha^{*}-\frac{\partial}{2 \partial \alpha}\right)\right. \\
&\quad \left.-\left(\frac{\partial}{2 \partial \alpha}+\alpha^{*}\right)\left(\frac{\partial}{2 \partial \alpha^{*}}+\alpha\right)+\left(\frac{\partial}{2 \partial \alpha}+\alpha^{*}\right)\left(\alpha-\frac{\partial}{2 \partial \alpha^{*}}\right)\right] W\left(\alpha, \alpha^{*}, t\right) \\
&= \kappa\left(-\frac{\partial^{2}}{\partial \alpha \partial \alpha^{*}}+\alpha \frac{\partial}{2 \partial \alpha}-\frac{\partial}{2 \partial \alpha} \alpha+\frac{\partial}{2 \partial \alpha^{*}} \alpha^{*}-\alpha^{*} \frac{\partial}{2 \partial \alpha^{*}}\right) W\left(\alpha, \alpha^{*}, t\right) \\
&= -\kappa \frac{\partial^{2}}{\partial \alpha \partial \alpha^{*}} W\left(\alpha, \alpha^{*}, t\right),
\end{aligned}
\tag{5.37}
$$

which is the same as Eq. (5.1).

On the other hand, in order to study the Wigner function evolutions of quantum states in the diffusion process, we can also start by discussing the evolution of the Wigner operator. Using Eq. (5.18), we have

$$
\begin{aligned}
W\left(\alpha, \alpha^{*}, t\right) &= \operatorname{tr}\left[\rho(t) \Delta\left(\alpha, \alpha^{*}\right)\right] \\
&= \operatorname{tr}\left[\sum_{m, n=0}^{\infty} M_{m, n} \rho(0) M_{m, n}^{\dagger} \Delta\left(\alpha, \alpha^{*}\right)\right] \\
&= \operatorname{tr}\left[\sum_{m, n=0}^{\infty} \rho(0) M_{m, n}^{\dagger} \Delta\left(\alpha, \alpha^{*}\right) M_{m, n}\right] \\
&= \operatorname{tr}\left[\rho(0) \Delta\left(\alpha, \alpha^{*}, t\right)\right],
\end{aligned}
\tag{5.38}
$$

where

$$
\Delta\left(\alpha, \alpha^{*}, t\right)=\sum_{m, n=0}^{\infty} M_{m, n}^{\dagger} \Delta\left(\alpha, \alpha^{*}\right) M_{m, n}. \tag{5.39}
$$

Therefore, finding the Wigner function $W\left(\alpha, \alpha^{*}, t\right)$ at time t comes down to calculating $\Delta\left(\alpha, \alpha^{*}, t\right)$. Noting that

$$
\Delta\left(\alpha, \alpha^{*}\right)=\frac{1}{\pi}: \mathrm{e}^{-2\left(a^{\dagger}-\alpha^{*}\right)(a-\alpha)}:=\frac{1}{\pi} \mathrm{e}^{-2 \alpha \alpha^{*}} \mathrm{e}^{2 a^{\dagger} \alpha}(-1)^{N} \mathrm{e}^{2 \alpha^{*} a}, \tag{5.40}
$$

and

$$
(-1)^{N} a=-a(-1)^{N}, \quad N=a^{\dagger} a, \tag{5.41}
$$

as well as

$$\begin{aligned}\exp(a^\dagger a \ln A) f(a^\dagger) \exp(-a^\dagger a \ln A) &= f(Aa^\dagger),\\ \exp(a^\dagger a \ln A) f(a) \exp(-a^\dagger a \ln A) &= f(a/A),\end{aligned} \tag{5.42}$$

thus we have

$$\begin{aligned}\Delta\left(\alpha, \alpha^*, t\right) = &\frac{e^{-2|\alpha|^2}}{\pi} \sum_{m,n=0}^{\infty} \frac{1}{m!n!} \frac{(\kappa t)^{m+n}}{(\kappa t+1)^{m+n+1}} a^{\dagger n} \left(\frac{1}{1+\kappa t}\right)^{a^\dagger a}\\ &\times a^m e^{2a^\dagger \alpha} (-1)^N e^{2\alpha^* a} a^{\dagger m} \left(\frac{1}{1+\kappa t}\right)^{a^\dagger a} a^n\\ = &\frac{e^{-2|\alpha|^2}}{\pi(1+\kappa t)} \sum_{m,n=0}^{\infty} \frac{1}{m!n!} \left(1 - \frac{1}{1+\kappa t}\right)^{m+n}\\ &\times a^{\dagger n} [(1+\kappa t) a]^m \left(\frac{1}{1+\kappa t}\right)^{a^\dagger a} e^{2a^\dagger \alpha}\\ &\times (-1)^N e^{2\alpha^* a} \left(\frac{1}{1+\kappa t}\right)^{a^\dagger a} [(1+\kappa t) a^\dagger]^m a^n.\end{aligned} \tag{5.43}$$

Letting

$$T_1 = 1 - \frac{1}{1+\kappa t}, \; T_2 = \frac{1}{1+\kappa t}, \tag{5.44}$$

thus Eq. (5.43) becomes

$$\begin{aligned}\Delta\left(\alpha, \alpha^*, t\right) = &\frac{T_2 e^{-2|\alpha|^2}}{\pi} \sum_{m,n=0}^{\infty} \frac{T_1^{m+n}}{m!n!T_2^{2m}} a^{\dagger n} a^m e^{a^\dagger a \ln T_2}\\ &\times e^{2a^\dagger \alpha} (-1)^N e^{2\alpha^* a} e^{a^\dagger a \ln T_2} a^{\dagger m} a^n\\ = &\frac{e^{-2|\alpha|^2} T_2}{\pi} \sum_{m,n=0}^{\infty} \frac{T_1^{m+n}}{m!n!T_2^{2m}} a^{\dagger n} a^m e^{2\alpha T_2 a^\dagger}\\ &\times e^{a^\dagger a \ln T_2} (-1)^N e^{a^\dagger a \ln T_2} e^{2\alpha^* T_2 a} a^{\dagger m} a^n\\ = &\frac{e^{-2|\alpha|^2} T_2}{\pi} \sum_{m,n=0}^{\infty} \frac{T_1^{m+n}}{m!n!T_2^{2m}} a^{\dagger n} a^m e^{2\alpha T_2 a^\dagger}\\ &\times e^{-2\alpha^* a/T_2} \exp\left[a^\dagger a \ln\left(-T_2^2\right)\right] a^{\dagger m} a^n\\ = &\frac{e^{-2|\alpha|^2} T_2}{\pi} \sum_{m,n=0}^{\infty} \frac{T_1^{m+n}}{m!n!T_2^{2m}} a^{\dagger n} a^m e^{2\alpha T_2 a^\dagger}\\ &\times e^{-2\alpha^* a/T_2} a^{\dagger m} a^n \exp\left[a^\dagger a \ln\left(-T_2^2\right)\right] \left(-T_2^2\right)^{m-n}.\end{aligned} \tag{5.45}$$

Further, using the following identities

$$\begin{aligned} e^{Aa} f\left(a^{\dagger}\right) e^{-Aa} &= \sum_{n=0}^{\infty} \frac{A^n}{n!} f^{(n)}\left(a^{\dagger}\right) = f\left(a^{\dagger} + A\right), \\ e^{Aa^{\dagger}} f(a) e^{-Aa^{\dagger}} &= \sum_{n=0}^{\infty} \frac{A^n}{n!} (-1)^n f^{(n)}\left(a^{\dagger}\right) = f(a - A) \end{aligned} \tag{5.46}$$

to rewrite Eq. (5.45) as

$$\begin{aligned} \Delta\left(\alpha, \alpha^*, t\right) = &\frac{e^{-2|\alpha|^2} T_2}{\pi} \sum_{m,n=0}^{\infty} \frac{T_1^{m+n}}{m!n!T_2^{2m}} a^{\dagger n} \exp\left(2\alpha T_2 a^{\dagger}\right)(a + 2\alpha T_2)^m \\ &\times \left(a^{\dagger} - \frac{2\alpha^*}{T_2}\right)^m \exp\left(\frac{-2\alpha^* a}{T_2}\right) a^n \exp\left[a^{\dagger} a \ln\left(-T_2^2\right)\right]\left(-T_2^2\right)^{m-n}. \end{aligned} \tag{5.47}$$

Inserting the completeness relation of coherent states, i.e.,

$$\int \frac{d^2 z}{\pi} |z\rangle \langle z| = \int \frac{d^2 z}{\pi} : e^{-|z|^2 + z a^{\dagger} + z^* a - a^{\dagger} a} : = 1 \tag{5.48}$$

into Eq. (5.47), and using $a|z\rangle = z|z\rangle$, we get

$$\begin{aligned} \Delta\left(\alpha, \alpha^*, t\right) = &\frac{e^{-2|\alpha|^2} T_2}{\pi} \sum_{n,m=0}^{\infty} \frac{T_1^{m+n}(-1)^{m-n}}{m!n!T_2^{2n}} e^{2\alpha T_2 a^{\dagger}} a^{\dagger n} \int \frac{d^2 z}{\pi} (z + 2\alpha T_2)^m \\ &|z\rangle\langle z| \left(z^* - \frac{2\alpha^*}{T_2}\right)^m a^n e^{-2\alpha^* a/T_2} \exp\left[a^{\dagger} a \ln\left(-T_2^2\right)\right] \\ = &\frac{\exp(-2|\alpha|^2) T_2}{\pi} \sum_{n=0}^{\infty} \frac{T_1^n}{n!\left(-T_2^2\right)^n} e^{2\alpha T_2 a^{\dagger}} a^{\dagger n} \int \frac{d^2 z}{\pi} \\ &\exp\left[-T_1 (z + 2\alpha T_2)\left(z^* - \frac{2\alpha^*}{T_2}\right)\right] |z\rangle\langle z| a^n e^{-2\alpha^* a/T_2} e^{a^{\dagger} a \ln(-T_2^2)} \\ = &\frac{\exp(-2|\alpha|^2) T_2}{\pi} \sum_{n=0}^{\infty} \frac{T_1^n}{n!\left(-T_2^2\right)^n} e^{2\alpha T_2 a^{\dagger}} : a^{\dagger n} \\ &\times \int \frac{d^2 z}{\pi} \exp[-(T_1 + 1)|z|^2 + 2z\alpha^* T_1/T_2 - 2z^* \alpha T_1 T_2 + z a^{\dagger} \\ &+ z^* a - a^{\dagger} a + 4|\alpha|^2 T_1] a^n : e^{-2\alpha^* a/T_2} \exp\left[a^{\dagger} a \ln\left(-T_2^2\right)\right]. \end{aligned} \tag{5.49}$$

Using the IWOP method to carry out the integral in Eq. (5.49), we have

$$
\begin{aligned}
\Delta\left(\alpha, \alpha^{*}, t\right) = & \frac{\mathrm{e}^{-2|\alpha|^{2}} T_{2}}{\pi\left(1+T_{1}\right)} \sum_{n=0}^{\infty} \frac{T_{1}^{n}}{n!\left(-T_{2}^{2}\right)^{n}} \mathrm{e}^{2 \alpha T_{2} a^{\dagger}}: a^{\dagger n} \\
& \times \exp \left[\frac{T_{1}\left(2 \alpha^{*} a / T_{2}-a^{\dagger} a-2 \alpha T_{2} a^{\dagger}+4 \alpha^{*} \alpha\right)}{1+T_{1}}\right] \\
& \times a^{n}: \mathrm{e}^{-2 \alpha^{*} a / T_{2}} \exp \left[a^{\dagger} a \ln \left(-T_{2}^{2}\right)\right] \\
= & \frac{T_{2}}{\pi\left(1+T_{1}\right)} \mathrm{e}^{\frac{\left(T_{1}-1\right) 2|\alpha|^{2}+2 \alpha T_{2} a^{\dagger}}{1+T_{1}}}: \exp \left[\frac{T_{1}\left(2 \alpha^{*} a / T_{2}-a^{\dagger} a-2 \alpha T_{2} a^{\dagger}\right)}{1+T_{1}}\right. \\
& \left.-\frac{T_{1}}{T_{2}^{2}} a^{\dagger} a\right]: \mathrm{e}^{-2 \alpha^{*} a / T_{2}} \exp \left[a^{\dagger} a \ln \left(-T_{2}^{2}\right)\right] \\
= & \frac{T_{2} \mathrm{e}^{\frac{\left(T_{1}-1\right) 2|\alpha|^{2}}{1+T_{1}}}}{\pi\left(1+T_{1}\right)} \mathrm{e}^{\frac{2 \alpha T_{2} a^{\dagger}}{1+T_{1}}}: \exp \left[\left(\frac{1}{1+T_{1}}-\frac{T_{1}}{T_{2}^{2}}-1\right) a^{\dagger} a\right]: \\
& \times \mathrm{e}^{-\frac{2 \alpha^{*} a}{T_{2}\left(1+T_{1}\right)}} \exp \left[a^{\dagger} a \ln \left(-T_{2}^{2}\right)\right] \\
= & \frac{T_{2} \mathrm{e}^{\frac{\left(T_{1}-1\right) 2|\alpha|^{2}}{1+T_{1}}}}{\pi\left(1+T_{1}\right)} \mathrm{e}^{\frac{2 \alpha T_{2} a^{\dagger}}{1+T_{1}}} \mathrm{e}^{\ln \left(\frac{1}{1+T_{1}}-\frac{T_{1}}{T_{2}^{2}}\right) a^{\dagger} a} \mathrm{e}^{a^{\dagger} a \ln \left(-T_{2}^{2}\right)} \mathrm{e}^{\frac{2 \alpha^{*} T_{2} a}{1+T_{1}}} \\
= & \frac{T_{2} \mathrm{e}^{\frac{\left(T_{1}-1\right) 2|\alpha|^{2}}{1+T_{1}}}}{\pi\left(1+T_{1}\right)} \mathrm{e}^{\frac{2 \alpha T_{2} a^{\dagger}}{1+T_{1}}} \exp \left[\ln \left(T_{1}-\frac{T_{2}^{2}}{1+T_{1}}\right) a^{\dagger} a\right] \mathrm{e}^{\frac{2 \alpha^{*} T_{2} a}{1+T_{1}}} .
\end{aligned} \tag{5.50}
$$

Further, Eq. (5.50) can be simplified as

$$
\begin{aligned}
\Delta\left(\alpha, \alpha^{*}, t\right) = & \frac{T_{2} \exp \left(2 \frac{T_{1}-1}{T_{1}+1}|\alpha|^{2}\right)}{\pi\left(T_{1}+1\right)} \exp \left[\frac{2 \alpha\left(1-T_{1}\right) a^{\dagger}}{T_{1}+1}\right] \\
& : \exp \left(2 \frac{T_{1}-1}{T_{1}+1} a^{\dagger} a\right): \exp \left[\frac{2 \alpha^{*}\left(1-T_{1}\right) a}{T_{1}+1}\right] \\
= & \frac{1-T_{1}}{\pi\left(T_{1}+1\right)}: \exp \left[2 \frac{T_{1}-1}{T_{1}+1}\left(a^{\dagger}-\alpha^{*}\right)(a-\alpha)\right]: \\
= & \frac{1}{\pi(2 \kappa t+1)}: \exp \left[\frac{-2}{2 \kappa t+1}\left(a^{\dagger}-\alpha^{*}\right)(a-\alpha)\right]: \\
= & \frac{\mathrm{e}^{\frac{2 \alpha a^{\dagger}}{2 \kappa t+1}}}{\pi(2 \kappa t+1)}: \exp \left(\frac{-2}{2 \kappa t+1} a^{\dagger} a\right): \mathrm{e}^{\frac{2 \alpha^{*} a}{2 \kappa t+1}} \mathrm{e}^{\frac{-2}{2 \kappa t+1}|\alpha|^{2}} \\
= & \frac{\mathrm{e}^{\frac{2 \alpha a^{\dagger}}{2 \kappa t+1}}}{\pi(2 \kappa t+1)} \exp \left(a^{\dagger} a \ln \frac{2 \kappa t-1}{2 \kappa t+1}\right) \mathrm{e}^{\frac{2 \alpha^{*} a}{2 \kappa t+1}} \mathrm{e}^{\frac{-2}{2 \kappa t+1}|\alpha|^{2}},
\end{aligned} \tag{5.51}
$$

which clearly reflects the evolution of the Wigner operator $\Delta\left(\alpha, \alpha^{*}\right)$ with time t. For $t=0$, Eq. (5.51) reduces to $\Delta\left(\alpha, \alpha^{*}, 0\right)=\pi^{-1}: \exp \left[-2\left(a^{\dagger}-\alpha^{*}\right)(a-\alpha)\right]:$.

For example, a pure number state $|n\rangle\langle n|$ acting as the initial state, thus its Wigner function is

$$\langle n|\,\Delta\left(\alpha,\alpha^*\right)|n\rangle = \frac{1}{\pi}\mathrm{e}^{-2|\alpha|^2}\mathrm{L}_n\left(4|\alpha|^2\right)(-1)^n\,, \tag{5.52}$$

where $\mathrm{L}_n(\cdot)$ is a Laguerre polynomial, using $a\,|n\rangle = \sqrt{n}\,|n-1\rangle$, and referring to Eq. (5.51), we have

$$\mathrm{e}^{\lambda a}\,|n\rangle = \sum_{l=0}^{n}\frac{(\lambda a)^l}{l!}\,|n\rangle = \sum_{l=0}^{n}\frac{\lambda^l}{l!}\sqrt{\frac{n!}{(n-l)!}}\,|n-l\rangle\,, \tag{5.53}$$

where $\lambda \equiv \frac{2\alpha^*}{2\kappa t+1}$, so the Wigner function evolution of the number state in the diffusion process is

$$\begin{aligned}
W\left(\alpha,\alpha^*,t\right) &= \mathrm{tr}\left[\rho(0)\Delta\left(\alpha,\alpha^*,t\right)\right] = \langle n|\,\Delta\left(\alpha,\alpha^*,t\right)|n\rangle \\
&= \frac{\mathrm{e}^{\frac{-2}{2\kappa t+1}|\alpha|^2}}{\pi(2\kappa t+1)}\sum_{m,l=0}^{n}\frac{\lambda^l\lambda^{*m}}{l!m!}\sqrt{\frac{n!n!}{(n-m)!\,(n-l)!}} \\
&\quad \langle n-m|\exp\left(a^\dagger a\ln\frac{2\kappa t-1}{2\kappa t+1}\right)|n-l\rangle \\
&= \frac{\mathrm{e}^{-\frac{2}{2\kappa t+1}|\alpha|^2}}{\pi(2\kappa t+1)}\sum_{m,l=0}^{n}\frac{\lambda^l\lambda^{*m}}{l!m!}\sqrt{\frac{n!n!}{(n-m)!\,(n-l)!}} \\
&\quad \times\exp\left[(n-l)\ln\frac{2\kappa t-1}{2\kappa t+1}\right]\delta_{m,l} \\
&= \frac{\mathrm{e}^{\frac{-2}{2\kappa t+1}|\alpha|^2}}{\pi(2\kappa t+1)}\sum_{l=0}^{n}\frac{\lambda^l\lambda^{*l}n!}{l!l!\,(n-l)!}\exp\left[(n-l)\ln\frac{2\kappa t-1}{2\kappa t+1}\right] \\
&= \frac{(2\kappa t-1)^n}{\pi(2\kappa t+1)^{n+1}}\mathrm{e}^{\frac{-2}{2\kappa t+1}|\alpha|^2}\sum_{l=0}^{n}\frac{|\lambda|^{2l}\,n!}{l!l!\,(n-l)!}\left(\frac{2\kappa t-1}{2\kappa t+1}\right)^l \\
&= \frac{1}{\pi}\frac{(2\kappa t-1)^n}{(2\kappa t+1)^{n+1}}\mathrm{e}^{\frac{-2}{2\kappa t+1}|\alpha|^2}\mathrm{L}_n\left(\frac{4\,|\alpha|^2}{1-4\kappa^2t^2}\right).
\end{aligned} \tag{5.54}$$

When $t=0$, Eq. (5.54) returns to Eq. (5.52).

5.3 Diffusion of Several Common Quantum Light Fields

5.3.1 Initial Coherent Light Field

When the initial state is a coherent light field, $\rho\,(0) = |z\rangle\langle z|$, whose P-representation is expressed as

$$\mathrm{P}_0 = \delta\left(z^*-\alpha^*\right)\delta\left(z-\alpha\right). \tag{5.55}$$

The solution of the diffusion equation $\frac{\partial \mathrm{P}(z,t)}{\partial t} = -\kappa \frac{\partial^2 \mathrm{P}(z,t)}{\partial z \partial z^*}$ is

$$\mathrm{P}_t = \frac{1}{\kappa t} \exp\left[-\frac{1}{\kappa t}\left(z^* - \alpha^*\right)(z - \alpha)\right], \tag{5.56}$$

which is the coherent state representation of the density operator, so its density operator corresponding to P_t is

$$\rho(t) = \frac{1}{\kappa t} \vdots \exp\left[-\frac{1}{\kappa t}\left(z^* - a^\dagger\right)(z - a)\right] \vdots . \tag{5.57}$$

Using the IWOP method to obtain the normal ordering product of $\rho(t)$, that is

$$\begin{aligned}
\rho(t) &= \frac{1}{\kappa t} \vdots \exp\left[-\frac{1}{\kappa t}\left(z^* - a^\dagger\right)(z - a)\right] \vdots \\
&= \frac{1}{\kappa t} \int \frac{\mathrm{d}^2\alpha}{\pi} |\alpha\rangle\langle\alpha| \exp\left[-\frac{1}{\kappa t}\left(z^* - a^\dagger\right)(z - a)\right] \\
&= \frac{1}{\kappa t} \int \frac{\mathrm{d}^2\alpha}{\pi} : \exp\left[-\frac{\left(z^* - \alpha^*\right)(z - \alpha)}{\kappa t} - |\alpha|^2 + \alpha a^\dagger + \alpha^* a - a^\dagger a\right] : \\
&= \frac{1}{1 + \kappa t} \mathrm{e}^{\frac{z}{1+\kappa t} a^\dagger} : \mathrm{e}^{\left(\frac{\kappa t}{1+\kappa t} - 1\right) a^\dagger a} : \mathrm{e}^{\frac{z^*}{1+\kappa t} a} \mathrm{e}^{-\frac{|z|^2}{1+\kappa t}} \\
&= \frac{1}{1 + \kappa t} \mathrm{e}^{\frac{z}{1+\kappa t} a^\dagger} \mathrm{e}^{aa \ln \frac{\kappa t}{1+\kappa t}} \mathrm{e}^{\frac{z^*}{1+\kappa t} a} \mathrm{e}^{-\frac{|z|^2}{1+\kappa t}},
\end{aligned} \tag{5.58}$$

which yields $\mathtt{tr}\rho(t) = 1$. Equation (5.58) shows that $\rho(t)$ is no longer a pure state, but a normalized generalized chaotic state. So, we say that the coherent light field becomes a generalized chaotic light field after diffusion [2].

Using the IWOP method, we can find the thermal vacuum state corresponding to the density operator $\rho(t)$. Since

$$\begin{aligned}
& \mathrm{e}^{\frac{z^* a}{\kappa t}} \mathrm{e}^{a^\dagger a \ln \frac{\kappa t}{1+\kappa t}} \mathrm{e}^{\frac{z a^\dagger}{\kappa t}} \\
&= \mathrm{e}^{\frac{z^* a}{\kappa t}} : \mathrm{e}^{\frac{\kappa t}{1+\kappa t} a^\dagger a - a^\dagger a} : \mathrm{e}^{\frac{z a^\dagger}{\kappa t}}
\end{aligned} \tag{5.59}$$

$$\begin{aligned}
&= \mathrm{e}^{\frac{z^* a}{\kappa t}} \int \frac{\mathrm{d}^2\beta}{\pi} : \mathrm{e}^{-|\beta|^2 + \beta^* \sqrt{\frac{\kappa t}{1+\kappa t}} a^\dagger + \beta \sqrt{\frac{\kappa t}{1+\kappa t}} a - a^\dagger a} : \mathrm{e}^{\frac{z a^\dagger}{\kappa t}} \\
&= \int \frac{\mathrm{d}^2\beta}{\pi} \left\langle\tilde{\beta}|\tilde{0}\right\rangle \mathrm{e}^{\frac{z^*\beta^*}{\kappa t}\sqrt{\frac{\kappa t}{1+\kappa t}}} \left\| \beta^* \sqrt{\frac{\kappa t}{1 + \kappa t}} \right\rangle \left\langle \beta^* \sqrt{\frac{\kappa t}{1 + \kappa t}} \right\| \mathrm{e}^{\frac{z\beta}{\kappa t}\sqrt{\frac{\kappa t}{1+\kappa t}}} \langle\tilde{0}|\tilde{\beta} \\
&= \int \frac{\mathrm{d}^2\beta}{\pi} \langle\tilde{\beta}| \mathrm{e}^{\frac{z^*\beta^*}{\kappa t}\sqrt{\frac{\kappa t}{1+\kappa t}} + \beta^* \sqrt{\frac{\kappa t}{1+\kappa t}} a^\dagger} \left|0, \tilde{0}\right\rangle \left\langle 0, \tilde{0}\right| \mathrm{e}^{\frac{z\beta}{\kappa t}\sqrt{\frac{\kappa t}{1+\kappa t}} + \beta \sqrt{\frac{\kappa t}{1+\kappa t}} a} \left|\tilde{\beta}\right\rangle \\
&= \int \frac{\mathrm{d}^2\beta}{\pi} \langle\tilde{\beta}| \mathrm{e}^{z^* \tilde{a}^\dagger \sqrt{\frac{1}{\kappa t(1+\kappa t)}} + \sqrt{\frac{\kappa t}{1+\kappa t}} \tilde{a}^\dagger a^\dagger} \left|0, \tilde{0}\right\rangle \left\langle 0, \tilde{0}\right| \mathrm{e}^{z\tilde{a}\sqrt{\frac{1}{\kappa t(1+\kappa t)}} + \sqrt{\frac{\kappa t}{1+\kappa t}} \tilde{a} a} \left|\tilde{\beta}\right\rangle \\
&= \widetilde{\mathtt{tr}} \int \frac{\mathrm{d}^2\beta}{\pi} \left|\tilde{\beta}\right\rangle \left(\mathrm{e}^{z^* \tilde{a}^\dagger \sqrt{\frac{1}{\kappa t(1+\kappa t)}} + \sqrt{\frac{\kappa t}{1+\kappa t}} \tilde{a}^\dagger a^\dagger} \left|0, \tilde{0}\right\rangle \left\langle 0, \tilde{0}\right| \mathrm{e}^{z\tilde{a}\sqrt{\frac{1}{\kappa t(1+\kappa t)}} + \sqrt{\frac{\kappa t}{1+\kappa t}} \tilde{a} a} \right),
\end{aligned} \tag{5.60}$$

by comparing Eq. (3.5), we find that the thermal vacuum state is

$$\frac{1}{\sqrt{1+\kappa t}}\mathrm{e}^{-\frac{|z|^2}{2\kappa t}}\mathrm{e}^{z^*\tilde{a}^\dagger\sqrt{\frac{1}{\kappa t(1+\kappa t)}}+\sqrt{\frac{\kappa t}{1+\kappa t}}\tilde{a}^\dagger a^\dagger}\left|0,\tilde{0}\right\rangle=|\psi\rangle\,. \tag{5.61}$$

Formally, it is a two-mode displaced squeezed state, which is normalized, namely

$$\begin{aligned}
\langle\psi\,|\psi\rangle &= \frac{1}{1+\kappa t}\mathrm{e}^{-\frac{|z|^2}{\kappa t}}\left\langle 0,\tilde{0}\right|\mathrm{e}^{z\tilde{a}\sqrt{\frac{1}{\kappa t(1+\kappa t)}}+\sqrt{\frac{\kappa t}{1+\kappa t}}\tilde{a}a}\mathrm{e}^{z^*\tilde{a}^\dagger\sqrt{\frac{1}{\kappa t(1+\kappa t)}}+\sqrt{\frac{\kappa t}{1+\kappa t}}\tilde{a}^\dagger a^\dagger}\left|0,\tilde{0}\right\rangle\\
&= \frac{1}{1+\kappa t}\mathrm{e}^{-\frac{|z|^2}{\kappa t}}\int\frac{\mathrm{d}^2z_1'\mathrm{d}^2z_2'}{\pi^2}\left\langle 0,\tilde{0}\right|\mathrm{e}^{z\tilde{a}\sqrt{\frac{1}{\kappa t(1+\kappa t)}}+\sqrt{\frac{\kappa t}{1+\kappa t}}\tilde{a}a}\\
&\quad |z_1',\tilde{z}_2'\rangle\langle z_1',\tilde{z}_2'|\mathrm{e}^{z^*\tilde{a}^\dagger\sqrt{\frac{1}{\kappa t(1+\kappa t)}}+\sqrt{\frac{\kappa t}{1+\kappa t}}\tilde{a}^\dagger a^\dagger}\left|0,\tilde{0}\right\rangle\\
&= \frac{1}{1+\kappa t}\mathrm{e}^{-\frac{|z|^2}{\kappa t}}\int\frac{\mathrm{d}^2z_1'\mathrm{d}^2z_2'}{\pi^2}\mathrm{e}^{-|z_1'|^2-|z_2'|^2+zz_2'\sqrt{\frac{1}{\kappa t(1+\kappa t)}}+\sqrt{\frac{\kappa t}{1+\kappa t}}z_2'z_1'}\\
&\quad\times\mathrm{e}^{z^*z_2'^*\sqrt{\frac{1}{\kappa t(1+\kappa t)}}+\sqrt{\frac{\kappa t}{1+\kappa t}}z_2'^*z_1'^*}\\
&= \frac{1}{1+\kappa t}\mathrm{e}^{-\frac{|z|^2}{\kappa t}}\int\frac{\mathrm{d}^2z_2'}{\pi}\mathrm{e}^{-\frac{1}{1+\kappa t}|z_2'|^2+zz_2'\sqrt{\frac{1}{\kappa t(1+\kappa t)}}+z^*z_2'^*\sqrt{\frac{1}{\kappa t(1+\kappa t)}}}=1.
\end{aligned}\tag{5.62}$$

Using the known thermal vacuum state $|\psi\rangle$, it is easy to calculate the average photon number evolution of the coherent state in the diffusion process. Let's first calculate $\langle\psi|\,\tilde{a}\tilde{a}^\dagger\,|\psi\rangle$, that is

$$\begin{aligned}
\langle\psi|\,\tilde{a}\tilde{a}^\dagger\,|\psi\rangle &= \frac{1}{1+\kappa t}\mathrm{e}^{-\frac{|z|^2}{\kappa t}}\left\langle 0,\tilde{0}\right|\mathrm{e}^{z\tilde{a}\sqrt{\frac{1}{\kappa t(1+\kappa t)}}+\sqrt{\frac{\kappa t}{1+\kappa t}}\tilde{a}a}\\
&\quad\times\int\frac{\mathrm{d}^2z_1'\mathrm{d}^2z_2'}{\pi^2}\tilde{a}|z_1',\tilde{z}_2'\rangle\langle z_1',\tilde{z}_2'|\tilde{a}^\dagger\mathrm{e}^{z^*\tilde{a}^\dagger\sqrt{\frac{1}{\kappa t(1+\kappa t)}}+\sqrt{\frac{\kappa t}{1+\kappa t}}\tilde{a}^\dagger a^\dagger}\left|0,\tilde{0}\right\rangle\\
&= \frac{1}{1+\kappa t}\mathrm{e}^{-\frac{|z|^2}{\kappa t}}\int\frac{\mathrm{d}^2z_2'}{\pi}\mathrm{e}^{-\frac{1}{1+\kappa t}|z_2'|^2+zz_2'\sqrt{\frac{1}{\kappa t(1+\kappa t)}}+z^*z_2'^*\sqrt{\frac{1}{\kappa t(1+\kappa t)}}}|z_2'|^2\\
&= (1+\kappa t)\,\mathrm{e}^{-\frac{|z|^2}{\kappa t}}\int\frac{\mathrm{d}^2z_2'}{\pi}\mathrm{e}^{-|z_2'|^2+zz_2'\sqrt{\frac{1}{\kappa t}}+z^*z_2'^*\sqrt{\frac{1}{\kappa t}}}|z_2'|^2\\
&= \mathrm{e}^{-\frac{|z|^2}{\kappa t}}\left(-\frac{\partial}{\partial f}\right)\int\frac{\mathrm{d}^2z_2'}{\pi}\,\mathrm{e}^{-f|z_2'|^2+zz_2'\sqrt{\frac{1}{\kappa t}}+z^*z_2'^*\sqrt{\frac{1}{\kappa t}}}\Bigg|_{f=1}\\
&= -(1+\kappa t)\,\mathrm{e}^{-\frac{|z|^2}{\kappa t}}\frac{\partial}{\partial f}\left(\frac{1}{f}\mathrm{e}^{\frac{|z|^2}{f\kappa t}}\right)\Bigg|_{f=1}\\
&= -\mathrm{e}^{-\frac{|z|^2}{\kappa t}}\left(-\frac{1}{f^2}\mathrm{e}^{\frac{|z|^2}{f\kappa t}}-\frac{1}{f^3}\frac{|z|^2}{\kappa t}\mathrm{e}^{\frac{|z|^2}{f\kappa t}}\right)\Bigg|_{f=1}\\
&= (1+\kappa t)\left(1+\frac{|z|^2}{\kappa t}\right).
\end{aligned}\tag{5.63}$$

According to the identity

$$a\left|\psi\right\rangle = \sqrt{\frac{\kappa t}{1+\kappa t}}\tilde{a}^{\dagger}\left|\psi\right\rangle, \tag{5.64}$$

we obtain

$$\left\langle\psi\right|a^{\dagger}a\left|\psi\right\rangle = \frac{\kappa t}{1+\kappa t}\left\langle\psi\right|\tilde{a}\tilde{a}^{\dagger}\left|\psi\right\rangle = \kappa t\left(1+\frac{|z|^{2}}{\kappa t}\right) = |z|^{2}+\kappa t. \tag{5.65}$$

Comparing with the average photon number $\mathtt{tr}\left(a^{\dagger}a\left|z\right\rangle\left\langle z\right|\right) = |z|^{2}$ of the initial coherent state, we can see that the average photon number of $\rho\left(t\right)$ increases by κt, which can be used for quantum control. When the diffusion coefficient κ is very small, it can reflect the original meaning of diffusion.

5.3.2 Initial Number State Light Field

This section studies the evolution of the number state $\rho(0) = \left|l\right\rangle\left\langle l\right|$ in the diffusion process, we find that the final-state density operator can be presented as a special function within normal ordering, and represents a new quantum light field [4]. Substituting the density operator $\rho\left(0\right)$ into Eq. (5.18), and considering the summation of n, and using the identities

$$\begin{aligned}\left(\frac{1}{1+\kappa t}\right)^{a^{\dagger}a}a^{n} &= \left[a(1+\kappa t)\right]^{n}\left(\frac{1}{1+\kappa t}\right)^{a^{\dagger}a},\\ a^{\dagger n}\left(\frac{1}{1+\kappa t}\right)^{a^{\dagger}a} &= \left(\frac{1}{1+\kappa t}\right)^{a^{\dagger}a}\left[a^{\dagger}(1+\kappa t)\right]^{n}\end{aligned} \tag{5.66}$$

and

$$a^{n}\left|l\right\rangle = \sqrt{\frac{l!}{n!\left(l-n\right)!}}\left|l-n\right\rangle, \tag{5.67}$$

as well as the series expansions of two-variable Hermite polynomials, i.e.,

$$\mathrm{H}_{m,n}\left(x,y\right) = \sum_{l=0}^{\min(m,n)}\frac{m!n!(-1)^{l}}{l!(m-l)!(n-l)!}x^{m-l}y^{n-l}, \tag{5.68}$$

we obtain

$$\sum_{n=0}^{l}\frac{(\kappa t)^{n}}{n!\left(\kappa t+1\right)^{n}}\left(\frac{1}{1+\kappa t}\right)^{a^{\dagger}a}a^{n}\left|l\right\rangle\left\langle l\right|a^{\dagger n}\left(\frac{1}{1+\kappa t}\right)^{a^{\dagger}a}$$

$$
\begin{aligned}
&= \sum_{n=0}^{l} \frac{(\kappa t)^n (\kappa t+1)^{n-2l}}{n!} a^n |l\rangle \langle l| a^{\dagger n} \\
&= \sum_{n=0}^{l} \frac{(\kappa t)^n (\kappa t+1)^{n-2l}\, l!}{n!\,(l-n)!} |l-n\rangle \langle l-n| \\
&= \sum_{n=0}^{l} \frac{l!\,(\kappa t)^n (\kappa t+1)^{n-2l}}{n!\,(l-n)!\,(l-n)!} a^{\dagger l-n} a^{l-n} \mathrm{e}^{-a^\dagger a} : \\
&= \frac{1}{l!}\left(\frac{-\kappa t}{\kappa t+1}\right)^l : \mathrm{H}_{l,l}\left(\frac{\mathrm{i}a^\dagger}{\sqrt{\kappa t(\kappa t+1)}}, \frac{\mathrm{i}a}{\sqrt{\kappa t(\kappa t+1)}}\right) \mathrm{e}^{-a^\dagger a} : . \qquad (5.69)
\end{aligned}
$$

Combining Eqs. (5.69) and (5.18) and using the summation method within normal ordering, we get

$$
\begin{aligned}
\rho(t) &= \sum_{m=0}^{\infty} \frac{(\kappa t)^m}{m!(\kappa t+1)^{m+1}} a^{\dagger m} \frac{1}{l!}\left(\frac{-\kappa t}{\kappa t+1}\right)^l \\
&\quad \times : \mathrm{H}_{l,l}\left(\frac{\mathrm{i}a^\dagger}{\sqrt{\kappa t(\kappa t+1)}}, \frac{\mathrm{i}a}{\sqrt{\kappa t(\kappa t+1)}}\right) \mathrm{e}^{-a^\dagger a} : a^m \\
&= \frac{(-\kappa t)^l}{l!\,(\kappa t+1)^{l+1}} : \mathrm{e}^{\frac{-1}{\kappa t+1}a^\dagger a} \mathrm{H}_{l,l}\left[\frac{\mathrm{i}a^\dagger}{\sqrt{\kappa t(\kappa t+1)}}, \frac{\mathrm{i}a}{\sqrt{\kappa t(\kappa t+1)}}\right] : \\
&= \frac{(\kappa t)^l}{(\kappa t+1)^{l+1}} : \mathrm{L}_l\left(\frac{-a^\dagger a}{\kappa t(\kappa t+1)}\right) \mathrm{e}^{\frac{-1}{\kappa t+1}a^\dagger a} : , \qquad (5.70)
\end{aligned}
$$

where we have used Eq. (2.81) and the definition of Laguerre polynomials in Eq. (3.195).

Letting

$$
\frac{1}{\kappa t+1} = \lambda, \quad \kappa t = \frac{1-\lambda}{\lambda}, \quad \frac{1}{\kappa t(\kappa t+1)} = \frac{\lambda^2}{1-\lambda}, \qquad (5.71)
$$

thus $\rho(t)$ becomes

$$
\rho(t) = \lambda(1-\lambda)^l : \mathrm{L}_l\left(\frac{-\lambda^2 a^\dagger a}{1-\lambda}\right) \mathrm{e}^{-\lambda a^\dagger a} : . \qquad (5.72)
$$

It can be seen that the number state $|l\rangle\langle l|$ evolves into the Laguerre-polynomial-weighted chaotic state within the normal ordering in the diffusion process. It is the ordered arrangement of operators that makes this density operator show its clue as a special function, i.e., forming a new light field that can be further studied. The ordered arrangement of density operators theoretically makes the entropy of the light field minimal, which makes it easier for people to find or study new light fields.

Using the coherent state representation and the integral formula

$$\int_0^\infty \mathrm{e}^{-bx}\mathrm{L}_l(x)\,\mathrm{d}x = \sum_{k=0}^{l}\binom{l}{l-k}\int_0^\infty \mathrm{e}^{-bx}\frac{(-x)^k}{k!}\mathrm{d}x = (b-1)^l b^{-l-1}, \quad (5.73)$$

it can be proved that $\rho(t)$ satisfies the basic condition of density operator, that is

$$\begin{aligned}\mathtt{tr}\rho(t) &= \int\frac{\mathrm{d}^2\alpha}{\pi}\langle\alpha|\,\lambda(1-\lambda)^l : \mathrm{L}_l\left(\frac{-\lambda^2 a^\dagger a}{1-\lambda}\right)\mathrm{e}^{-\lambda a^\dagger a} : |\alpha\rangle \\ &= \lambda(1-\lambda)^l\int\frac{\mathrm{d}^2\alpha}{\pi}\mathrm{e}^{-\lambda|\alpha|^2}\mathrm{L}_l\left(\frac{-\lambda^2|\alpha|^2}{1-\lambda}\right) \\ &= \lambda(1-\lambda)^l\int_0^\infty \mathrm{d}\left(\frac{\lambda-1}{\lambda^2}x\right)\mathrm{e}^{-\frac{\lambda-1}{\lambda}x}\mathrm{L}_l(x) = 1, \end{aligned} \quad (5.74)$$

so it is qualified to be a new light field.

Next, we investigate the average photon number in the state $\rho(t)$. Using the completeness relation of coherent states and the integral of Laguerre polynomials

$$\begin{aligned}\int_0^\infty \mathrm{e}^{-bx}x\mathrm{L}_l(x)\,\mathrm{d}x &= -\frac{\partial}{\partial b}\left[(b-1)^l b^{-l-1}\right] \\ &= -l(b-1)^{l-1}b^{-l-1} + (b-1)^l(l+1)b^{-l-2} \\ &= \frac{1}{b^{l+2}}(b-1)^{l-1}(b-l-1), \end{aligned} \quad (5.75)$$

we can calculate

$$\begin{aligned}\mathtt{tr}\left[\rho(t)aa^\dagger\right] &= \lambda(1-\lambda)^l\,\mathtt{tr}\left[\int\frac{\mathrm{d}^2\alpha}{\pi}|\alpha\rangle\langle\alpha|\,a^\dagger : \mathrm{L}_l\left(\frac{-\lambda^2 a^\dagger a}{1-\lambda}\right)\mathrm{e}^{-\lambda a^\dagger a} : a\right] \\ &= \lambda(1-\lambda)^l\int\frac{\mathrm{d}^2\alpha}{\pi}|\alpha|^2\mathrm{e}^{-\lambda|\alpha|^2}\mathrm{L}_l\left(\frac{-\lambda^2|\alpha|^2}{1-\lambda}\right) \\ &= \lambda(1-\lambda)^l\int_0^\infty \mathrm{d}\left(\frac{\lambda-1}{\lambda^2}x\right)\frac{\lambda-1}{\lambda^2}x\mathrm{e}^{-\frac{\lambda-1}{\lambda}x}\mathrm{L}_l(x) \\ &= \lambda(1-\lambda)^l\left(\frac{\lambda-1}{\lambda^2}\right)^2\int_0^\infty \mathrm{d}x x\mathrm{e}^{-\frac{\lambda-1}{\lambda}x}\mathrm{L}_l(x) = \frac{1}{\lambda}+l, \end{aligned} \quad (5.76)$$

so we have

$$\mathtt{tr}\left[\rho(t)a^\dagger a\right] = \mathtt{tr}\left[\rho(t)aa^\dagger\right] - 1 = \frac{1}{\lambda}+l-1 = l+\kappa t. \quad (5.77)$$

That is, after passing through the diffusion process, the average photon number l of the initial state becomes $l+\kappa t$. When κ is very small, the increment κt is obvious only after a long time t. So, the diffusion process can be regarded as the quantum control of photon addition.

Besides, how to determine that a light field is a chaotic light field weighted by Laguerre polynomials? Generally, the properties of a quantum light field are verified by photon counting of the field. The photon counting distribution formula in quantum mechanics was first proposed by Kelley and Kleiner. For a single-mode light field with the density operator ρ, the probability distribution $p\left(m, \Delta\tau\right)$ of m photoelectrons measured within the time interval $\Delta\tau$ is

$$p\left(m, \Delta\tau\right) = \mathtt{tr}\left[\rho\colon \frac{\left(\xi a^{\dagger}a\right)^{m}}{m!}\mathrm{e}^{-\xi a^{\dagger}a}\colon\right], \tag{5.78}$$

where ξ is called the detection efficiency of the detector. Using the IWOP method, the formula (5.78) has been converted into the following form [Details can be seen in Sect. 9.1]:

$$p\left(m, \Delta\tau\right) = \left(\frac{\xi}{\xi-1}\right)^{m}\int\frac{\mathrm{d}^2z}{\pi}\mathrm{e}^{-\xi|z|^2}\mathrm{L}_m\left(|z|^2\right)\left\langle\sqrt{1-\xi}z\right|\rho\left|\sqrt{1-\xi}z\right\rangle, \tag{5.79}$$

where $\left|\sqrt{1-\xi}z\right\rangle$ is a coherent state, and $\mathrm{L}_m\left(\cdot\right)$ are the Laguerre polynomials of order m. When the normal ordering product of the density operator ρ is known, it is very convenient to calculate the photon counting distribution by using Eq. (5.79). Substituting Eq. (5.72) into Eq. (5.79), its photon counting distribution can be calculated as:

$$\begin{aligned} p\left(m, \Delta\tau\right) = {}& \lambda\left(1-\lambda\right)^{l}\left(\frac{\xi}{\xi-1}\right)^{m}\int\frac{\mathrm{d}^2z}{\pi}\mathrm{e}^{-\left(\lambda-\lambda\xi+\xi\right)|z|^2} \\ & \times\mathrm{L}_m\left(|z|^2\right)\mathrm{L}_l\left(\frac{-\lambda^2\left(1-\xi\right)|z|^2}{1-\lambda}\right). \end{aligned} \tag{5.80}$$

Using the integral formula of the product of two different Laguerre polynomials

$$\begin{aligned} \int_0^{\infty}\mathrm{e}^{-bx}\mathrm{L}_m\left(\nu x\right)\mathrm{L}_l\left(\mu x\right)\mathrm{d}x = {}& \frac{(m+l)!\left(b-\nu\right)^{m}\left(b-\mu\right)^{l}}{m!k!b^{m+l+1}} \\ & \times\mathrm{F}\left(-m, -l; -m-l, \frac{b\left(b-\nu-\mu\right)}{\left(b-\nu\right)\left(b-\mu\right)}\right), \end{aligned} \tag{5.81}$$

where $\mathrm{F}(\cdot)$ are the confluent hypergeometric functions, and

$$b = \lambda - \lambda\xi + \xi, \nu = 1, \mu = \frac{-\lambda^2}{1-\lambda}\left(1-\xi\right). \tag{5.82}$$

It follows from Eqs. (5.80) and (5.81) that

$$p(m,\Delta\tau) = \lambda(1-\lambda)^l\left(\frac{\xi(b-\nu)}{\xi-1}\right)^m\frac{(m+l)!(b-\mu)^l}{m!k!b^{m+l+1}}$$
$$\times\mathrm{F}\left(-m,-l;-m-l,\frac{b(b-\nu-\mu)}{(b-\nu)(b-\mu)}\right). \tag{5.83}$$

Therefore, experimental physicists can explore the Laguerre-polynomial-weighted chaotic light field according to Eq. (5.83).

5.3.3 Initial Chaotic Light Field

This section discusses the evolution of the chaotic light field in the diffusion process [8, 9]. Substituting the density operator $\rho(0) = \gamma e^{a^\dagger a\ln(1-\gamma)}$ into the solution of the diffusion equation in Eq. (5.18), we obtain

$$\rho(t) = \gamma\sum_{m,n=0}^{\infty}\frac{1}{m!n!}\frac{(\kappa t)^{m+n}}{(\kappa t+1)^{m+n+1}}$$
$$\times a^{\dagger m}\left(\frac{1}{1+\kappa t}\right)^{a^\dagger a}a^n e^{a^\dagger a\ln(1-\gamma)}a^{\dagger n}\left(\frac{1}{1+\kappa t}\right)^{a^\dagger a}a^m. \tag{5.84}$$

Let's first examine the summation of n in Eq. (5.84). Using the operator identities

$$\left(\frac{1}{1+\kappa t}\right)^{a^\dagger a} = e^{-a^\dagger a\ln(1+\kappa t)},\ e^{fa^\dagger a}ae^{-fa^\dagger a} = ae^{-f}, \tag{5.85}$$

we get

$$\sum_{n=0}^{\infty}\frac{1}{n!}\frac{(\kappa t)^n}{(\kappa t+1)^n}e^{-a^\dagger a\ln(1+\kappa t)}a^n e^{a^\dagger a\ln(1-\gamma)}a^{\dagger n}e^{-a^\dagger a\ln(1+\kappa t)}$$
$$= \sum_{n=0}^{\infty}\frac{(\kappa t)^n(1+\kappa t)^n}{n!}a^n e^{a^\dagger a\ln\frac{1-\gamma}{(1+\kappa t)^2}}a^{\dagger n}. \tag{5.86}$$

On the other hand, using the completeness relation of coherent states, that is $\pi^{-1}\int d^2z\,|z\rangle\langle z| = 1$, $|z\rangle = e^{-|z|^2/2}e^{za^\dagger}|0\rangle$, and the IWOP method, we obtain the operator identity

$$a^n e^{ga^\dagger a}a^{\dagger n} = \int\frac{d^2z}{\pi}a^n e^{ga^\dagger a}|z\rangle\langle z|a^{\dagger n}$$
$$= \int\frac{d^2z}{\pi}e^{\frac{-|z|^2}{2}}a^n e^{ga^\dagger a}e^{za^\dagger}e^{-ga^\dagger a}|0\rangle\langle z|z^{*n}$$
$$= \int\frac{d^2z}{\pi}e^{\frac{-|z|^2}{2}}a^n e^{za^\dagger e^g}|0\rangle\langle z|z^{*n}$$

$$
\begin{aligned}
&= \int \frac{d^2 z}{\pi} \left(z e^{g}\right)^n z^{*n} : e^{-|z|^2 + z a^\dagger e^{g} + z^* a - a^\dagger a} : \\
&= e^{gn} : \sum_{l=0}^{n} e^{(e^{g}-1)a^\dagger a} \frac{(n!)^2 \left(a^\dagger a e^{g}\right)^{n-l}}{l!\,[(n-l)!]^2} : \\
&= n! e^{gn} : e^{(e^{g}-1)a^\dagger a} \mathrm{L}_s \left(-a^\dagger a e^{g}\right) : . \qquad (5.87)
\end{aligned}
$$

Thus, using Eq. (5.87) to rewrite the term $a^n e^{a^\dagger a \ln[(1-\gamma)/(1+\kappa t)^2]} a^{\dagger n}$ in Eq. (5.86) as

$$
\begin{aligned}
&a^n e^{a^\dagger a[\ln(1-\gamma)-2\ln(1+\kappa t)]} a^{\dagger n} \\
&= n! e^{n \ln \frac{1-\gamma}{(1+\kappa t)^2}} : e^{\left[\frac{1-\gamma}{(1+\kappa t)^2}-1\right]a^\dagger a} \mathrm{L}_n \left(a^\dagger a \frac{\gamma-1}{(1+\kappa t)^2}\right) : . \qquad (5.88)
\end{aligned}
$$

Further, inserting Eq. (5.88) into Eq. (5.86) leads to

$$
\begin{aligned}
&\sum_{n=0}^{\infty} (\kappa t+1)^n (\kappa t)^n e^{n \ln \frac{1-\gamma}{(1+\kappa t)^2}} : e^{\left[\frac{1-\gamma}{(1+\kappa t)^2}-1\right]a^\dagger a} \mathrm{L}_n \left(a^\dagger a \frac{\gamma-1}{(1+\kappa t)^2}\right) : \\
&= \sum_{n=0}^{\infty} \frac{[\kappa t (1-\gamma)]^n}{(\kappa t+1)^n} : e^{\left[\frac{1-\gamma}{(1+\kappa t)^2}-1\right]a^\dagger a} \mathrm{L}_n \left(a^\dagger a \frac{\gamma-1}{(1+\kappa t)^2}\right) : . \qquad (5.89)
\end{aligned}
$$

On the other hand, using the generating function for the Laguerre polynomial in Eq. (2.34), thus we have

$$
\begin{aligned}
\text{Equation (5.86)} &= \frac{\kappa t+1}{1+\kappa t\gamma} : e^{a^\dagger a \frac{\kappa t(1-\gamma)^2}{(1+\kappa t\gamma)(1+\kappa t)^2}} e^{\left[\frac{1-\gamma}{(1+\kappa t)^2}-1\right]a^\dagger a} : \\
&= \frac{\kappa t+1}{1+\kappa t\gamma} : e^{\left[\frac{1-\gamma}{(t\kappa+1)(t\kappa\gamma+1)}-1\right]a^\dagger a} : , \qquad (5.90)
\end{aligned}
$$

substituting Eq. (5.90) into $\rho(t)$ in Eq. (5.84) and completing the sum of n within normal ordering, thus we have

$$
\begin{aligned}
\rho(t) &= \frac{\gamma(\kappa t+1)}{1+\kappa t\gamma} \sum_{m=0}^{\infty} \frac{(\kappa t)^m}{m!\,(\kappa t+1)^{m+1}} : a^{\dagger m} e^{\left[\frac{1-\gamma}{(t\kappa+1)(t\kappa\gamma+1)}-1\right]a^\dagger a} a^m : \\
&= \frac{\gamma}{1+\kappa t\gamma} : e^{-\frac{\gamma}{\kappa t\gamma+1}a^\dagger a} : , \qquad (5.91)
\end{aligned}
$$

which is the evolution of the chaotic light field in the diffusion process. Compared with the density operator $\rho(0) = \gamma : e^{-\gamma a^\dagger a} :$ at the initial time, it is easy to find that chaotic light field still retains the chaotic characteristics in the diffusion process, and we only needs to make the substitution

$$\gamma \to \gamma' \equiv \frac{\gamma}{1+\kappa t\gamma} \tag{5.92}$$

in $\rho(0) = \gamma e^{a^\dagger a \ln(1-\gamma)}$, thus the chaotic light field in the diffusion process can be obtained.

Obviously, the state $\rho(t)$ is trace preserving, that is

$$\begin{aligned} \mathrm{tr}\rho(t) &= \frac{\gamma}{1+\kappa t\gamma}\mathrm{tr}\left(: e^{-\frac{\gamma}{\kappa t\gamma+1}a^\dagger a} : \int \frac{d^2 z}{\pi} |z\rangle \langle z|\right) \\ &= \frac{\gamma}{1+\kappa t\gamma}\int \frac{d^2 z}{\pi} e^{-\frac{\gamma}{t\kappa\gamma+1}|z|^2} = 1, \end{aligned} \tag{5.93}$$

and the energy change of the chaotic light field caused by the diffusion is

$$\begin{aligned} \mathrm{tr}(\rho(t) H) &= \omega \mathrm{tr}\left[\rho(t)\left(a^\dagger a + \frac{1}{2}\right)\right] \\ &= \omega\left(\frac{1}{\gamma'} - \frac{1}{2}\right) \\ &= \omega\left(\frac{1}{\gamma} + \kappa t - \frac{1}{2}\right) \\ &= \omega \frac{1+e^{-\beta\omega\hbar}}{2\left(1-e^{-\beta\omega\hbar}\right)} + \omega\kappa t \end{aligned} \tag{5.94}$$

which shows that after the diffusion, the energy of the chaotic light field increases $\omega\kappa t$, which is consistent with Eq. (5.65). By introducing an equivalent frequency

$$\gamma' = 1 - e^{-\beta\omega'\hbar}, \tag{5.95}$$

that is

$$\omega' = \frac{1}{\beta\hbar}\ln\frac{1}{1-\gamma'} = \frac{1}{\beta\hbar}\ln\frac{1}{1-\frac{\gamma}{1+\kappa t\gamma}} = \frac{1}{\beta\hbar}\ln\frac{1+\kappa t\gamma}{1+\kappa t\gamma-\gamma}, \tag{5.96}$$

thus we get the entropy change after the diffusion, that is

$$-k_B \mathrm{tr}[\rho(t)\ln\rho(t)] = -k_B\left[\ln\left(1-e^{-\beta\omega'\hbar}\right) + \frac{\beta\omega' e^{-\beta\omega'\hbar}}{e^{-\beta\omega'\hbar}-1}\right]. \tag{5.97}$$

5.3.4 Initial Squeezed Chaotic Light Field

In this section, we discuss the evolution of the initial squeezed chaotic light field in the diffusion process [10]. The initial density operator $\rho(0)$ is

$$\begin{aligned}\rho(0) =& \left(1-\mathrm{e}^{k}\right) S^{\dagger}(r)\, \mathrm{e}^{k a^{\dagger} a} S(r)\\ =& \left(1-\mathrm{e}^{k}\right) \sum_{n=0}^{\infty} \mathrm{e}^{k n} S^{-1}(r)\,|n\rangle\langle n|\, S(r),\end{aligned} \tag{5.98}$$

where $|n\rangle=\frac{a^{\dagger n}}{\sqrt{n!}}|0\rangle$ is a number state, $\left(1-\mathrm{e}^{k}\right)\mathrm{e}^{k a^{\dagger} a}$ is a chaotic light field, $S^{-1}|n\rangle$ is a squeezed number state. In the coordinate representation, the squeezed operator is defined as

$$S(r)=\exp\left[\frac{r}{2}\left(a^{2}-a^{\dagger 2}\right)\right]=\int_{-\infty}^{+\infty} \frac{\mathrm{d}x}{\sqrt{\mu}}\left|\frac{x}{\mu}\right\rangle\langle x|, \tag{5.99}$$

where $\mu=\mathrm{e}^{r}$, $S^{\dagger}(r)=S^{-1}(r)$, and the state $|x\rangle$ is the coordinate eigen-state, that is

$$|x\rangle=\pi^{-\frac{1}{4}} \exp\left(-\frac{x^{2}}{2}+\sqrt{2} x a^{\dagger}-\frac{a^{\dagger 2}}{2}\right)|0\rangle, \tag{5.100}$$

its inner product with the state $\langle n|$ is

$$\langle n| x\rangle=\langle x| n\rangle=\frac{1}{\sqrt{\sqrt{\pi} 2^{n} n!}} \mathrm{e}^{-\frac{x^{2}}{2}} \mathrm{H}_{n}(x), \tag{5.101}$$

where $\mathrm{H}_{n}(x)$ are the Hermite polynomials. Using the integral formula of $\mathrm{H}_{n}(x)$, i.e.,

$$\int_{-\infty}^{+\infty} \mathrm{e}^{-\frac{(x-y)^{2}}{2d}} \mathrm{H}_{n}(x)\, \mathrm{d}x=\sqrt{2\pi d}\,(1-2d)^{n/2}\, \mathrm{H}_{n}\left[y(1-2d)^{-1/2}\right], \tag{5.102}$$

we can change the squeezed number state into the form

$$\begin{aligned}S(r)|n\rangle =& \int_{-\infty}^{+\infty} \frac{\mathrm{d}x}{\sqrt{\mu}}\left|\frac{x}{\mu}\right\rangle\langle x| n\rangle\\ =& \int_{-\infty}^{+\infty} \frac{\mathrm{d}x}{\sqrt{\mu\pi 2^{n} n!}} \mathrm{H}_{n}(x) \exp\left[-\frac{x^{2}}{2}\left(1+\frac{1}{\mu^{2}}\right)\right.\\ &\left.+\sqrt{2}\left(\frac{x}{\mu}\right) a^{\dagger}-\frac{a^{\dagger 2}}{2}\right]|0\rangle\\ =& \int_{-\infty}^{+\infty} \frac{\mathrm{d}x}{\sqrt{\mu\pi 2^{n} n!}} \mathrm{H}_{n}(x) \exp\left[\left(-\frac{\mu^{2}+1}{2\mu^{2}}\right)\left(x-\frac{\sqrt{2}\mu a^{\dagger}}{\mu^{2}+1}\right)^{2}\right.\\ &\left.-\frac{2\mu^{2}}{\left(\mu^{2}+1\right)^{2}} a^{\dagger 2}+\frac{\mu^{2}}{\mu^{2}+1} a^{\dagger 2}\right]|0\rangle\end{aligned}$$

$$= \frac{\sqrt{\mathrm{sech} r}}{\sqrt{2^n n!}} (-\tanh r)^{\frac{n}{2}} \mathrm{H}_n \left(\frac{a^\dagger}{\sqrt{-2 \sin 2r}} \right) \exp\left(-\frac{\tanh r}{2} a^{\dagger 2} \right) |0\rangle . \tag{5.103}$$

It can be seen that the squeezed number state can actually be regarded as a Hermite polynomial squeezed vacuum state.

Making the substitutions $r \to -r$, $S \to S^{-1}$ in Eq. (5.103) leads to

$$S^{-1}(r)|n\rangle = \frac{\sqrt{\mathrm{sech} r}}{\sqrt{2^n n!}} (\tanh r)^{n/2} \mathrm{H}_n \left(\frac{a^\dagger}{\sqrt{2 \sin (2r)}} \right) \exp\left(\frac{\tanh r}{2} a^{\dagger 2} \right) |0\rangle \tag{5.104}$$

Substituting Eq. (5.104) into Eq. (5.98) and using the operator identity $|0\rangle\langle 0| = :\mathrm{e}^{-a^\dagger a}:$, we get

$$\begin{aligned}\left(1-\mathrm{e}^k\right) S^\dagger(r) \mathrm{e}^{k a^\dagger a} S(r) = & \frac{1-\mathrm{e}^k}{\sqrt{\cosh^2 r - \mathrm{e}^{2k} \sinh^2 r}} \exp\left[\frac{-a^{\dagger 2}\left(\mathrm{e}^{2k}-1\right)\tanh r}{2\left(1-\mathrm{e}^{2k}\tanh^2 r\right)} \right] \\ & \times : \exp\left[a^\dagger a \left(\frac{\mathrm{e}^k}{\cosh^2 r - \mathrm{e}^{2k}\sinh^2 r} - 1 \right) \right] : \\ & \times \exp\left[\frac{-a^2\left(\mathrm{e}^{2k}-1\right)\tanh r}{2\left(1-\mathrm{e}^{2k}\tanh^2 r\right)} \right] . \end{aligned} \tag{5.105}$$

Considering that the parameter $k = -\omega\hbar/(k_B T)$ of the chaotic light field, and the average photon number is

$$\frac{1}{\mathrm{e}^{-k}-1} = \frac{1}{\mathrm{e}^{\frac{\omega\hbar}{k_B T}}-1} \equiv \bar{n}, \tag{5.106}$$

we introduce two characteristic parameters τ and θ

$$\tau = \frac{1}{(2\bar{n}+1)\,\mathrm{e}^{-2r}+1}, \quad \theta = \frac{1}{(2\bar{n}+1)\,\mathrm{e}^{2r}+1}, \tag{5.107}$$

thus combining Eqs. (5.98) and (5.105) leads to

$$\rho(0) = 2\sqrt{\tau\theta} : \exp\left[\frac{1}{2}(\tau-\theta)\left(a^{\dagger 2}+a^2\right) - (\tau+\theta)\, a^\dagger a \right] : , \tag{5.108}$$

which is the normal ordering product of the density operator of a squeezed chaotic state.

Next, we will see that this diffusion evolution process is reflected in the changes of τ and θ with time t, while the form of the squeezed chaotic light field remains unchanged. Using the anti-normal ordering formula in Eq. (3.190) and the integral formula in Eq. (1.57), we obtain the anti-normal ordering product of $\rho(0)$ as

$$\rho(0) = 2\sqrt{\tau\theta}\int\frac{d^2\beta}{\pi}\vdots\exp\left[-|\beta|^2-\frac{\theta}{2}(\beta-\beta^*)^2\right.$$
$$\left.+\frac{\tau}{2}(\beta+\beta^*)^2+\beta^*a-\beta a^\dagger+a^\dagger a\right]\vdots$$
$$=2\sqrt{\tau\theta}\int\frac{d^2\beta}{\pi}\vdots\exp\left[(\tau+\theta-1)|\beta|^2-a^\dagger\beta+a\beta^*\right.$$
$$\left.+\frac{1}{2}(\tau-\theta)(\beta^2+\beta^{*2})+a^\dagger a\right]\vdots$$
$$=2\sqrt{\frac{\tau\theta}{\Gamma}}\vdots\exp\left[\frac{\tau-\theta}{2\Gamma}\left(a^{\dagger 2}+a^2\right)+\frac{4\tau\theta-\tau-\theta}{\Gamma}a^\dagger a\right]\vdots, \quad (5.109)$$

where

$$\Gamma=(2\tau-1)(2\theta-1). \quad (5.110)$$

Using the operator identity

$$e^{\beta a^\dagger a}=e^{-\beta}\vdots\exp\left[\left(1-e^{-\beta}\right)aa^\dagger\right]\vdots, \quad (5.111)$$

Equation (5.109) becomes

$$\rho(0)=\frac{2\sqrt{\tau\theta\Gamma}}{1-\tau-\theta}\exp\left(\frac{\tau-\theta}{2\Gamma}a^2\right)\exp\left(a^\dagger a\ln\frac{\Gamma}{1-\tau-\theta}\right)\exp\left(\frac{\tau-\theta}{2\Gamma}a^{\dagger 2}\right). \quad (5.112)$$

Inserting Eq. (5.112) into Eq. (5.18) leads to

$$\rho(t)=\sum_{m,n=0}^{\infty}\frac{(\kappa t)^{m+n}}{m!n!(\kappa t+1)^{m+n+1}}a^{\dagger m}\left(\frac{1}{1+\kappa t}\right)^{a^\dagger a}a^n$$
$$\times\frac{2\sqrt{\tau\theta\Gamma}}{1-\tau-\theta}\exp\left(\frac{\tau-\theta}{2\Gamma}a^2\right)\exp\left(a^\dagger a\ln\frac{\Gamma}{1-\tau-\theta}\right)$$
$$\times\exp\left(\frac{\tau-\theta}{2\Gamma}a^{\dagger 2}\right)a^{\dagger n}\left(\frac{1}{1+\kappa t}\right)^{a^\dagger a}a^m. \quad (5.113)$$

Letting the sum of n be

$$\mathrm{R}\equiv\sum_{n=0}^{\infty}\frac{(\kappa t)^n}{n!(\kappa t+1)^n}\left(\frac{1}{1+\kappa t}\right)^{a^\dagger a}a^n\exp\left(\frac{\tau-\theta}{2\Gamma}a^2\right)$$
$$\times\exp\left(a^\dagger a\ln\frac{\Gamma}{1-\tau-\theta}\right)\exp\left(\frac{\tau-\theta}{2\Gamma}a^{\dagger 2}\right)a^{\dagger n}\left(\frac{1}{1+\kappa t}\right)^{a^\dagger a}, \quad (5.114)$$

and then using the operator identity

$$\left(\frac{1}{1+\kappa t}\right)^{a^\dagger a} a^n = (\kappa t+1)^n a^n \exp\left[a^\dagger a \ln\left(\frac{1}{1+\kappa t}\right)\right], \tag{5.115}$$

we obtain

$$\begin{aligned}
\mathrm{R} = & \sum_{n=0}^{\infty} \frac{[\kappa t(\kappa t+1)]^n}{n!} a^n \exp\left[a^\dagger a \ln\left(\frac{1}{1+\kappa t}\right)\right] \exp\left(\frac{\tau-\theta}{2\Gamma} a^2\right) \\
& \times \exp\left(a^\dagger a \ln\frac{\Gamma}{1-\tau-\theta}\right) \exp\left(\frac{\tau-\theta}{2\Gamma} a^{\dagger 2}\right) \exp\left[a^\dagger a \ln\left(\frac{1}{1+\kappa t}\right)\right] a^{\dagger n} \\
= & \sum_{n=0}^{\infty} \frac{[\kappa t(1+\kappa t)]^n}{n!} a^n \exp\left[\frac{(\tau-\theta)(1+\kappa t)^2}{2\Gamma} a^2\right] \\
& \times \exp\left[a^\dagger a \ln\frac{\Gamma}{(1+\kappa t)^2(1-\tau-\theta)}\right] \exp\left[\frac{(\tau-\theta)(1+\kappa t)^2}{2\Gamma} a^{\dagger 2}\right] a^{\dagger n}.
\end{aligned} \tag{5.116}$$

It follows from Eq. (5.111) that

$$\begin{aligned}
\mathrm{R} = & \frac{(1+\kappa t)^2(1-\tau-\theta)}{\Gamma} \sum_{n=0}^{\infty} \frac{[\kappa t(1+\kappa t)]^n}{n!} \colon a^n \exp\left[\frac{(\tau-\theta)(1+\kappa t)^2 a^2}{2\Gamma}\right] \\
& \times \exp\left[\frac{\Gamma-(1+\kappa t)^2(1-\tau-\theta)}{\Gamma} a^\dagger a\right] \exp\left[\frac{(\tau-\theta)(1+\kappa t)^2 a^{\dagger 2}}{2\Gamma}\right] a^{\dagger n} \colon \\
= & \frac{(1+\kappa t)^2(1-\tau-\theta)}{\Gamma} \colon \exp\left[\kappa t(1+\kappa t) a a^\dagger\right] \exp\left[\frac{(\tau-\theta)(1+\kappa t)^2}{2\Gamma} a^2\right] \\
& \times \exp\left[\frac{\Gamma-(1+\kappa t)^2(1-\tau-\theta)}{\Gamma} a^\dagger a\right] \exp\left[\frac{(\tau-\theta)(1+\kappa t)^2}{2\Gamma} a^{\dagger 2}\right] \colon .
\end{aligned} \tag{5.117}$$

Using the completeness relation of coherent states, and the normal ordering product of coherent state projection operator, that is

$$|z\rangle\langle z| = \colon \exp\left(-|z|^2 + z a^\dagger + z^* a - a^\dagger a\right) \colon , \tag{5.118}$$

as well as the integral formula in Eq. (1.57), we get the normal ordering product of R, that is

$$\begin{aligned}
\mathrm{R} &= \int \frac{d^2 z}{\pi} \mathrm{R} \left|z\right\rangle \left\langle z\right| \\
&= \frac{(1+\kappa t)^2 (1-\tau-\theta)}{\Gamma} \int \frac{d^2 z}{\pi} \\
&\quad : \exp\left[\frac{\kappa t (1+\kappa t) \Gamma - (1+\kappa t)^2 (1-\tau-\theta)}{\Gamma} |z|^2 \right. \\
&\quad \left. + z a^\dagger + z^* a + \frac{(\tau-\theta)(1+\kappa t)^2}{2\Gamma} \left(z^2 + z^{*2}\right) - a^\dagger a : \right] \\
&= \frac{(1+\kappa t)(1-\tau-\theta)}{\sqrt{D (1+2\kappa t \tau)(1+2\kappa t \theta)}} : \exp\left[\frac{\left(a^{\dagger 2} + a^2\right)(\tau-\theta)}{2(1+2\kappa t \tau)(1+2\kappa t \theta)} \right. \\
&\quad \left. + \frac{(1-\tau-\theta) - \kappa t (4\tau\theta - \tau - \theta)}{(1+\kappa t)(1+2\kappa t \tau)(1+2\kappa t \theta)} a^\dagger a - a^\dagger a \right] : . \qquad (5.119)
\end{aligned}$$

Further, substituting Eq. (5.119) into Eq. (5.113) and completing the sum of m, thus we have

$$\begin{aligned}
\rho(t) &= \frac{2\sqrt{\tau\theta\Gamma}}{1-\theta-\tau} \sum_{m=0}^{\infty} \frac{(\kappa t)^m}{m!\,(\kappa t+1)^{m+1}} a^{\dagger m} \mathrm{R} a^m \\
&= \frac{2\sqrt{\tau\theta}}{\sqrt{(1+2\kappa t \tau)(1+2\kappa t \theta)}} : \exp\left\{ \frac{(\tau-\theta)\left(a^{\dagger 2} + a^2\right)}{2(1+2\kappa t \tau)(1+2\kappa t \theta)} + \right. \\
&\quad \left. \left[\frac{(1-\tau-\theta) - \kappa t (4\tau\theta - \tau - \theta)}{(1+\kappa t)(1+2\kappa t \tau)(1+2\kappa t \theta)} - \frac{1}{(\kappa t+1)} \right] a^\dagger a \right\} : \\
&= \frac{2\sqrt{\tau\theta}}{\sqrt{(1+2\kappa t \tau)(1+2\kappa t \theta)}} : \exp\left[\frac{(\tau-\theta)\left(a^{\dagger 2} + a^2\right)}{2(1+2\kappa t \tau)(1+2\kappa t \theta)} \right. \\
&\quad \left. - \left(\frac{\tau}{1+2\kappa t \tau} + \frac{\theta}{1+2\kappa t \theta} \right) a^\dagger a \right] : . \qquad (5.120)
\end{aligned}$$

Denoting

$$\tau' = \frac{\tau}{1+2\kappa t \tau}, \qquad \theta' = \frac{\theta}{1+2\kappa t \theta} \qquad (5.121)$$

thus we have

$$\begin{aligned}
\rho(t) &= 2\sqrt{\tau'\theta'} : \exp\left[\frac{1}{2} \left(\tau' - \theta'\right) \left(a^{\dagger 2} + a^2\right) - \left(\tau' + \theta'\right) a^\dagger a \right] : \\
&= \left(1 - e^{k'}\right) S^\dagger\left(r'\right) e^{k' a^\dagger a} S\left(r'\right), \qquad (5.122)
\end{aligned}$$

which is the same as the density operator $\rho(0)$ in Eq. (5.108) because making the substitutions $\tau \to \tau' = \tau/(1+2\kappa t \tau)$, $\theta \to \theta' = \theta/(1+2\kappa t \theta)$ and taking

$$k' = \frac{1}{4}\ln\frac{\tau'(1-\theta')}{\theta'(1-\tau')},\quad r' = \ln\left(1-\frac{2\sqrt{\tau'\theta'(1-\tau')(1-\theta')}}{1-\tau'-\theta'}\right) \tag{5.123}$$

in Eq. (5.108) leads to the evolved density operator $\rho(t)$ in Eq. (5.122), which shows that the initial squeezed chaotic state is still squeezed and chaotic after undergoing the diffusion process.

Letting

$$\delta = -(\tau'+\theta'),\quad \gamma = \frac{1}{2}(\tau'-\theta'), \tag{5.124}$$

thus we obtain

$$\delta^2 - 4\gamma^2 = 4\tau'\theta', \tag{5.125}$$

so Eq. (5.122) can be simplified as

$$\rho(t) = \sqrt{\delta^2-4\gamma^2} : \exp\left[\delta a^\dagger a + \gamma\left(a^{\dagger 2}+a^2\right)\right] : . \tag{5.126}$$

For $t=0$, it is clearly that $\rho(t) \to \rho(0)$. For the density operator $\rho(t)$ in Eq. (5.126), we can prove that $\mathtt{tr}\rho(t) = 1$. In fact, using the completeness relation of coherent states, we obtain

$$\begin{aligned}\mathtt{tr}\rho(t) &= \sqrt{\delta^2-4\gamma^2}\,\mathtt{tr}\left(: \exp\left[\delta a^\dagger a + \gamma\left(a^{\dagger 2}+a^2\right)\right] :\right) \\ &= \sqrt{\delta^2-4\gamma^2}\int\frac{d^2\alpha}{\pi}\langle\alpha| : \exp\left[\delta a^\dagger a + \gamma\left(a^{\dagger 2}+a^2\right)\right] : |\alpha\rangle \\ &= 1.\end{aligned} \tag{5.127}$$

Finally, we consider the evolution of the average photon number of the squeezed chaotic light field in the diffusion process. It follows from Eq. (5.108) that the average photon number at initial time reads

$$\begin{aligned}\langle N\rangle_0 &= \mathtt{tr}\left(\rho(0)a^\dagger a\right) = \mathtt{tr}\left(\rho(0)aa^\dagger\right) - 1 \\ &= 2\sqrt{\tau\theta}\,\mathtt{tr}\left\{: a^\dagger a \exp\left[\frac{1}{2}(\tau-\theta)\left(a^{\dagger 2}+a^2\right) - (\tau+\theta)a^\dagger a\right] :\right\} - 1 \\ &= 2\sqrt{\tau\theta}\int\frac{d^2z}{\pi}|z|^2\exp\left[\frac{1}{2}(\tau-\theta)\left(z^{*2}+z^2\right) - (\tau+\theta)|z|^2\right] - 1,\end{aligned} \tag{5.128}$$

taking $\sigma = (\tau-\theta)/2$ and $\epsilon = -(\tau+\theta)$, thus Eq. (5.128) becomes

$$\begin{aligned}\langle N\rangle_0 &= 2\sqrt{\tau\theta}\int\frac{d^2z}{\pi}|z|^2\exp\left[\sigma\left(z^{*2}+z^2\right)+\epsilon|z|^2\right] - 1 \\ &= 2\sqrt{\tau\theta}\int\frac{d^2z}{\pi}\frac{\partial}{\partial\epsilon}\exp\left[\sigma\left(z^{*2}+z^2\right)+\epsilon|z|^2\right] - 1\end{aligned}$$

$$
\begin{aligned}
&= 2\sqrt{\tau\theta}\frac{\partial}{\partial\epsilon}\frac{1}{\sqrt{\epsilon^2-4\sigma^2}}-1 \\
&= -\frac{2\epsilon\sqrt{\tau\theta}}{\sqrt{\left(\epsilon^2-4\sigma^2\right)^3}}-1=\frac{\tau+\theta}{4\tau\theta}-1.
\end{aligned} \tag{5.129}
$$

Noting that the density operator form of the squeezed chaotic state remains unchanged in the diffusion process, so the average photon number at time t can be directly given by

$$
\langle N\rangle_t=\mathrm{tr}\left[\rho\left(t\right)a^{\dagger}a\right]=\frac{\tau+\theta}{4\tau\theta}-1+\kappa t=\langle N\rangle_0+\kappa t, \tag{5.130}
$$

which shows that, compared with the average photon number at the initial time in Eq. (5.129), the average photon number of the state $\rho\left(t\right)$ at time t has increased by κt after passing through the diffusion process. Therefore, the diffusion process can be seen as a quantum control scheme to increase the average photon number, and κ is the control parameter.

5.3.5 Initial Thermal Vacuum Light Field

This section discusses the evolution of the initial thermal vacuum light field with the density operator

$$
\rho\left(0\right)=\mathrm{sech}^2\lambda \mathrm{e}^{a^{\dagger}\tilde{a}^{\dagger}\tanh\lambda}\left|0\tilde{0}\right\rangle\left\langle 0\tilde{0}\right|\mathrm{e}^{a\tilde{a}\tanh\lambda} \tag{5.131}
$$

in the single-mode diffusion process, where $\tilde{a}^{\dagger}$ is the creation operator of fictitious mode. Substituting Eq. (5.131) into Eq. (5.18) leads to

$$
\begin{aligned}
\rho\left(t\right)= &\ \mathrm{sech}^2\lambda\sum_{m,n=0}^{\infty}\frac{(\kappa t)^{m+n}}{m!n!\left(\kappa t+1\right)^{m+n+1}}a^{\dagger m}\left(\frac{1}{1+\kappa t}\right)^{a^{\dagger}a} \\
&\times a^n\mathrm{e}^{a^{\dagger}\tilde{a}^{\dagger}\tanh\lambda}\left|0\tilde{0}\right\rangle\left\langle 0\tilde{0}\right|\mathrm{e}^{a\tilde{a}\tanh\lambda}a^{\dagger n}\left(\frac{1}{1+\kappa t}\right)^{a^{\dagger}a}a^m.
\end{aligned} \tag{5.132}
$$

Noting that

$$
a^n\mathrm{e}^{a^{\dagger}\tilde{a}^{\dagger}\tanh\lambda}\left|0\tilde{0}\right\rangle=\left(\tilde{a}^{\dagger}\tanh\lambda\right)^m\mathrm{e}^{a^{\dagger}\tilde{a}^{\dagger}\tanh\lambda}\left|0\tilde{0}\right\rangle, \tag{5.133}
$$

so we get

$$\left(\frac{1}{1+\kappa t}\right)^{a^\dagger a} a^n \mathrm{e}^{a^\dagger \tilde{a}^\dagger \tanh\lambda} \left|0\tilde{0}\right\rangle = \left(\tilde{a}^\dagger \tanh\lambda\right)^n \mathrm{e}^{a^\dagger a \ln\frac{1}{1+\kappa t}} \mathrm{e}^{a^\dagger \tilde{a}^\dagger \tanh\lambda} \left|0\tilde{0}\right\rangle$$
$$= \left(\tilde{a}^\dagger \tanh\lambda\right)^n \mathrm{e}^{a^\dagger \tilde{a}^\dagger \frac{\tanh\lambda}{1+\kappa t}} \left|0\tilde{0}\right\rangle. \tag{5.134}$$

Using the normal ordering product of two-mode vacuum projector, that is $\left|0\tilde{0}\right\rangle\left\langle 0\tilde{0}\right| =: \mathrm{e}^{-a^\dagger a-\tilde{a}^\dagger \tilde{a}}:$, thus Eq. (5.132) becomes

$$\rho(t) = \operatorname{sech}^2\lambda \sum_{m,n=0}^{\infty} \frac{(\kappa t)^{m+n}}{m!n!\,(\kappa t+1)^{m+n+1}} : a^{\dagger m} \left(\tilde{a}^\dagger \tanh\lambda\right)^n$$
$$\times \mathrm{e}^{\left(a^\dagger \tilde{a}^\dagger + a\tilde{a}\right)\frac{\tanh\lambda}{1+\kappa t} - a^\dagger a - \tilde{a}^\dagger \tilde{a}} \left(\tilde{a}\tanh\lambda\right)^n a^m :$$
$$= \frac{\operatorname{sech}^2\lambda}{\kappa t+1} : \mathrm{e}^{\left(a^\dagger \tilde{a}^\dagger + a\tilde{a}\right)\frac{\tanh\lambda}{1+\kappa t} + \left(\frac{\kappa t}{1+\kappa t}-1\right)a^\dagger a + \left(\frac{\kappa t}{1+\kappa t}\tanh^2\lambda - 1\right)\tilde{a}^\dagger \tilde{a}} :$$
$$= \frac{\operatorname{sech}^2\lambda}{\kappa t+1} \mathrm{e}^{a^\dagger \tilde{a}^\dagger \frac{\tanh\lambda}{1+\kappa t}} : \mathrm{e}^{\left(\frac{\kappa t}{1+\kappa t}-1\right)a^\dagger a + \left(\frac{\kappa t \tanh^2\lambda}{1+\kappa t}-1\right)\tilde{a}^\dagger \tilde{a}} : \mathrm{e}^{a\tilde{a}\frac{\tanh\lambda}{1+\kappa t}}$$
$$= \frac{\operatorname{sech}^2\lambda}{\kappa t+1} \mathrm{e}^{a^\dagger \tilde{a}^\dagger \frac{\tanh\lambda}{1+\kappa t}} \mathrm{e}^{a^\dagger a \ln\frac{\kappa t}{1+\kappa t} + \tilde{a}^\dagger \tilde{a} \ln\frac{\kappa t \tanh^2\lambda}{1+\kappa t}} \mathrm{e}^{a\tilde{a}\frac{\tanh\lambda}{1+\kappa t}}, \tag{5.135}$$

which is a mixed state light filed. The partial tracing of the real field mode a leads to

$$\mathtt{tr}_a\rho(t) = \frac{\operatorname{sech}^2\lambda}{\kappa t+1} \int \frac{\mathrm{d}^2 z_1}{\pi} \langle z| \, \mathrm{e}^{a^\dagger \tilde{a}^\dagger \frac{\tanh\lambda}{1+\kappa t}}$$
$$\times : \mathrm{e}^{\frac{-1}{1+\kappa t}a^\dagger a + \left(\frac{\kappa t \tanh^2\lambda}{1+\kappa t}-1\right)\tilde{a}^\dagger \tilde{a}} : \mathrm{e}^{a\tilde{a}\frac{\tanh\lambda}{1+\kappa t}} |z\rangle$$
$$= \frac{\operatorname{sech}^2\lambda}{\kappa t+1} \int \frac{\mathrm{d}^2 z}{\pi} \mathrm{e}^{\frac{-1}{1+\kappa t}|z|^2} \mathrm{e}^{z^* \tilde{a}^\dagger \frac{\tanh\lambda}{1+\kappa t}}$$
$$\times : \mathrm{e}^{\left(\frac{\kappa t \tanh^2\lambda}{1+\kappa t}-1\right)\tilde{a}^\dagger \tilde{a}} : \mathrm{e}^{z\tilde{a}\frac{\tanh\lambda}{1+\kappa t}}$$
$$= \operatorname{sech}^2\lambda : \mathrm{e}^{\tilde{a}^\dagger \tilde{a}\frac{\tanh^2\lambda}{1+\kappa t} + \left(\frac{\kappa t \tanh^2\lambda}{1+\kappa t}-1\right)\tilde{a}^\dagger \tilde{a}} :$$
$$= \operatorname{sech}^2\lambda : \mathrm{e}^{(\tanh^2\lambda - 1)\tilde{a}^\dagger \tilde{a}} : , \tag{5.136}$$

thus we have $\mathtt{tr}\rho(t) = \operatorname{sech}^2\lambda \mathtt{tr}_{\tilde{a}}\left[: \mathrm{e}^{(\tanh^2\lambda-1)\tilde{a}^\dagger \tilde{a}} :\right] = 1$, while the partial tracing of the fictitious field mode $\tilde{a}$ leads to

$$\mathtt{tr}_{\tilde{a}}\rho(t) = \frac{\operatorname{sech}^2\lambda}{\kappa t+1} \int \frac{\mathrm{d}^2 z_2}{\pi} \langle \tilde{z}| \, \mathrm{e}^{a^\dagger \tilde{a}^\dagger \frac{\tanh\lambda}{1+\kappa t}} : \mathrm{e}^{\frac{-1}{1+\kappa t}a^\dagger a + \left(\frac{\kappa t \tanh^2\lambda}{1+\kappa t}-1\right)\tilde{a}^\dagger \tilde{a}} : \mathrm{e}^{a\tilde{a}\frac{\tanh\lambda}{1+\kappa t}} |\tilde{z}\rangle$$
$$= \frac{\operatorname{sech}^2\lambda}{\kappa t+1} \int \frac{\mathrm{d}^2 z_2}{\pi} \mathrm{e}^{-\left(1-\frac{\kappa t \tanh^2\lambda}{1+\kappa t}\right)|z_2|^2} \mathrm{e}^{z^* a^\dagger \frac{\tanh\lambda}{1+\kappa t}} : \mathrm{e}^{\frac{-1}{1+\kappa t}a^\dagger a} : \mathrm{e}^{za\frac{\tanh\lambda}{1+\kappa t}}$$
$$= f : \exp\left(-f a^\dagger a\right) : = \rho_a(t), \tag{5.137}$$

where

$$f = \frac{\mathrm{sech}^2\lambda}{1 + \kappa t \mathrm{sech}^2\lambda}. \tag{5.138}$$

Therefore, the average photon number of the thermal vacuum light field in the diffusion process is obtained as

$$\begin{aligned}
\mathtt{tr}\left[\rho_a(t)\, a^\dagger a\right] &= \mathtt{tr}\left[\rho_a(t)\left(aa^\dagger - 1\right)\right] \\
&= \int \frac{\mathrm{d}^2 z}{\pi} \langle z|\, a^\dagger \rho_a(t)\, a\, |z\rangle - 1 \\
&= f \int \frac{\mathrm{d}^2 z}{\pi} |z|^2 \exp\left(-f|z|^2\right) \\
&= f \frac{\partial}{\partial(-f)} \int \frac{\mathrm{d}^2 z}{\pi} \exp\left(-f|z|^2\right) \\
&= -f \frac{\partial}{\partial f} \frac{1}{f} = \frac{1}{f} \\
&= \cosh^2\lambda + \kappa t,
\end{aligned} \tag{5.139}$$

which means that the average number of photons in the thermal vacuum light field is increased by κt as a result of the diffusion.

5.3.6 Initial Forced Oscillator System

When a quantum harmonic oscillator interacts with a time-dependent external field, the Hamiltonian of the system is [11]

$$H_s = \hbar\Omega a^\dagger a + \hbar f(t)\left(a + a^\dagger\right). \tag{5.140}$$

Here, Ω is a harmonic oscillator with a frequency, and $f(t)$ is a real number. If a time-dependent harmonic oscillator under external force undergoes a diffusion process, the corresponding master equation is

$$\frac{\mathrm{d}\rho_s(t)}{\mathrm{d}t} = \frac{1}{\mathrm{i}\hbar}\left[H_s, \rho_s(t)\right] - \kappa\left(a^\dagger a \rho_s(t) - a\rho_s(t)\, a^\dagger - a^\dagger \rho_s(t)\, a + \rho_s(t)\, aa^\dagger\right), \tag{5.141}$$

where κ is the decay constant. Using the transformation relation $\rho(t) = \mathrm{e}^{\mathrm{i}\Omega t a^\dagger a}\rho_s(t)\mathrm{e}^{-\mathrm{i}\Omega t a^\dagger a}$, the master equation (5.141) can be changed into a simpler form, that is

$$\begin{aligned}
\frac{\mathrm{d}\rho(t)}{\mathrm{d}t} &= -\mathrm{i}f(t)\left[\left(a\mathrm{e}^{-\mathrm{i}\Omega t} + a^\dagger \mathrm{e}^{\mathrm{i}\Omega t}\right), \rho(t)\right] \\
&\quad -\kappa\left(a^\dagger a\rho(t) - a\rho(t)\, a^\dagger - a^\dagger \rho(t)\, a + \rho(t)\, aa^\dagger\right).
\end{aligned} \tag{5.142}$$

Applying the above equation to the entangled state $|I\rangle \equiv |\tau = 0\rangle$, and use Eq. (5.10) to obtain a like Schrödinger equation for the state vector $|\rho(t)\rangle \equiv \rho(t)|I\rangle$:

$$\begin{aligned}\frac{\mathrm{d}}{\mathrm{d}t}|\rho(t)\rangle &= -\mathrm{i}f(t)\left[\left(a-\tilde{a}^{\dagger}\right)\mathrm{e}^{-\mathrm{i}\Omega t}+\left(a^{\dagger}-\tilde{a}\right)\mathrm{e}^{\mathrm{i}\Omega t}\right]|\rho(t)\rangle \\ &\quad+\kappa\left(a^{\dagger}\tilde{a}^{\dagger}+a\tilde{a}-a^{\dagger}a-\tilde{a}\tilde{a}^{\dagger}\right)|\rho(t)\rangle \\ &= \left[g(t)+G_0\right]|\rho(t)\rangle,\end{aligned} \tag{5.143}$$

where

$$\begin{aligned}G_0 &= \kappa\left(a^{\dagger}\tilde{a}^{\dagger}+a\tilde{a}-a^{\dagger}a-\tilde{a}\tilde{a}^{\dagger}\right) = -\kappa\hat{\eta}^{\dagger}\hat{\eta} \\ g(t) &= -\mathrm{i}f(t)\left[\left(a-\tilde{a}^{\dagger}\right)\mathrm{e}^{-\mathrm{i}\Omega t}+\left(a^{\dagger}-\tilde{a}\right)\mathrm{e}^{\mathrm{i}\Omega t}\right] \\ \hat{\eta}^{\dagger} &= a^{\dagger}-\tilde{a},\ \hat{\eta}=a-\tilde{a}^{\dagger}.\end{aligned} \tag{5.144}$$

Using the following commutative relations

$$\begin{aligned}\left[G_0, g(t)\right] &= -\mathrm{i}\kappa f(t)\mathrm{e}^{-\mathrm{i}\Omega t}\left[a^{\dagger}\tilde{a}^{\dagger}+a\tilde{a}-a^{\dagger}a-\tilde{a}\tilde{a}^{\dagger}, a-\tilde{a}^{\dagger}\right] \\ &\quad-\mathrm{i}\kappa f(t)\mathrm{e}^{\mathrm{i}\Omega t}\left[a^{\dagger}\tilde{a}^{\dagger}+a\tilde{a}-a^{\dagger}a-\tilde{a}\tilde{a}^{\dagger}, a^{\dagger}-\tilde{a}\right] \\ &= -\mathrm{i}\kappa f(t)\mathrm{e}^{-\mathrm{i}\Omega t}\left\{\left[a^{\dagger}\tilde{a}^{\dagger}, a\right]+\left[a\tilde{a}, -\tilde{a}^{\dagger}\right]+\left[-a^{\dagger}a, a\right]+\left[-\tilde{a}\tilde{a}^{\dagger}, -\tilde{a}^{\dagger}\right]\right\} \\ &\quad-\mathrm{i}\kappa f(t)\mathrm{e}^{\mathrm{i}\Omega t}\left\{\left[a^{\dagger}\tilde{a}^{\dagger}, -\tilde{a}\right]+\left[a\tilde{a}, a^{\dagger}\right]+\left[-a^{\dagger}a, a^{\dagger}\right]+\left[-\tilde{a}\tilde{a}^{\dagger}, -\tilde{a}\right]\right\} \\ &= -\mathrm{i}\kappa f(t)\left[\mathrm{e}^{-\mathrm{i}\Omega t}\left(-\tilde{a}^{\dagger}-a+a+\tilde{a}^{\dagger}\right)+\mathrm{e}^{\mathrm{i}\Omega t}\left(a^{\dagger}+\tilde{a}-a^{\dagger}-\tilde{a}\right)\right] \\ &= 0\end{aligned} \tag{5.145}$$

and

$$\begin{aligned}\left[g(t), g(t')\right] &= \left[-\mathrm{i}f(t)\mathrm{e}^{-\mathrm{i}\Omega t}\left(a-\tilde{a}^{\dagger}\right)-\mathrm{i}f(t)\mathrm{e}^{\mathrm{i}\Omega t}\left(a^{\dagger}-\tilde{a}\right),\right. \\ &\quad\left.-\mathrm{i}f(t')\mathrm{e}^{-\mathrm{i}\Omega t'}\left(a-\tilde{a}^{\dagger}\right)-\mathrm{i}f(t')\mathrm{e}^{\mathrm{i}\Omega t'}\left(a^{\dagger}-\tilde{a}\right)\right] \\ &= 0,\end{aligned} \tag{5.146}$$

as well as

$$\left[g(t)+G_0, g\left(t'\right)+G_0\right] = 0, \tag{5.147}$$

we can obtain the formal solution of Eq. (5.143) as

$$\begin{aligned}|\rho(t)\rangle &= \exp\left\{\int_0^t\left[g(t')+G_0\right]\mathrm{d}t'\right\}|\rho(0)\rangle \\ &= \exp\left[\int_0^t g(t')\,\mathrm{d}t'\right]\exp(G_0t)|\rho(0)\rangle \\ &= \exp\left[\lambda(t)\hat{\eta}^{\dagger}-\lambda^*(t)\hat{\eta}\right]\exp\left(-\kappa t\hat{\eta}^{\dagger}\hat{\eta}\right)|\rho(0)\rangle \\ &= \exp\left[\lambda(t)\hat{\eta}^{\dagger}\right]\exp\left[-\lambda^*(t)\hat{\eta}\right]\exp\left(-\kappa t\hat{\eta}^{\dagger}\hat{\eta}\right)|\rho(0)\rangle,\end{aligned} \tag{5.148}$$

where $\lambda(t)$ is related to the external force, that is

$$\lambda(t) \equiv -\mathrm{i}\int_0^t \mathrm{d}t' f(t')\, \mathrm{e}^{\mathrm{i}\Omega t'}. \tag{5.149}$$

Projecting Eq. (5.148) onto the thermal entangled state $\langle\eta|$, we have

$$\begin{aligned}\langle\eta|\rho(t)\rangle &= \langle\eta|\exp\left[\lambda(t)\hat{\eta}^\dagger\right]\exp\left[-\lambda^*(t)\hat{\eta}\right]\exp\left(-\kappa t\hat{\eta}^\dagger\hat{\eta}\right)|\rho(0)\rangle\\ &= \mathrm{e}^{\lambda(t)\eta^*-\lambda^*(t)\eta-\kappa t|\eta|^2}\langle\eta|\rho(0)\rangle,\end{aligned} \tag{5.150}$$

and using the completeness relation of the entangled states $|\eta\rangle$, we get

$$\begin{aligned}|\rho(t)\rangle &= \int\frac{\mathrm{d}^2\eta}{\pi}\mathrm{e}^{\lambda(t)\eta^*-\lambda^*(t)\eta-\kappa t|\eta|^2}|\eta\rangle\langle\eta|\rho(0)\rangle\\ &= \mathrm{e}^{-\frac{|\lambda(t)|^2}{1+\kappa t}}\sum_{m,n=0}^{\infty}\frac{1}{m!n!}\frac{(\kappa t)^{m+n}}{(1+\kappa t)^{m+n+1}}a^{\dagger m}\mathrm{e}^{\frac{\lambda(t)a^\dagger}{1+\kappa t}}\left(\frac{1}{1+\kappa t}\right)^{a^\dagger a}\\ &\quad\times a^n\mathrm{e}^{-\frac{\lambda^*(t)a}{1+\kappa t}}\rho(0)\,\mathrm{e}^{-\frac{\lambda(t)a^\dagger}{1+\kappa t}}a^{\dagger n}\left(\frac{1}{1+\kappa t}\right)^{a^\dagger a}\mathrm{e}^{\frac{\lambda^*(t)a}{1+\kappa t}}a^m|I\rangle.\end{aligned} \tag{5.151}$$

Therefore, the Kraus operator-sum representation of the density operator $\rho(t)$ is

$$\begin{aligned}\rho(t) &= \sum_{m,n=0}^{\infty}\frac{\mathrm{e}^{-\frac{|\lambda(t)|^2}{1+\kappa t}}}{m!n!}\frac{(\kappa t)^{m+n}}{(1+\kappa t)^{m+n+1}}a^{\dagger m}\mathrm{e}^{\frac{\lambda(t)a^\dagger}{1+\kappa t}}\left(\frac{1}{1+\kappa t}\right)^{a^\dagger a}\\ &\quad\times a^n\mathrm{e}^{-\frac{\lambda^*(t)a}{1+\kappa t}}\rho(0)\,\mathrm{e}^{-\frac{\lambda(t)a^\dagger}{1+\kappa t}}a^{\dagger n}\left(\frac{1}{1+\kappa t}\right)^{a^\dagger a}\mathrm{e}^{\frac{\lambda^*(t)a}{1+\kappa t}}a^m\\ &= \sum_{m,n=0}^{\infty}M_{m,n}\rho(0)\,M_{m,n}^\dagger,\end{aligned} \tag{5.152}$$

where the Kraus operator $M_{m,n}$ is

$$M_{m,n} = \sqrt{\frac{\mathrm{e}^{-\frac{|\lambda(t)|^2}{1+\kappa t}}}{m!n!}\frac{(\kappa t)^{m+n}}{(1+\kappa t)^{m+n+1}}}a^{\dagger m}\mathrm{e}^{\frac{\lambda(t)a^\dagger}{1+\kappa t}}\left(\frac{1}{1+\kappa t}\right)^{a^\dagger a}a^n\mathrm{e}^{-\frac{\lambda^*(t)a}{1+\kappa t}}. \tag{5.153}$$

Since

$$\sum_{m,n=0}^{\infty}M_{m,n}^\dagger M_{m,n} = 1, \tag{5.154}$$

thus we obtain

$$\mathtt{tr}\rho(t)=\mathtt{tr}\sum_{m,n=0}^{\infty}M_{m,n}^{\dagger}M_{m,n}\rho(0)=\mathtt{tr}\rho(0)=1, \tag{5.155}$$

so $M_{m,n}$ is a trace preserving quantum operator.

5.3.7 Initial Coherent State of Forced Oscillator

Assuming that the initial state is a coherent state, i.e. $\rho(0)=|z\rangle\langle z|$ and $a|z\rangle=z|z\rangle$, $|z\rangle=\exp\left(za^{\dagger}-z^{*}a\right)|0\rangle$, the density operator $\rho(t)$ of the final state can be obtained from Eq. (5.152) as follows:

$$\rho(t)=GV\left(\frac{\kappa t}{1+\kappa t}\right)^{a^{\dagger}a}V^{\dagger}, \tag{5.156}$$

where

$$G=\frac{\mathrm{e}^{\frac{|\lambda(t)|^{2}\left(|z|^{2}-1\right)+\kappa t|z|^{2}}{1+\kappa t}-|z|^{2}}}{1+\kappa t},\quad V=\mathrm{e}^{\frac{[\lambda(t)+z]a^{\dagger}}{1+\kappa t}}, \tag{5.157}$$

i.e., V is related to both $\lambda(t)$ and the decay factor κ. Due to the fact that $\left(\frac{\kappa t}{1+\kappa t}\right)^{a^{\dagger}a}$ is a chaotic field and the operator V acts as a displaced operator, thus the evolution of $\rho(0)$ evolves into a displaced chaotic state, which clearly exhibits decoherence.

To sum up, using the thermal entangled state representation and the IWOP method, we derived the exact solution of the master equation for the density operator of the harmonic oscillator system under classical time-dependent external forces in the diffusion process, known as the Kraus operator-sum representation. The results indicate that if the system is initially in a coherent state, it evolves into a displaced chaotic state.

Appendix: Proof of Eqs. (5.151), (5.154) and (5.156).

Proof of Eq. (5.151):

$$\begin{aligned}
|\rho(t)\rangle &= \int\frac{\mathrm{d}^{2}\eta}{\pi}\mathrm{e}^{\lambda(t)\eta^{*}-\lambda^{*}(t)\eta-\kappa t|\eta|^{2}}|\eta\rangle\langle\eta|\rho(0)\rangle\\
&= \int\frac{\mathrm{d}^{2}\eta}{\pi}:\mathrm{e}^{-(1+\kappa t)|\eta|^{2}+\eta\left(a^{\dagger}-\tilde{a}+\lambda^{*}(t)\right)+\eta^{*}\left(a-\tilde{a}^{\dagger}+\lambda(t)\right)+a^{\dagger}\tilde{a}^{\dagger}+a\tilde{a}-a^{\dagger}a-\tilde{a}^{\dagger}\tilde{a}}:|\rho(0)\rangle\\
&= \frac{1}{1+\kappa t}:\mathrm{e}^{\frac{1}{1+\kappa t}\left[\kappa t\left(a^{\dagger}\tilde{a}^{\dagger}+a\tilde{a}-a^{\dagger}a-\tilde{a}^{\dagger}\tilde{a}\right)+\lambda(t)\left(a^{\dagger}-\tilde{a}\right)-\lambda^{*}(t)\left(a-\tilde{a}^{\dagger}\right)-|\lambda(t)|^{2}\right]}:|\rho(0)\rangle\\
&= \frac{\mathrm{e}^{\frac{-|\lambda(t)|^{2}}{1+\kappa t}}}{1+\kappa t}\mathrm{e}^{\frac{\kappa t a^{\dagger}\tilde{a}^{\dagger}+\lambda(t)a^{\dagger}+\lambda^{*}(t)\tilde{a}^{\dagger}}{1+\kappa t}}:\mathrm{e}^{\frac{-\kappa t}{1+\kappa t}\left(a^{\dagger}a+\tilde{a}^{\dagger}\tilde{a}\right)}:\mathrm{e}^{\frac{\kappa t a\tilde{a}-\lambda^{*}(t)a-\lambda(t)\tilde{a}}{1+\kappa t}}|\rho(0)\rangle\\
&= \frac{\mathrm{e}^{\frac{-|\lambda(t)|^{2}}{1+\kappa t}}}{1+\kappa t}\mathrm{e}^{\frac{\kappa t a^{\dagger}\tilde{a}^{\dagger}+\lambda(t)a^{\dagger}+\lambda^{*}(t)\tilde{a}^{\dagger}}{1+\kappa t}}\left(\frac{1}{1+\kappa t}\right)^{a^{\dagger}a+\tilde{a}^{\dagger}\tilde{a}}\mathrm{e}^{\frac{\kappa t a\tilde{a}-\lambda(t)\tilde{a}-\lambda^{*}(t)a}{1+\kappa t}}|\rho(0)\rangle
\end{aligned}$$

$$
\begin{aligned}
&= \frac{e^{\frac{-|\lambda(t)|^2}{1+\kappa t}}}{1+\kappa t} e^{\frac{\kappa t a^\dagger \tilde{a}^\dagger}{1+\kappa t}} e^{\frac{\lambda(t)a^\dagger}{1+\kappa t}} e^{\frac{\lambda^*(t)\tilde{a}^\dagger}{1+\kappa t}} \left(\frac{1}{1+\kappa t}\right)^{a^\dagger a+\tilde{a}^\dagger \tilde{a}} e^{\frac{\kappa t a\tilde{a}}{1+\kappa t}} e^{\frac{-\lambda(t)\tilde{a}}{1+\kappa t}} e^{\frac{-\lambda^*(t)a}{1+\kappa t}} \rho(0)|I\rangle \\
&= \frac{e^{\frac{-|\lambda(t)|^2}{1+\kappa t}}}{1+\kappa t} \sum_{m=0}^{\infty} \frac{1}{m!} \left(\frac{\kappa t a^\dagger}{1+\kappa t}\right)^m e^{\frac{\lambda(t)a^\dagger}{1+\kappa t}} \left(\frac{1}{1+\kappa t}\right)^{a^\dagger a} \sum_{n=0}^{\infty} \frac{1}{n!} \\
&\quad \times \left(\frac{\kappa t a}{1+\kappa t}\right)^n e^{\frac{-\lambda^*(t)a}{1+\kappa t}} \rho(0) e^{\frac{-\lambda(t)a^\dagger}{1+\kappa t}} a^{\dagger n} \left(\frac{1}{1+\kappa t}\right)^{a^\dagger a} e^{\frac{\lambda^*(t)a}{1+\kappa t}} a^m |I\rangle \\
&= e^{\frac{-|\lambda(t)|^2}{1+\kappa t}} \sum_{m,n=0}^{\infty} \frac{1}{m!n!} \frac{(\kappa t)^{m+n}}{(1+\kappa t)^{m+n+1}} a^{\dagger m} e^{\frac{\lambda(t)a^\dagger}{1+\kappa t}} \left(\frac{1}{1+\kappa t}\right)^{a^\dagger a} a^n e^{\frac{-\lambda^*(t)a}{1+\kappa t}} \rho(0) \\
&\quad \times e^{\frac{-\lambda(t)a^\dagger}{1+\kappa t}} a^{\dagger n} \left(\frac{1}{1+\kappa t}\right)^{a^\dagger a} e^{\frac{\lambda^*(t)a}{1+\kappa t}} a^m |I\rangle ,
\end{aligned}
$$

where we have used the operator identities $\left|0\tilde{0}\right\rangle\left\langle 0\tilde{0}\right| =: e^{-a^\dagger a-\tilde{a}^\dagger \tilde{a}}:$, $:\exp\left[\frac{-\kappa t}{1+\kappa t}\left(a^\dagger a+\tilde{a}^\dagger \tilde{a}\right)\right]: = \left(\frac{1}{1+\kappa t}\right)^{a^\dagger a+\tilde{a}^\dagger \tilde{a}}$, $\left[\tilde{a}, \rho(0)\right] = 0$ and $\tilde{a}|I\rangle = a^\dagger |I\rangle$, as well as the commutation relations of real and fictitious modes.

Proof of Eq. (5.154):

$$
\begin{aligned}
\sum_{m,n=0}^{\infty} M_{m,n}^\dagger M_{m,n} &= \sum_{m,n=0}^{\infty} \frac{e^{-\frac{|\lambda(t)|^2}{1+\kappa t}}}{m!n!} \frac{(\kappa t)^{m+n}}{(1+\kappa t)^{m+n+1}} e^{-\frac{\lambda(t)a^\dagger}{1+\kappa t}} a^{\dagger n} \left(\frac{1}{1+\kappa t}\right)^{a^\dagger a} \\
&\quad \times e^{\frac{\lambda^*(t)a}{1+\kappa t}} a^m a^{\dagger m} e^{\frac{\lambda(t)a^\dagger}{1+\kappa t}} \left(\frac{1}{1+\kappa t}\right)^{a^\dagger a} a^n e^{-\frac{\lambda^*(t)a}{1+\kappa t}} \\
&= \sum_{n=0}^{\infty} \frac{e^{-\frac{|\lambda(t)|^2}{1+\kappa t}}}{n!} \frac{(\kappa t)^n}{(1+\kappa t)^{n+1}} e^{-\frac{\lambda(t)a^\dagger}{1+\kappa t}} a^{\dagger n} \left(\frac{1}{1+\kappa t}\right)^{a^\dagger a} e^{\frac{\lambda^*(t)a}{1+\kappa t}} \\
&\quad \times \vdots \sum_{m=0}^{\infty} \frac{1}{m!} \left(\frac{\kappa t}{1+\kappa t} a a^\dagger\right)^m \vdots e^{\frac{\lambda(t)a^\dagger}{1+\kappa t}} \left(\frac{1}{1+\kappa t}\right)^{a^\dagger a} a^n e^{-\frac{\lambda^*(t)a}{1+\kappa t}} \\
&= \sum_{n=0}^{\infty} \frac{e^{-\frac{|\lambda(t)|^2}{1+\kappa t}}}{n!} \frac{(\kappa t)^n}{(1+\kappa t)^{n+1}} e^{-\frac{\lambda(t)a^\dagger}{1+\kappa t}} a^{\dagger n} \left(\frac{1}{1+\kappa t}\right)^{a^\dagger a} \\
&\quad \times e^{\frac{\lambda^*(t)a}{1+\kappa t}} \vdots e^{\frac{\kappa t a a^\dagger}{1+\kappa t}} \vdots e^{\frac{\lambda(t)a^\dagger}{1+\kappa t}} \left(\frac{1}{1+\kappa t}\right)^{a^\dagger a} a^n e^{\frac{-\lambda^*(t)a}{1+\kappa t}} \\
&= \sum_{n=0}^{\infty} \frac{e^{-\frac{|\lambda(t)|^2}{1+\kappa t}}}{n!} \frac{(\kappa t)^n}{(1+\kappa t)^n} e^{-\frac{\lambda(t)a^\dagger}{1+\kappa t}} a^{\dagger n} \left(\frac{1}{1+\kappa t}\right)^{a^\dagger a} e^{\frac{\lambda^*(t)a}{1+\kappa t}} \\
&\quad \times \frac{1}{1+\kappa t} \vdots e^{\left(1-\frac{1}{1+\kappa t}\right)a a^\dagger} \vdots e^{\frac{\lambda(t)a^\dagger}{1+\kappa t}} \left(\frac{1}{1+\kappa t}\right)^{a^\dagger a} a^n e^{-\frac{\lambda^*(t)a}{1+\kappa t}} \\
&= \sum_{n=0}^{\infty} \frac{e^{-\frac{|\lambda(t)|^2}{1+\kappa t}}}{n!} \frac{(\kappa t)^n}{(1+\kappa t)^n} e^{-\frac{\lambda(t)a^\dagger}{1+\kappa t}} a^{\dagger n} \left(\frac{1}{1+\kappa t}\right)^{a^\dagger a} e^{\frac{\lambda^*(t)a}{1+\kappa t}}
\end{aligned}
$$

$$
\begin{aligned}
&\times\left(\frac{1}{1+\kappa t}\right)^{-a^{\dagger}a}\mathrm{e}^{\frac{\lambda(t)a^{\dagger}}{1+\kappa t}}\left(\frac{1}{1+\kappa t}\right)^{a^{\dagger}a}a^{n}\mathrm{e}^{-\frac{\lambda^{*}(t)a}{1+\kappa t}}\\
&=\sum_{n=0}^{\infty}\frac{\mathrm{e}^{-\frac{|\lambda(t)|^{2}}{1+\kappa t}}}{n!}\frac{(\kappa t)^{n}}{(1+\kappa t)^{n}}a^{\dagger n}\mathrm{e}^{-\frac{\lambda(t)a^{\dagger}}{1+\kappa t}}\mathrm{e}^{\lambda^{*}(t)a}\mathrm{e}^{\frac{\lambda(t)a^{\dagger}}{1+\kappa t}}\left(\frac{1}{1+\kappa t}\right)^{a^{\dagger}a}a^{n}\mathrm{e}^{-\frac{\lambda^{*}(t)a}{1+\kappa t}}\\
&=\sum_{n=0}^{\infty}\frac{1}{n!}\frac{(\kappa t)^{n}}{(1+\kappa t)^{n}}a^{\dagger n}\mathrm{e}^{\lambda^{*}(t)a}\left(\frac{1}{1+\kappa t}\right)^{a^{\dagger}a}\mathrm{e}^{-\frac{\lambda^{*}(t)a}{1+\kappa t}}a^{n}\\
&=\sum_{n=0}^{\infty}\frac{1}{n!}\frac{(\kappa t)^{n}}{(1+\kappa t)^{n}}a^{\dagger n}\left(\frac{1}{1+\kappa t}\right)^{a^{\dagger}a}\left(\frac{1}{1+\kappa t}\right)^{-a^{\dagger}a}\\
&\quad\times\mathrm{e}^{\lambda^{*}(t)a}\left(\frac{1}{1+\kappa t}\right)^{a^{\dagger}a}\mathrm{e}^{-\frac{\lambda^{*}(t)a}{1+\kappa t}}a^{n}\\
&=\sum_{n=0}^{\infty}\frac{1}{n!}\frac{(\kappa t)^{n}}{(1+\kappa t)^{n}}a^{\dagger n}\left(\frac{1}{1+\kappa t}\right)^{a^{\dagger}a}a^{n}\\
&=\,:\sum_{n=0}^{\infty}\frac{1}{n!}\frac{\left(\kappa t a^{\dagger}a\right)^{n}}{(1+\kappa t)^{n}}\mathrm{e}^{-\frac{\kappa t}{1+\kappa t}a^{\dagger}a}:\\
&=\,:\mathrm{e}^{\frac{\kappa t}{1+\kappa t}a^{\dagger}a}\mathrm{e}^{-\frac{\kappa t}{1+\kappa t}a^{\dagger}a}:\\
&=1
\end{aligned}
$$

Proof of Eq. (5.156):

$$
\begin{aligned}
\rho(t)&=\sum_{m,n=0}^{\infty}M_{m,n}\left|z\right\rangle\left\langle z\right|M_{m,n}^{\dagger}\\
&=\sum_{m,n=0}^{\infty}\frac{\mathrm{e}^{-\frac{|\lambda(t)|^{2}}{1+\kappa t}}(\kappa t)^{m+n}a^{\dagger m}\mathrm{e}^{\frac{\lambda(t)a^{\dagger}}{1+\kappa t}}}{m!n!(1+\kappa t)^{m+n+1}}\left(\frac{1}{1+\kappa t}\right)^{a^{\dagger}a}a^{n}\mathrm{e}^{-\frac{\lambda^{*}(t)a}{1+\kappa t}}\\
&\quad\left|z\right\rangle\left\langle z\right|\mathrm{e}^{-\frac{\lambda(t)a^{\dagger}}{1+\kappa t}}a^{\dagger n}\left(\frac{1}{1+\kappa t}\right)^{a^{\dagger}a}\mathrm{e}^{\frac{\lambda^{*}(t)a}{1+\kappa t}}a^{m}\\
&=\sum_{m,n=0}^{\infty}\frac{\mathrm{e}^{-\frac{|\lambda(t)|^{2}}{1+\kappa t}}(\kappa t)^{m+n}}{m!n!(1+\kappa t)^{m+n+1}}a^{\dagger m}\mathrm{e}^{\frac{\lambda(t)a^{\dagger}}{1+\kappa t}}\left(\frac{1}{1+\kappa t}\right)^{a^{\dagger}a}\\
&\quad\times z^{n}\mathrm{e}^{-\frac{\lambda^{*}(t)z}{1+\kappa t}}\left|z\right\rangle\left\langle z\right|\mathrm{e}^{-\frac{\lambda(t)z^{*}}{1+\kappa t}}(z)^{*n}\left(\frac{1}{1+\kappa t}\right)^{a^{\dagger}a}\mathrm{e}^{\frac{\lambda^{*}(t)a}{1+\kappa t}}a^{m}\\
&=\sum_{m,n=0}^{\infty}\frac{|z|^{2n}\mathrm{e}^{\frac{|\lambda(t)|^{2}\left(|z|^{2}-1\right)}{1+\kappa t}}(\kappa t)^{m+n}}{m!n!(1+\kappa t)^{m+n+1}}a^{\dagger m}\mathrm{e}^{\frac{\lambda(t)a^{\dagger}}{1+\kappa t}}\left(\frac{1}{1+\kappa t}\right)^{a^{\dagger}a}\\
&\quad\times\mathrm{e}^{-\frac{1}{2}|z|^{2}+za^{\dagger}}\left|0\right\rangle\left\langle 0\right|\mathrm{e}^{-\frac{1}{2}|z|^{2}+z^{*}a}\left(\frac{1}{1+\kappa t}\right)^{a^{\dagger}a}\mathrm{e}^{\frac{\lambda^{*}(t)a}{1+\kappa t}}a^{m}
\end{aligned}
$$

$$
\begin{aligned}
&= \sum_{m,n=0}^{\infty} \frac{|z|^{2n} \mathrm{e}^{\frac{|\lambda(t)|^2\left(|z|^2-1\right)}{1+\kappa t}-|z|^2} (\kappa t)^{m+n}}{m!n!\,(1+\kappa t)^{m+n+1}} a^{\dagger m} \mathrm{e}^{\frac{\lambda(t) a^{\dagger}}{1+\kappa t}} \left(\frac{1}{1+\kappa t}\right)^{a^{\dagger} a} \mathrm{e}^{z a^{\dagger}} \\
&\left(\frac{1}{1+\kappa t}\right)^{-a^{\dagger} a} |0\rangle\langle 0| \left(\frac{1}{1+\kappa t}\right)^{-a^{\dagger} a} \mathrm{e}^{z^* a} \left(\frac{1}{1+\kappa t}\right)^{a^{\dagger} a} \mathrm{e}^{\frac{\lambda^*(t) a}{1+\kappa t}} a^m \\
&= \frac{\mathrm{e}^{\frac{|\lambda(t)|^2\left(|z|^2-1\right)}{1+\kappa t}-|z|^2}}{1+\kappa t} \sum_{n=0}^{\infty} \frac{1}{n!} \left(\frac{\kappa t\,|z|^2}{1+\kappa t}\right)^n \sum_{m=0}^{\infty} \frac{1}{m!} \\
&\left(\frac{\kappa t}{1+\kappa t}\right)^m \mathrm{e}^{\frac{[\lambda(t)+z] a^{\dagger}}{1+\kappa t}} a^{\dagger m} |0\rangle\langle 0| a^m \mathrm{e}^{\frac{[\lambda^*(t)+z^*] a}{1+\kappa t}} \\
&= \frac{\mathrm{e}^{\frac{|\lambda(t)|^2\left(|z|^2-1\right)+\kappa t |z|^2}{1+\kappa t}-|z|^2}}{1+\kappa t} \mathrm{e}^{\frac{[\lambda(t)+z] a^{\dagger}}{1+\kappa t}} : \sum_{m=0}^{\infty} \frac{1}{m!} \left(\frac{\kappa t}{1+\kappa t} a^{\dagger} a\right)^m \mathrm{e}^{-a^{\dagger} a} : \mathrm{e}^{\frac{[\lambda^*(t)+z^*] a}{1+\kappa t}} \\
&= \frac{\mathrm{e}^{\frac{|\lambda(t)|^2\left(|z|^2-1\right)+\kappa t |z|^2}{1+\kappa t}-|z|^2}}{1+\kappa t} \mathrm{e}^{\frac{[\lambda(t)+z] a^{\dagger}}{1+\kappa t}} : \mathrm{e}^{\left(\frac{\kappa t}{1+\kappa t}-1\right) a^{\dagger} a} : \mathrm{e}^{\frac{[\lambda^*(t)+z^*] a}{1+\kappa t}} \\
&= \frac{\mathrm{e}^{\frac{|\lambda(t)|^2\left(|z|^2-1\right)+\kappa t |z|^2}{1+\kappa t}-|z|^2}}{1+\kappa t} \mathrm{e}^{\frac{[\lambda(t)+z] a^{\dagger}}{1+\kappa t}} \left(\frac{\kappa t}{1+\kappa t}\right)^{a^{\dagger} a} \mathrm{e}^{\frac{[\lambda^*(t)+z^*] a}{1+\kappa t}},
\end{aligned}
$$

where we have used the operator identities $|0\rangle\langle 0| =: \mathrm{e}^{-a^{\dagger} a}:$ and $: \exp\left[\left(\frac{\kappa t}{1+\kappa t}-1\right) a^{\dagger} a\right]: = \left(\frac{1}{1+\kappa t}\right)^{a^{\dagger} a}$.

References

1. Zhou J, Fan HY (2018) From the classical diffusion equation to quantum master equation for describing optical diffusion. Optik 164:455–459
2. Liu TK, Shan CJ, Liu JB, Fan HY (2014) Master equation describing the diffusion process for a coherent state. Chin Phys B 23(3):030303
3. Meng XG, Fan HY, Wang JS (2018) Generation of a kind of displaced thermal states in the diffusion process and its statistical properties. Int J Theor Phys 57:1202–1209
4. Fan HY, Lou SY, Pan XY, Hu LY (2014) A new optical field state as an output of diffusion channel when the input being number state. Sci China Phys Mech Astron 57(9):1649–1653
5. Fan HY, Hu LY (2008) Operator-sum representation of density operators as solution to master equations obtained via the entangled states approach. Mod Phys Lett B 22(25):2435–2468
6. Meng XG, Wang JS, Liang BL (2023) Entangled state representations in quantum optics. Springer, Singapore
7. Du JM, He R, Ren G, Yu HJ (2015) Time evolution of Wigner function in diffusion channel. Int J Theor Phys 54(4):1225–1232
8. Zhang CZ, Fan HY (2020) Temperature variation of chaotic optical field in quantum dissipation channel and diffusion channel. Int J Theor Phys 59(7):2137–2146
9. Chen F, Xiong W, Fang BL, Fan HY (2020) Chaotic state as an output of vacuum state evolving in diffusion channel and generation of displaced chaotic state for quantum controlling. Chin Phys B 29(12):124202

10. Da C, Fan HY (2016) Time evolution and characteristic quantities of squeezed chaotic field in diffusion channel. Int J Theorl Phys 55(10):4521–4531
11. Fan HY, Chen QF (2014) Density operator for describing driven damped harmonic oscillator in the diffusion-limited channel. Can J Phys 92(10):1069–1073

Chapter 6
Light Fields Caused by Quantum Amplitude Decay

6.1 Evolution Equation Followed by Quantum Decay

Since no system in nature can be completely isolated from its environment, the coupling between the system and the environment always causes noise. So, what is the dynamic evolution equation describing the amplitude decay channel of the system?

Let's start with the amplitude decay of a coherent state $|\alpha\rangle\langle\alpha|$. For

$$|\alpha\rangle\langle\alpha| \to \left|\alpha e^{-\kappa t}\right\rangle\left\langle\alpha e^{-\kappa t}\right|, \tag{6.1}$$

it means that what equation governs this evolution (κ is the decay rate)? Using the normal ordering product

$$\left|\alpha e^{-\kappa t}\right\rangle\left\langle\alpha e^{-\kappa t}\right| =: \exp\left(-|\alpha|^2 e^{-2\kappa t} + \alpha e^{-\kappa t} a^\dagger + \alpha^* e^{-\kappa t} a - a^\dagger a\right):, \tag{6.2}$$

we have

$$\begin{aligned}
&\frac{d}{dt}\left|\alpha e^{-\kappa t}\right\rangle\left\langle\alpha e^{-\kappa t}\right| \\
=& \frac{d}{dt}: \exp\left(-|\alpha|^2 e^{-2\kappa t} + \alpha e^{-\kappa t} a^\dagger + \alpha^* e^{-\kappa t} a - a^\dagger a\right): \\
=& 2\kappa|\alpha|^2 e^{-2\kappa t}\left|\alpha e^{-\kappa t}\right\rangle\left\langle\alpha e^{-\kappa t}\right| - \kappa a^\dagger \alpha e^{-\kappa t}\left|\alpha e^{-\kappa t}\right\rangle\left\langle\alpha e^{-\kappa t}\right| \\
&-\kappa\left|\alpha e^{-\kappa t}\right\rangle\left\langle\alpha e^{-\kappa t}\right|\alpha^* e^{-\kappa t} a \\
=& 2\kappa a\left|\alpha e^{-\kappa t}\right\rangle\left\langle\alpha e^{-\kappa t}\right| a^\dagger - \kappa a^\dagger a\left|\alpha e^{-\kappa t}\right\rangle\left\langle\alpha e^{-\kappa t}\right| - \kappa\left|\alpha e^{-\kappa t}\right\rangle\left\langle\alpha e^{-\kappa t}\right| a^\dagger a.
\end{aligned} \tag{6.3}$$

Letting $\left|\alpha e^{-\kappa t}\right\rangle\left\langle\alpha e^{-\kappa t}\right| = \rho(t)$, thus Eq. (6.3) is equivalent to

$$\frac{d}{dt}\rho(t) = \kappa\left(2a\rho(t)a^\dagger - a^\dagger a\rho(t) - \rho(t)a^\dagger a\right), \tag{6.4}$$

H. Fan and L. Hu, *Dissipation and Diffusion of Quantum Light Field*,
https://doi.org/10.1007/978-981-95-1528-8_6

which is the evolution equation followed by quantum amplitude decay.

Next, we will examine how the Wigner function evolves in the amplitude decay channel described by the master equation in Eq. (6.4) [1]. Noting $W(\alpha, \alpha^*) = \mathtt{tr}(\Delta(\alpha, \alpha^*)\rho)$, the time-evolution of the Wigner function can be expressed as

$$\begin{aligned}\frac{\partial W(\alpha, \alpha^*, t)}{\partial t} &= \mathtt{tr}\left[\Delta(\alpha, \alpha^*)\frac{\partial}{\partial t}\rho(t)\right] \\ &= \kappa\mathtt{tr}[(2a^\dagger\Delta(\alpha, \alpha^*)a\rho(t) - \Delta(\alpha, \alpha^*) \\ &\quad \times a^\dagger a\rho(t) - a^\dagger a\Delta(\alpha, \alpha^*)\rho(t))].\end{aligned} \tag{6.5}$$

Using Eqs. (5.33) and (5.34) to obtain

$$\begin{aligned}\frac{\partial W(\alpha, \alpha^*, t)}{\partial t} &= \kappa\left[2\left(\frac{\partial}{2\partial\alpha} + \alpha^*\right)\left(\frac{\partial}{2\partial\alpha^*} + \alpha\right) - \left(\alpha^* - \frac{\partial}{2\partial\alpha}\right)\right. \\ &\quad \left.\times\left(\frac{\partial}{2\partial\alpha^*} + \alpha\right) - \left(\alpha - \frac{\partial}{2\partial\alpha^*}\right)\left(\frac{\partial}{2\partial\alpha} + \alpha^*\right)\right]W(\alpha, \alpha^*, t) \\ &= \kappa\left(\frac{\partial^2}{\partial\alpha\partial\alpha^*} + 2\alpha^*\frac{\partial}{2\partial\alpha^*} + 2\frac{\partial}{2\partial\alpha}\alpha - \alpha^*\frac{\partial}{2\partial\alpha^*}\right. \\ &\quad \left.+\frac{\partial}{2\partial\alpha^*}\alpha^* + \frac{\partial}{2\partial\alpha}\alpha - \alpha\frac{\partial}{2\partial\alpha}\right)W(\alpha, \alpha^*, t) \\ &= \kappa\left(\frac{\partial^2}{\partial\alpha\partial\alpha^*} + \alpha^*\frac{\partial}{2\partial\alpha^*} + \frac{\partial}{\partial\alpha}\alpha + \frac{\partial}{2\partial\alpha^*}\alpha^* + \frac{1}{2}\right)W(\alpha, \alpha^*, t) \\ &= \kappa\left(\frac{\partial^2}{\partial\alpha\partial\alpha^*} + \alpha^*\frac{\partial}{\partial\alpha^*} + \frac{\partial}{\partial\alpha}\alpha + 1\right)W(\alpha, \alpha^*, t) \\ &= \kappa\left(\frac{\partial^2}{\partial\alpha\partial\alpha^*} + \alpha^*\frac{\partial}{\partial\alpha^*} + \alpha\frac{\partial}{\partial\alpha} + 2\right)W(\alpha, \alpha^*, t),\end{aligned} \tag{6.6}$$

which is the time-evolution equation of the Wigner function $W(\alpha, \alpha^*)$ in the amplitude decay channel.

6.2 Converting the Decay Master Equation into a Like Schrödinger Equation

To understand the quantum master equation in Eq. (6.4), we can also avoid using its Wigner function correspondence or P-representation. Introducing the thermal entangled state [2, 3], that is

$$|\eta\rangle = \exp\left(-\frac{1}{2}|\eta|^2 + \eta a^\dagger - \eta^*\tilde{a}^\dagger + a^\dagger\tilde{a}^\dagger\right)\left|0, \tilde{0}\right\rangle \tag{6.7}$$

where $\tilde{a}^{\dagger}$ is the creation operator of fictitious mode that accompanies the creation operator $a^{\dagger}$ of real mode, $\left|\tilde{0}\right\rangle$ is vacuum of fictitious mode, $\tilde{a}\left|\tilde{0}\right\rangle = 0$, $\left[\tilde{a}, \tilde{a}^{\dagger}\right] = 1$, $|n, \tilde{n}\rangle = \left(a^{\dagger}\tilde{a}^{\dagger}\right)^{n}\left|0, \tilde{0}\right\rangle / n!$, $\left[a, \tilde{a}^{\dagger}\right] = 0$. The states $|\eta\rangle$ have completeness, that is $\pi^{-1}\int d^{2}\eta\,|\eta\rangle\langle\eta| = 1$. Acting the density operator ρ on the state $|I\rangle = |\eta = 0\rangle$, and defining a pure state $|\rho\rangle$ as

$$|\rho\rangle = \rho\,|I\rangle\,,\;|I\rangle = \exp\left(a^{\dagger}\tilde{a}^{\dagger}\right)\left|0, \tilde{0}\right\rangle = \sum_{n=0}^{\infty}|n, \tilde{n}\rangle\,. \tag{6.8}$$

It follows from Eq. (6.8) that

$$a\,|I\rangle = \tilde{a}^{\dagger}\,|I\rangle\,,\; a^{\dagger}\,|I\rangle = \tilde{a}\,|I\rangle\,,\; \left(a^{\dagger}a\right)^{n}|I\rangle = \left(\tilde{a}^{\dagger}\tilde{a}\right)^{n}|I\rangle\,. \tag{6.9}$$

Acting Eq. (6.4) on the state $|I\rangle$ leads to

$$\frac{d}{dt}\,|\rho(t)\rangle = \kappa\left(2a\rho(t)\,a^{\dagger} - a^{\dagger}a\rho(t) - \rho(t)\,a^{\dagger}a\right)|I\rangle\,. \tag{6.10}$$

Using the operator exchanges between real and fictitious modes realized by Eq. (6.9), and noting that $\rho(t)$ is only a density operator of real mode and is commutative with all the operators of fictitious mode, thus Eq. (6.10) becomes

$$\frac{d}{dt}\,|\rho(t)\rangle = \kappa\left(2a\tilde{a} - a^{\dagger}a - \tilde{a}^{\dagger}\tilde{a}\right)|\rho(t)\rangle\,, \tag{6.11}$$

its formal solution is expressed as

$$|\rho(t)\rangle = \exp\left[\kappa t\left(2a\tilde{a} - a^{\dagger}a - \tilde{a}^{\dagger}\tilde{a}\right)\right]|\rho(0)\rangle \tag{6.12}$$

where $|\rho(0)\rangle \equiv \rho(0)\,|I\rangle$ denotes as the pure state at initial time $t = 0$. Therefore, by introducing the fictitious mode corresponding to the real mode of the system, the master equation of the density operator can be transformed into the evolution equation of a pure state, which is formally similar to a Schrödinger equation, and its formal solution is given.

Noting the operator identity

$$\left[\frac{a^{\dagger}a + \tilde{a}^{\dagger}\tilde{a}}{2}, a\tilde{a}\right] = -a\tilde{a}, \tag{6.13}$$

and using the operator formula

$$e^{\lambda(A+\sigma B)} = e^{\lambda A}\exp\left[\frac{\sigma\left(1 - e^{-\lambda\tau}\right)B}{\tau}\right], \tag{6.14}$$

which holds for $[A, B] = \tau B$, we can decompose the exponential operator in Eq. (6.12) into

$$\exp\left[-2\kappa t\left(\frac{a^{\dagger}a+\tilde{a}^{\dagger}\tilde{a}}{2}-a\tilde{a}\right)\right]=\exp\left[-\kappa t\left(a^{\dagger}a+\tilde{a}^{\dagger}\tilde{a}\right)\right]\exp\left(Ta\tilde{a}\right), \quad (6.15)$$

where $T = 1-\mathrm{e}^{-2\kappa t}$. Therefore, Eq. (6.12) can be expressed as

$$\begin{aligned}|\rho(t)\rangle &= \rho(t)|\tau=0\rangle=\exp\left[-\kappa t\left(a^{\dagger}a+\tilde{a}^{\dagger}\tilde{a}\right)\right]\exp\left(Ta\tilde{a}\right)|\rho(0)\rangle\\ &= \exp\left[-\kappa t\left(a^{\dagger}a+\tilde{a}^{\dagger}\tilde{a}\right)\right]\sum_{n=0}^{\infty}\frac{T^n}{n!}a^n\tilde{a}^n\rho(0)|I\rangle. \quad (6.16)\end{aligned}$$

Using the identities

$$\tilde{a}^n\rho(0)|I\rangle=\rho(0)\tilde{a}^n|I\rangle=\rho(0)a^{\dagger n}|I\rangle, \quad (6.17)$$

and

$$\begin{aligned}\exp\left(-\kappa t\tilde{a}^{\dagger}\tilde{a}\right)a^n\rho(0)a^{\dagger n}|I\rangle &= a^n\rho(0)a^{\dagger n}\exp\left(-\kappa t\tilde{a}^{\dagger}\tilde{a}\right)|I\rangle\\ &= a^n\rho(0)a^{\dagger n}\exp\left(-\kappa ta^{\dagger}a\right)|I\rangle, \quad (6.18)\end{aligned}$$

we thus obtain

$$\begin{aligned}\rho(t)|I\rangle &= \exp\left(-\kappa ta^{\dagger}a\right)\sum_{n=0}^{\infty}\frac{T^n}{n!}a^n\rho(0)a^{\dagger n}\exp\left(-\kappa t\tilde{a}^{\dagger}\tilde{a}\right)|I\rangle\\ &= \sum_{n=0}^{\infty}\frac{T^n}{n!}\exp\left(-\kappa ta^{\dagger}a\right)a^n\rho(0)a^{\dagger n}\exp\left(-\kappa ta^{\dagger}a\right)|I\rangle. \quad (6.19)\end{aligned}$$

Therefore, the solution of the master equation in Eq. (6.4) reads

$$\rho(t)=\sum_{n=0}^{\infty}\frac{T^n}{n!}\exp\left(-\kappa ta^{\dagger}a\right)a^n\rho(0)a^{\dagger n}\exp\left(-\kappa ta^{\dagger}a\right). \quad (6.20)$$

Compared with the standard operator-sum representations, it can be concluded that

$$\rho(t)=\sum_{m=0}^{\infty}M_m\rho(0)M_m^{\dagger}, \quad (6.21)$$

where M_m is the Kraus operator, that is

$$M_m = \sqrt{\frac{T^m}{m!}} e^{-\kappa t a^\dagger a} a^m. \tag{6.22}$$

It can be proven that the Kraus operator M_m satisfies the normalization condition:

$$\sum_{m=0}^{\infty} M_m^\dagger M_m = 1. \tag{6.23}$$

6.3 Description of Decay Mechanism in Phase Space

This section discusses how to describe the quantum decay in the phase space [4]. Assuming that the Weyl correspondence of the initial density operator of the light field in phase space is $h_0(\alpha, \alpha^*)$, thus the density operator $\rho(0)$ can be expressed as

$$\rho(0) = 2 \int d^2\alpha \Delta\left(\alpha, \alpha^*\right) h_0\left(\alpha, \alpha^*\right). \tag{6.24}$$

After passing through the amplitude decay channel, $\rho(0) \to \rho(t)$, the Weyl correspondence of $\rho(t)$ in the phase space is denotes as $h_t(\alpha, \alpha^*)$, thus we have

$$\rho(t) = 2 \int d^2\alpha \Delta\left(\alpha, \alpha^*\right) h_t\left(\alpha, \alpha^*\right). \tag{6.25}$$

What transformation can we use to connect $h_t(\alpha, \alpha^*)$ and $h_0(\alpha, \alpha^*)$?

Using Eq. (1.92), we get

$$h_t\left(\alpha, \alpha^*\right) = 2\pi \mathrm{tr}\left[\rho(t) \Delta\left(\alpha, \alpha^*\right)\right], \tag{6.26}$$

thus substituting Eq. (6.20) into Eq. (6.26) leads to

$$\begin{aligned} h_t\left(\alpha, \alpha^*\right) &= 2\pi \mathrm{tr}\left(\sum_{n=0}^{\infty} M_n \rho(0) M_n^\dagger \Delta\left(\alpha, \alpha^*\right)\right) \\ &= 2\pi \mathrm{tr}\left(\rho(0) \sum_{n=0}^{\infty} M_n^\dagger \Delta\left(\alpha, \alpha^*\right) M_n\right) \\ &= 2\pi \mathrm{tr}\left(\rho(0) \Delta\left(\alpha, \alpha^*, t\right)\right), \end{aligned} \tag{6.27}$$

where $\Delta(\alpha, \alpha^*, t)$ is defined as

$$\Delta\left(\alpha, \alpha^*, t\right) = \sum_{n=0}^{\infty} M_n^\dagger \Delta\left(\alpha, \alpha^*\right) M_n. \tag{6.28}$$

Using Eq. (6.22), we can write $\Delta(\alpha, \alpha^*, t)$ as

$$
\begin{aligned}
\Delta\left(\alpha, \alpha^*, t\right) &= \frac{1}{\pi}\mathrm{e}^{-2\alpha^*\alpha}\sum_{n=0}^{\infty}\frac{T^n}{n!}a^{\dagger n}\mathrm{e}^{-\kappa t a^\dagger a}\mathrm{e}^{2a^\dagger\alpha}\mathrm{e}^{\mathrm{i}\pi a^\dagger a}\mathrm{e}^{2\alpha^* a}\mathrm{e}^{-\kappa t a^\dagger a}a^n \\
&= \frac{1}{\pi}\mathrm{e}^{-2\alpha^*\alpha}\sum_{n=0}^{\infty}\frac{T^n}{n!}a^{\dagger n}\mathrm{e}^{2a^\dagger\alpha\mathrm{e}^{-\kappa t}}\mathrm{e}^{(\mathrm{i}\pi-2\kappa t)a^\dagger a}\mathrm{e}^{2\alpha^* a\mathrm{e}^{-\kappa t}}a^n \\
&= \frac{1}{\pi}\mathrm{e}^{-2\alpha^*\alpha}\sum_{n=0}^{\infty}\frac{T^n}{n!}a^{\dagger n}\mathrm{e}^{2a^\dagger\alpha\mathrm{e}^{-\kappa t}} \\
&\quad\times :\exp\left\{\left[\mathrm{e}^{(\mathrm{i}\pi-2\kappa t)}-1\right]a^\dagger a\right\}: \mathrm{e}^{2\alpha^* a\mathrm{e}^{-\kappa t}}a^n \\
&= \frac{1}{\pi}\mathrm{e}^{-2\alpha^*\alpha}\sum_{n=0}^{\infty}\frac{T^n}{n!}: a^{\dagger n}a^n\mathrm{e}^{2a^\dagger\alpha\mathrm{e}^{-\kappa t}} \\
&\quad\times\exp\left\{\left[\mathrm{e}^{(\mathrm{i}\pi-2\kappa t)}-1\right]a^\dagger a\right\}\mathrm{e}^{2\alpha^* a\mathrm{e}^{-\kappa t}}: \\
&= \frac{1}{\pi}\mathrm{e}^{-2\alpha^*\alpha}: \mathrm{e}^{2a^\dagger\alpha\mathrm{e}^{-\kappa t}}\exp\left\{\left[T+\mathrm{e}^{(\mathrm{i}\pi-2\kappa t)}-1\right]a^\dagger a\right\}\mathrm{e}^{2\alpha^* a\mathrm{e}^{-\kappa t}}: \\
&= \frac{1}{\pi}\mathrm{e}^{-2\alpha^*\alpha}: \mathrm{e}^{2a^\dagger\alpha\mathrm{e}^{-\kappa t}}\exp\left(\mathrm{e}^{-2\kappa t}a^\dagger a\right)\mathrm{e}^{2\alpha^* a\mathrm{e}^{-\kappa t}}: \\
&= \frac{1}{\pi}: \mathrm{e}^{-2\left(\alpha^*-a^\dagger\mathrm{e}^{-\kappa t}\right)\left(\alpha-a\mathrm{e}^{-\kappa t}\right)}:,
\end{aligned}
\tag{6.29}
$$

thus Eq. (6.27) becomes

$$
h_t\left(\alpha, \alpha^*\right) = 2\mathtt{tr}\left(\rho(0): \mathrm{e}^{-2\left(\alpha^*-a^\dagger\mathrm{e}^{-\kappa t}\right)\left(\alpha-a\mathrm{e}^{-\kappa t}\right)}:\right), \tag{6.30}
$$

which is the classical Weyl correspondence of the density operator $\rho(t)$.

Further, in order to find the transformation relation between $h_t(\alpha, \alpha^*)$ and $h_0(\alpha, \alpha^*)$, we need to convert the operator $\Delta(\alpha, \alpha^*, t)$ into its Weyl ordering product using the following integral formula

$$
\rho = 2 \vdots \int\frac{\mathrm{d}^2\beta}{\pi}\langle-\beta|\,\rho\,|\beta\rangle\,\mathrm{e}^{2\left(a^\dagger a+a\beta^*-\beta a^\dagger\right)} \vdots, \tag{6.31}
$$

where $|\beta\rangle = \exp\left(-|\beta|^2/2+\beta a^\dagger\right)|0\rangle$ is a coherent state. Inserting the normal ordering product of $\Delta(\alpha, \alpha^*, t)$ in Eq. (6.29) into Eq. (6.31) leads to

$$
\begin{aligned}
\Delta\left(\alpha, \alpha^*, t\right) = & \frac{2}{\pi} \vdots \int \frac{\mathrm{d}^2\beta}{\pi} \langle-\beta| : \mathrm{e}^{-2\left(\alpha^*-a^\dagger \mathrm{e}^{-\kappa t}\right)\left(\alpha-a\mathrm{e}^{-\kappa t}\right)} : \\
& |\beta\rangle\, \mathrm{e}^{2\left(a^\dagger a+a\beta^*-\beta a^\dagger\right)} \vdots \\
= & \frac{2}{\pi} \vdots \exp\left(-2|\alpha|^2+2a^\dagger a\right) \int \frac{\mathrm{d}^2\beta}{\pi} \exp\left[-2|\beta|^2\left(1-\mathrm{e}^{-2\kappa t}\right)\right. \\
& \left.+2\beta\left(\alpha^*\mathrm{e}^{-\kappa t}-a^\dagger\right)+2\beta^*\left(\alpha-a\mathrm{e}^{-\kappa t}\right)\right] \vdots \\
= & \frac{1}{\pi T} \vdots \exp\left[-\frac{2}{T}\left(\alpha^*-a^\dagger \mathrm{e}^{-\kappa t}\right)\left(\alpha-a\mathrm{e}^{-\kappa t}\right)\right] \vdots,
\end{aligned} \tag{6.32}
$$

where $T = 1 - \mathrm{e}^{-2\kappa t}$. Therefore, Eq. (6.30) becomes

$$
h_t\left(\alpha, \alpha^*\right) = \frac{2}{T}\mathrm{tr}\left\{\rho(0) \vdots \exp\left[-\frac{2}{T}\left(\alpha^*-a^\dagger \mathrm{e}^{-\kappa t}\right)\left(\alpha-a\mathrm{e}^{-\kappa t}\right)\right] \vdots\right\}. \tag{6.33}
$$

Using Eqs. (6.24) and (6.33), we obtain

$$
\begin{aligned}
h_t\left(\alpha, \alpha^*\right) = & \frac{2}{T}\left\{\mathrm{tr}\rho(0) \int \mathrm{d}^2\alpha' \vdots \delta\left(\alpha'^*-a^\dagger\right)\delta\left(\alpha'-a\right) \vdots\right. \\
& \left.\times \exp\left[-\frac{2}{T}\left(\alpha^*-\alpha'^*\mathrm{e}^{-\kappa t}\right)\left(\alpha-\alpha'\mathrm{e}^{-\kappa t}\right)\right]\right\} \\
= & \frac{2}{T}\int \mathrm{d}^2\beta \mathrm{d}^2\alpha' h_0\left(\beta, \beta^*\right) \mathrm{tr}\left[4\Delta\left(\beta, \beta^*\right)\Delta\left(\alpha', \alpha'^*\right)\right] \\
& \times \exp\left[-\frac{2}{T}\left(\alpha^*-\alpha'^*\mathrm{e}^{-\kappa t}\right)\left(\alpha-\alpha'\mathrm{e}^{-\kappa t}\right)\right].
\end{aligned} \tag{6.34}
$$

Noting that the trace of the product of two different Wigner operators is the delta δ function (see also Eq. (1.91)), thus Eq. (6.34) can further be written as

$$
\begin{aligned}
h_t\left(\alpha, \alpha^*\right) = & \frac{2}{\pi T}\int \mathrm{d}^2\beta \int \mathrm{d}^2\alpha' h_0\left(\beta, \beta^*\right)\delta\left(\alpha'^*-\beta^*\right)\delta\left(\alpha'-\beta\right) \\
& \times \exp\left[-\frac{2}{T}\left(\alpha^*-\alpha'^*\mathrm{e}^{-\kappa t}\right)\left(\alpha-\alpha'\mathrm{e}^{-\kappa t}\right)\right] \\
= & \frac{2}{T}\int \frac{\mathrm{d}^2\beta}{\pi} h_0\left(\beta, \beta^*\right)\exp\left[-\frac{2}{T}\left(\alpha^*-\beta^*\mathrm{e}^{-\kappa t}\right)\left(\alpha-\beta\mathrm{e}^{-\kappa t}\right)\right],
\end{aligned} \tag{6.35}
$$

which is the relation between the classical correspondence of the density operator at any given time t and the classical correspondence of the initial density operator. The decay of $h_0\left(\alpha, \alpha^*\right)$ caused by the channel is described by the Gaussian decay factor.

By virtue of Eqs. (6.25) and (6.35), the Weyl corresponding rule can also be represented as

$$\begin{aligned}\rho(t) &= 2\int \mathrm{d}^2\alpha\Delta(\alpha,\alpha^*)\,h_t(\alpha,\alpha^*)\\&= \frac{2}{T}\int \mathrm{d}^2\alpha \,\vdots\, \delta(\alpha^*-a^\dagger)(\alpha-a)\,\vdots\\&\times\int\frac{\mathrm{d}^2\beta}{\pi}h_0(\beta,\beta^*)\exp\left[-\frac{2}{T}(\alpha^*-\beta^*\mathrm{e}^{-\kappa t})(\alpha-\beta\mathrm{e}^{-\kappa t})\right]\\&= \frac{2}{T}\int\frac{\mathrm{d}^2\beta}{\pi}h_0(\beta,\beta^*)\,\vdots\,\exp\left[-\frac{2}{T}(a^\dagger-\beta^*\mathrm{e}^{-\kappa t})(a-\beta\mathrm{e}^{-\kappa t})\right]\vdots. \end{aligned} \tag{6.36}$$

Especially, when $\rho(0)=|z\rangle\langle z|$ is a coherent state, its classical Weyl correspondence is

$$\begin{aligned}h_0(\alpha,\alpha^*) &= 2\pi\mathrm{tr}\left[|z\rangle\langle z|\,\Delta(\alpha,\alpha^*)\right]\\&= 2\langle z|:\mathrm{e}^{-2(\alpha^*-a^\dagger)(\alpha-a)}:|z\rangle\\&= 2\mathrm{e}^{-2(\alpha^*-z^*)(\alpha-z)},\end{aligned} \tag{6.37}$$

substituting it into Eq. (6.35) leads to

$$\begin{aligned}h_t(\alpha,\alpha^*) &= \frac{4}{T}\mathrm{e}^{-2|z|^2-\frac{2}{T}|\alpha|^2}\int\frac{\mathrm{d}^2\beta}{\pi}\exp\left[-\frac{2}{T}|\beta|^2\right.\\&\left.+2\beta^*\left(\frac{\mathrm{e}^{-\kappa t}}{T}\alpha+z\right)+2\beta\left(\frac{\mathrm{e}^{-\kappa t}}{T}\alpha^*+z^*\right)\right]\\&= 2\exp\left[-2(\alpha^*-z^*\mathrm{e}^{-\kappa t})(\alpha-z\mathrm{e}^{-\kappa t})\right],\end{aligned} \tag{6.38}$$

which presents the decay of quantum channels in a classical way.

In short, we established the integration transformation relation between the classical correspondence $h_0(\alpha,\alpha^*)$ of the initial density operator $\rho(0)$ and the classical correspondence $h_t(\alpha,\alpha^*)$ of the evolved density operator $\rho(t)$ in the quantum amplitude decay channel. This method is useful for examining the amplitude decay case within the classical framework and can also be extended to other quantum decay cases.

6.4 Laguerre-Polynomial-Weighted Coherent States Generated via Displaced Fock States for Amplitude Decay

This section considers the evolution of the displaced Fock state $\rho(0)=D(\alpha)|m\rangle\langle m|D^\dagger(\alpha)$ in the amplitude decay channel [5]. Using the identities

$$a^n|m\rangle=\sqrt{\frac{m!}{(m-n)!}}|m-n\rangle \tag{6.39}$$

and

$$D(\alpha)|m\rangle = \frac{1}{\sqrt{m!}}\left(a^\dagger - \alpha^*\right)^m D(\alpha)|0\rangle$$
$$= \frac{1}{\sqrt{m!}}\sum_{l=0}^{m}\binom{m}{l}\left(-\alpha^*\right)^{m-l} a^{\dagger l}|\alpha\rangle, \tag{6.40}$$

as well as the operator identity

$$a^n a^{\dagger l} = (-\mathrm{i})^{n+l} : \mathrm{H}_{l,n}\left(\mathrm{i}a^\dagger, \mathrm{i}a\right) : , \tag{6.41}$$

where $\mathrm{H}_{l,n}(\cdot)$ is a two-mode Hermite polynomial, we have

$$a^n D(\alpha)|m\rangle = \frac{1}{\sqrt{m!}}\sum_{l=0}^{m}\binom{m}{l}\left(-\alpha^*\right)^{m-l} a^n a^{\dagger l}|\alpha\rangle$$
$$= \frac{1}{\sqrt{m!}}\sum_{l=0}^{m}(-\mathrm{i})^{n+l}\binom{m}{l}\left(-\alpha^*\right)^{m-l} : \mathrm{H}_{l,n}\left(\mathrm{i}a^\dagger, \mathrm{i}a\right) : |\alpha\rangle$$
$$= \frac{1}{\sqrt{m!}}\sum_{l=0}^{m}(-\mathrm{i})^{n+l}\binom{m}{l}\left(-\alpha^*\right)^{m-l}\mathrm{H}_{l,n}\left(\mathrm{i}a^\dagger, \mathrm{i}\alpha\right)|\alpha\rangle. \tag{6.42}$$

Noting that $\mathrm{e}^{-\kappa t a^\dagger a} a^\dagger \mathrm{e}^{\kappa t a^\dagger a} = a^\dagger \mathrm{e}^{-\kappa t}$, thus

$$\mathrm{e}^{-\kappa t a^\dagger a} a^n D(\alpha)|m\rangle = \frac{1}{\sqrt{m!}}(-1)^m(-\mathrm{i})^n\sum_{l=0}^{m}\mathrm{i}^l\binom{m}{l}$$
$$\times\alpha^{*m-l}\mathrm{H}_{l,n}\left(\mathrm{i}a^\dagger \mathrm{e}^{-\kappa t}, \mathrm{i}\alpha\right)\mathrm{e}^{-\kappa t a^\dagger a}|\alpha\rangle. \tag{6.43}$$

Inserting the density operator $\rho(0)$ into Eq. (6.20), and using the following identity

$$\mathrm{e}^{-\kappa t a^\dagger a}|\alpha\rangle = \mathrm{e}^{-|\alpha|^2/2+\alpha a^\dagger \mathrm{e}^{-\kappa t}}|0\rangle \tag{6.44}$$

and the operator identity $|0\rangle\langle 0| =: \mathrm{e}^{-a^\dagger a} :$, we obtain

$$\rho(t) = \sum_{n=0}^{\infty}\frac{T^n}{n!}\mathrm{e}^{-\kappa t a^\dagger a} a^n D(\alpha)|m\rangle\langle m| D^\dagger(\alpha) a^{\dagger n}\mathrm{e}^{-\kappa t a^\dagger a}$$
$$= \frac{1}{m!}\sum_{n=0}^{\infty}\frac{T^n}{n!}\sum_{l=0}^{m}\mathrm{i}^l\binom{m}{l}\alpha^{*m-l}\mathrm{H}_{l,n}\left(\mathrm{i}a^\dagger \mathrm{e}^{-\kappa t}, \mathrm{i}\alpha\right)$$
$$\times\,\mathrm{e}^{-\kappa t a^\dagger a}|\alpha\rangle\langle\alpha|\,\mathrm{e}^{-\kappa t a^\dagger a}\sum_{j=0}^{m}(-\mathrm{i})^j\binom{m}{j}\mathrm{H}_{j,n}\left(-\mathrm{i}a\mathrm{e}^{-\kappa t}, -\mathrm{i}\alpha^*\right)\alpha^{m-j}$$

$$
\begin{aligned}
&= \frac{1}{m!}\sum_{j,l=0}^{m} \mathrm{i}^l (-\mathrm{i})^j \binom{m}{j}\binom{m}{l}\alpha^{m-j}\alpha^{*m-l}\sum_{n=0}^{\infty}\frac{T^n}{n!} \\
&\quad \times :\mathrm{H}_{l,n}\left(\mathrm{i}a^{\dagger}\mathrm{e}^{-\kappa t}, \mathrm{i}\alpha\right)\mathrm{H}_{j,n}\left(-\mathrm{i}a\mathrm{e}^{-\kappa t}, -\mathrm{i}\alpha^*\right) \\
&\quad \times \mathrm{e}^{-|\alpha|^2+\alpha a^{\dagger}\mathrm{e}^{-\kappa t}+\alpha^* a\mathrm{e}^{-\kappa t}-a^{\dagger}a}: . \qquad (6.45)
\end{aligned}
$$

To further calculate Eq. (6.45), we need to introduce a new summation formula for two-variable Hermite polynomials, that is

$$
\begin{aligned}
&\sum_{n=0}^{\infty}\frac{T^n}{n!}\mathrm{H}_{l,n}(x,y)\,\mathrm{H}_{j,n}\left(x',y'\right) \\
&= \mathrm{i}^{l+j}\sqrt{T^{l+j}}\mathrm{e}^{Tyy'}\mathrm{H}_{l,j}\left[\mathrm{i}\left(\sqrt{T}y'-x/\sqrt{T}\right), \mathrm{i}(\sqrt{T}y-x'/\sqrt{T}\right)\right]. \qquad (6.46)
\end{aligned}
$$

Proof of Eq. (6.46) as follows. Using the generating function of $\mathrm{H}_{m,n}(x,y)$, i.e.,

$$
\mathrm{H}_{m,n}(x,y) = \frac{\partial^{n+m}}{\partial t^m \partial \tau^n}\exp\left(-t\tau+tx+\tau y\right)|_{t,\tau=0}, \qquad (6.47)
$$

we have

$$
\begin{aligned}
&\sum_{n=0}^{\infty}\frac{z^n}{n!}\mathrm{H}_{m,n}(x,y)\mathrm{H}_{m',n}(x',y') \\
&= \sum_{n=0}^{\infty}\frac{z^n}{n!}\frac{\partial^{n+m}}{\partial t^m\partial\tau^n}\mathrm{e}^{-t\tau+tx+\tau y}\frac{\partial^{n+m'}}{\partial t'^{m'}\partial\tau'^{n}}\mathrm{e}^{-t'\tau'+t'x'+\tau'y'}\Big|_{t,\tau,t',\tau'=0} \\
&= \frac{\partial^{m+m'}}{\partial t^m\partial t'^{m'}}\mathrm{e}^{tx+t'x'}\sum_{n=0}^{\infty}\frac{z^n}{n!}\frac{\partial^{2n}}{\partial\tau'^n\partial\tau^n}\mathrm{e}^{\tau(y-t)+\tau'(y'-t')}\Big|_{t,\tau,t',\tau'=0} \\
&= \frac{\partial^{m+m'}}{\partial t^m\partial t'^{m'}}\mathrm{e}^{tx+t'x'}\exp\left(z\frac{\partial^2}{\partial\tau'\partial\tau}\right)\mathrm{e}^{\tau(y-t)+\tau'(y'-t')}\Big|_{t,\tau,t',\tau'=0} \\
&= \frac{\partial^{m+m'}}{\partial t^m\partial t'^{m'}}\mathrm{e}^{tx+t'x'}\exp\left[z\left(y'-t'\right)\left(y-t\right)\right]\Big|_{t,t'=0} \\
&= \exp\left(zy'y\right)\frac{\partial^{m+m'}}{\partial t^m\partial t'^{m'}}\mathrm{e}^{zt't+tx+t'x'-t'zy-tzy'}\Big|_{t,t'=0} \\
&= \mathrm{i}^{m+m'}\sqrt{z^{m+m'}}\exp\left(zy'y\right)\mathrm{H}_{m,m'}\left[\mathrm{i}(\sqrt{z}y'-x/\sqrt{z}), \mathrm{i}(\sqrt{z}y-x'/\sqrt{z})\right]. \qquad (6.48)
\end{aligned}
$$

Thus Eq. (6.45) can be written as

$$\rho(t) = \frac{1}{m!} \sum_{j,l=0}^{m} \mathrm{i}^l (-\mathrm{i})^j \binom{m}{j} \binom{m}{l} \alpha^{m-j} \alpha^{*m-l} (\mathrm{i}\sqrt{T})^{l+j}$$
$$\times : \mathrm{H}_{l,j} \left[\left(\sqrt{T}\alpha^* + \frac{a^\dagger \mathrm{e}^{-\kappa t}}{\sqrt{T}} \right), \left(-\sqrt{T}\alpha - \frac{a\mathrm{e}^{-\kappa t}}{\sqrt{T}} \right) \right]$$
$$\times \mathrm{e}^{-|\mathrm{e}^{-\kappa t}\alpha|^2 + \alpha a^\dagger \mathrm{e}^{-\kappa t} + \alpha^* a \mathrm{e}^{-\kappa t} - a^\dagger a} : . \tag{6.49}$$

Next, we derive the new binomial theorem involving two-mode Hermite polynomials:

$$\sum_{l=0}^{m} \binom{m}{l} \tau^l q^{m-l} \sum_{j=0}^{n} \binom{n}{j} \sigma^j p^{n-j} \mathrm{H}_{l,j}(x, y)$$
$$= \tau^m \sigma^n \mathrm{H}_{m,n} \left(\frac{q}{\tau} + x, \frac{p}{\sigma} + y \right). \tag{6.50}$$

Proof. Noting that

$$\sum_{n,m=0}^{\infty} \frac{\tau^n t^m}{n!m!} a^{\dagger n} a^m = \mathrm{e}^{\tau a^\dagger} \mathrm{e}^{ta} = \mathrm{e}^{-t\tau} \mathrm{e}^{ta} \mathrm{e}^{\tau a^\dagger}$$
$$= \vdots \exp\left(-t\tau + ta + \tau a^\dagger\right) \vdots$$
$$= \sum_{n,m=0}^{\infty} \frac{\tau^n t^m}{n!m!} \vdots \mathrm{H}_{m,n}\left(a, a^\dagger\right) \vdots, \tag{6.51}$$

and then comparing the same coefficients on both sides of Eq. (6.51), we obtain the anti-normal ordering product of the operator $a^{\dagger n} a^m$ as

$$a^{\dagger n} a^m = \vdots \mathrm{H}_{m,n}\left(a, a^\dagger\right) \vdots = \vdots \mathrm{H}_{n,m}\left(a^\dagger, a\right) \vdots. \tag{6.52}$$

Using Eq. (6.52), we can calculate the summation

$$\sum_{l=0}^{m} \binom{m}{l} \tau^{m-l} q^l \sum_{k=0}^{n} \binom{n}{k} \sigma^{n-k} p^k \vdots \mathrm{H}_{m-l,n-k}\left(a^\dagger, a\right) \vdots$$
$$= \sum_{l=0}^{m} \binom{m}{l} \tau^{m-l} q^l \sum_{k=0}^{n} \binom{n}{k} \sigma^{n-k} p^k : a^{\dagger m-l} a^{n-k} :$$
$$=: \left(q + \tau a^\dagger\right)^m (p + \sigma a)^n : . \tag{6.53}$$

When $\sum_{m,n=0}^{\infty} \frac{t^m s^n}{m!n!}$ is summed on the right side of Eq. (6.53), we obtain

$$
\begin{aligned}
&\sum_{m,n=0}^{\infty}\frac{t^m s^n}{m!n!}\colon\left(q+\tau a^{\dagger}\right)^m(p+\sigma a)^n\colon\\
&=\mathrm{e}^{t\left(q+\tau a^{\dagger}\right)}\mathrm{e}^{s(p+\sigma a)}\\
&=\vdots\mathrm{e}^{s(p+\sigma a)}\mathrm{e}^{t\left(q+\tau a^{\dagger}\right)-t\tau s\sigma}\vdots\\
&=\vdots\mathrm{e}^{\sigma s\left(\frac{p}{\sigma}+a\right)+t\tau\left(\frac{q}{\tau}+a^{\dagger}\right)-(t\tau)(s\sigma)}\vdots\\
&=\sum_{m,n=0}^{\infty}\frac{(t\tau)^m(s\sigma)^n}{m!n!}\vdots\mathrm{H}_{m,n}\left(\frac{q}{\tau}+a^{\dagger},\frac{p}{\sigma}+a\right)\vdots.
\end{aligned}
\tag{6.54}
$$

Considering the same summation $\sum_{m,n}^{\infty}\frac{t^m s^n}{m!n!}$ on the both sides of Eq. (6.54) and using Eq. (6.53), we have

$$
\begin{aligned}
&\sum_{l=0}^{m}\binom{m}{l}\tau^{m-l}q^l\sum_{k=0}^{n}\binom{n}{k}\sigma^{n-k}p^k\vdots\mathrm{H}_{m-l,n-k}\left(a^{\dagger},a\right)\vdots\\
=\;&\tau^m\sigma^n\vdots\mathrm{H}_{m,n}\left(\frac{q}{\tau}+a^{\dagger},\frac{p}{\sigma}+a\right)\vdots.
\end{aligned}
\tag{6.55}
$$

Since the operators on both sides of Eq. (6.55) are in anti-normal ordering, thus Eq. (6.50) has been proven. Obviously, it is a new binomial theorem involving two-mode Hermite polynomials.

Therefore, using Eq. (6.50), the summation in Eq. (6.45) is simplified as

$$
\begin{aligned}
\rho(t)=\;&\frac{1}{m!}\left(\mathrm{i}\sqrt{T}\right)^{2m}\sum_{j=0}^{m}\mathrm{i}^{m-j}\binom{m}{j}\left(\frac{1}{\mathrm{i}\sqrt{T}}\right)^{m-j}\\
&\times\alpha^{m-j}\sum_{l=0}^{m}(-\mathrm{i})^{m-l}\binom{m}{l}\alpha^{*m-l}\left(\frac{1}{\mathrm{i}\sqrt{T}}\right)^{m-l}\\
&\times\colon\mathrm{H}_{l,j}\left[\left(\sqrt{T}\alpha^*+\frac{a^{\dagger}\mathrm{e}^{-\kappa t}}{\sqrt{T}}\right),\left(-\sqrt{T}\alpha-\frac{a\mathrm{e}^{-\kappa t}}{\sqrt{T}}\right)\right]\\
&\times\mathrm{e}^{-\mathrm{e}^{-2\kappa t}|\alpha|^2+\alpha a^{\dagger}\mathrm{e}^{-\kappa t}+\alpha^* a\mathrm{e}^{-\kappa t}-a^{\dagger}a}\colon\\
=\;&\frac{1}{m!}\left(\mathrm{i}\sqrt{T}\right)^{2m}\colon\mathrm{e}^{-\mathrm{e}^{-2\kappa t}|\alpha|^2+\alpha a^{\dagger}\mathrm{e}^{-\kappa t}+\alpha^* a\mathrm{e}^{-\kappa t}-a^{\dagger}a}\\
&\times\mathrm{H}_{m,m}\left[\left(\sqrt{T}\alpha^*+\frac{a^{\dagger}\mathrm{e}^{-\kappa t}}{\sqrt{T}}-\frac{1}{\sqrt{T}}\alpha^*\right),\left(-\sqrt{T}\alpha-\frac{a\mathrm{e}^{-\kappa t}}{\sqrt{T}}+\frac{1}{\sqrt{T}}\alpha\right)\right]\colon\\
=\;&\frac{1}{m!}\left(\mathrm{i}\sqrt{T}\right)^{2m}\colon\mathrm{H}_{m,m}\left(\frac{a^{\dagger}\mathrm{e}^{-\kappa t}-\mathrm{e}^{-2\kappa t}\alpha^*}{\sqrt{T}},-\frac{a\mathrm{e}^{-\kappa t}-\mathrm{e}^{-2\kappa t}\alpha}{\sqrt{T}}\right)\\
&\times\mathrm{e}^{-\mathrm{e}^{-2\kappa t}|\alpha|^2+\alpha a^{\dagger}\mathrm{e}^{-kt}+\alpha^* a\mathrm{e}^{-kt}-a^{\dagger}a}\colon\\
=\;&T^m\colon\mathrm{L}_m\left[-\frac{\mathrm{e}^{-2kt}}{T}\left(a^{\dagger}-\mathrm{e}^{-\kappa t}\alpha^*\right)\left(a-\mathrm{e}^{-\kappa t}\alpha\right)\right]\left|\alpha\mathrm{e}^{-\kappa t}\right\rangle\left\langle\alpha\mathrm{e}^{-\kappa t}\right|\colon.
\end{aligned}
\tag{6.56}
$$

In the last step of Eq. (6.56), we have used the relations between Laguerre polynomials and two-mode Hermite polynomials in Eq. (2.81). Equation (6.56) represents a new light field (Laguerre-polynomial-weighted coherent state) generated by the displaced Fock state in the amplitude decay channel. The decay process clearly reflects the quantum decoherence effect.

For the new density operator $\rho(t)$, we must verify whether $\mathtt{tr}\rho(t) = 1$ holds. Actually, using Eq. (6.56) and the completeness relation of coherent states, we thus have

$$\begin{aligned}
\mathtt{tr}\rho(t) &= T^m \int \frac{d^2 z}{\pi} \langle z| : \mathrm{L}_m \left[-\frac{e^{-2\kappa t}}{T} \left(a^\dagger - e^{-\kappa t}\alpha^*\right)\left(a - e^{-\kappa t}\alpha\right)\right] \\
&\left|\alpha e^{-\kappa t}\right\rangle\left\langle \alpha e^{-\kappa t}\right| : |z\rangle \\
&= T^m \int \frac{d^2 z}{\pi} \mathrm{L}_m \left[-\frac{e^{-2\kappa t}}{T} \left(z^* - e^{-\kappa t}\alpha^*\right)\left(z - e^{-\kappa t}\alpha\right)\right] \\
&\times e^{-\left(z^* - e^{-\kappa t}\alpha^*\right)\left(z - e^{-\kappa t}\alpha\right)} \\
&= T^m \int \frac{d^2 z}{\pi} \mathrm{L}_m \left(-\frac{e^{-2\kappa t}}{T} |z|^2 \right) e^{-|z|^2} \\
&= T^{m+1} e^{2\kappa t} \int \frac{d^2 z}{\pi} \mathrm{L}_m \left(-|z|^2\right) e^{-T e^{2\kappa t}|z|^2} \\
&= -T^{m+1} e^{2\kappa t} \left(-T e^{2\kappa t} - 1\right)^m \left(-T e^{2\kappa t}\right)^{-m-1} = 1,
\end{aligned} \tag{6.57}$$

where we have used the integral formula in Eq. (5.73). Obviously, $\mathtt{tr}\rho(t) = 1$ is valid.

In particular, for $m = 0, \mathrm{L}_0(x) = 1$, thus Eq. (6.56) reduces to : $\left|\alpha e^{-\kappa t}\right\rangle\left\langle \alpha e^{-\kappa t}\right| : = \left|\alpha e^{-\kappa t}\right\rangle\left\langle \alpha e^{-\kappa t}\right|$. It shows that the initial coherent state remains coherent in the amplitude decay channel, but the amplitude decays over time t. On the other hand, when $\alpha = 0$, Eq. (6.56) reduces to

$$\begin{aligned}
\rho(t)_{\alpha=0} &= T^m : \mathrm{L}_m \left[-\frac{1}{T} \left(a^\dagger e^{-\kappa t}\right)\left(a e^{-\kappa t}\right)\right] |0\rangle\langle 0| : \\
&= \sum_{n'=0}^{m} \binom{m}{m-n'} \left(e^{-2\kappa t}\right)^{n'} \left(1 - e^{-2\kappa t}\right)^{m-n'} : \frac{a^{\dagger n'} a^{n'}}{n'!} e^{-a^\dagger a} : \\
&= \sum_{n'=0}^{m} \binom{m}{m-n'} \left(e^{-2\kappa t}\right)^{n'} \left(1 - e^{-2\kappa t}\right)^{m-n'} \left|n'\right\rangle\left\langle n'\right|,
\end{aligned} \tag{6.58}$$

which is a binomial state. This indicates that the number state $|m\rangle$ as the initial state becomes a binomial state after passing through the amplitude decay channel.

Besides, we use Eq. (6.56) and the integral formula

$$\int_0^\infty x\mathrm{e}^{-bx}\mathrm{L}_l\left(x\right)\mathrm{d}x = (b-1)^{l-1}b^{-l-2}\left(b-l-1\right) \tag{6.59}$$

to derive the average photon number of the state $\rho(t)$, that is

$$\begin{aligned}
\mathtt{tr}\left(\rho(t)a^\dagger a\right) &= \mathtt{tr}\left(\rho(t)aa^\dagger\right) - 1 \\
&= T^m \int \frac{\mathrm{d}^2 z}{\pi} \langle z| : a^\dagger \mathrm{L}_m\left[-\frac{\mathrm{e}^{-2\kappa t}}{T}\left(a^\dagger - \mathrm{e}^{-\kappa t}\alpha^*\right)\left(a - \mathrm{e}^{-\kappa t}\alpha\right)\right] \\
&\left|\alpha\mathrm{e}^{-\kappa t}\right\rangle\left\langle\alpha\mathrm{e}^{-\kappa t}\right| a : |z\rangle - 1 \\
&= T^m \int \frac{\mathrm{d}^2 z}{\pi} |z|^2 \mathrm{L}_m\left(-\frac{\mathrm{e}^{-2\kappa t}}{T}|z - \mathrm{e}^{-\kappa t}\alpha|^2\right)\mathrm{e}^{-|z-\mathrm{e}^{-\kappa t}\alpha|^2} - 1 \\
&= T^m \int \frac{\mathrm{d}^2 z}{\pi} |z + \mathrm{e}^{-\kappa t}\alpha|^2 \mathrm{L}_m\left(-\frac{\mathrm{e}^{-2\kappa t}}{T}|z|^2\right)\mathrm{e}^{-|z|^2} - 1 \\
&= T^m \int \frac{\mathrm{d}^2 z}{\pi}\left(|z|^2 + \mathrm{e}^{-2\kappa t}|\alpha|^2\right)\mathrm{L}_m\left(-\frac{\mathrm{e}^{-2\kappa t}}{T}|z|^2\right)\mathrm{e}^{-|z|^2} - 1 \\
&= T^{m+2}\mathrm{e}^{4\kappa t}\int \frac{\mathrm{d}^2 z}{\pi}|z|^2 \mathrm{L}_m\left(-|z|^2\right)\mathrm{e}^{-T\mathrm{e}^{2\kappa t}|z|^2} + \mathrm{e}^{-2\kappa t}|\alpha|^2 - 1 \\
&= \left(m + |\alpha|^2\right)\mathrm{e}^{-2\kappa t}.
\end{aligned} \tag{6.60}$$

Obviously, Eq. (6.60) indicates that the average photon number decays exponentially with time t. Just as many characteristics of the light field can be determined by photon counting, we can use this approach to study the noise light field of this new Laguerre-polynomial-weighted coherent state.

To sum up, we have obtained a new type of light field in Eq. (6.56) for the first time, and also revealed the time-evolution law of the displaced number state in the amplitude decay channel, and the photon number distribution $\left(m + |\alpha|^2\right)\mathrm{e}^{-2\kappa t}$ indicates that photon number control can be achieved by controlling the parameter m or $|\alpha|^2$.

6.5 Laguerre-Polynomial-Weighted Chaotic Fields Generated via Decay of Photon-Added Chaotic Fields

There is a cavity with a chaotic light field stored inside, and an external light source injects photons into the cavity. The density operator is

$$\rho(0) = Ca^{\dagger s}\mathrm{e}^{\lambda a^\dagger a}a^s, \tag{6.61}$$

where C is an undetermined normalization coefficient, this light field is called the photon-added chaotic light field. This section discusses how this light field dissipates in the amplitude decay channel [6]. We will point out that it evolves into a Laguerre-polynomial-weighted chaotic light field.

Let's first determine the normalization coefficient. Using the completeness of coherent states to obtain

$$\begin{aligned}
1 = & \ \mathtt{tr}\rho(0) = C\mathtt{tr}\left(\int \frac{d^2 z}{\pi} |z\rangle \langle z| a^{\dagger s} e^{\lambda a^\dagger a} a^s\right) \\
= & \ C \int \frac{d^2 z}{\pi} \langle z| a^{\dagger s} : \exp\left[\left(e^{\lambda} - 1\right) a^\dagger a\right] : a^s |z\rangle \\
= & \ C \int \frac{d^2 z}{\pi} |z|^{2s} \exp\left[-\left(1 - e^{\lambda}\right) |z|^2\right] \\
= & \ Cs! \left(\frac{1}{1 - e^{\lambda}}\right)^{s+1}, \qquad (6.62)
\end{aligned}$$

where $C = \frac{1}{s!}\left(1 - e^{\lambda}\right)^{s+1}$. Thus, we have

$$\rho(0) = \frac{1}{s!}\left(1 - e^{\lambda}\right)^{s+1} a^{\dagger s} e^{\lambda a^\dagger a} a^s. \qquad (6.63)$$

Obviously, when $s = 0$, $\rho(0) \to \rho_c = \left(1 - e^{\lambda}\right) e^{\lambda a^\dagger a}$, just a chaotic light field. The average photon number of the state $\rho(0)$ is

$$\begin{aligned}
\mathtt{tr}\left(\rho(0) a^\dagger a\right) = & \ C \int \frac{d^2 z}{\pi} \langle z| a^{\dagger s} e^{\lambda a^\dagger a} a^s a^\dagger a |z\rangle \\
= & \ C \int \frac{d^2 z}{\pi} z^{*s} z \langle z| e^{\lambda a^\dagger a} \left(a^\dagger a^s + s a^{s-1}\right) |z\rangle \\
= & \ C \int \frac{d^2 z}{\pi} z^{*s} z^{s+1} \langle z| e^{\lambda a^\dagger a} a^\dagger |z\rangle \\
& + Cs \int \frac{d^2 z}{\pi} z^{*s} z^s \langle z| : e^{\left(e^{\lambda} - 1\right) a a^\dagger} : |z\rangle \\
= & \ \frac{1}{s!}\left(1 - e^{\lambda}\right)^{s+1} \int \frac{d^2 z}{\pi} \left(z^{*s+1} z^{s+1} e^{\lambda} + s z^{*s} z^s\right) e^{\left(e^{\lambda} - 1\right)|z|^2} \\
= & \ \frac{s + e^{\lambda}}{1 - e^{\lambda}}. \qquad (6.64)
\end{aligned}$$

For $s = 0$, Eq. (6.64) reduces to $\mathtt{tr}\left(\rho_c a^\dagger a\right) = \left(e^{-\lambda} - 1\right)^{-1}$, as expected.

Substituting the density operator $\rho(0)$ into Eq. (6.20) leads to

$$\begin{aligned}
\rho(t) = & \ \frac{1}{s!}\left(1 - e^{\lambda}\right)^{s+1} \sum_{n=0}^{\infty} \frac{T^n}{n!} e^{-\kappa t a^\dagger a} a^n a^{\dagger s} e^{\lambda a^\dagger a} a^s a^{\dagger n} e^{-\kappa t a^\dagger a} \\
= & \ \frac{1}{s!}\left(1 - e^{\lambda}\right)^{s+1} \sum_{n=0}^{\infty} \frac{T^n}{n!} e^{2\kappa t(n-s)} a^n a^{\dagger s} e^{\lambda' a^\dagger a} a^s a^{\dagger n}, \qquad (6.65)
\end{aligned}$$

where $\lambda' = \lambda - 2\kappa t$. At this stage, we need to convert the operator $a^{\dagger s} e^{\lambda' a a^{\dagger}} a^s$ into its anti-normal product so that summation can be performed. Acorrding to the anti-normal ordering formula in Eq. (3.190), we get

$$\begin{aligned} a^{\dagger s} e^{\lambda' a^{\dagger} a} a^s &= \vdots \int \frac{d^2\beta}{\pi} \langle -\beta| : a^{\dagger s} \exp\left[\left(e^{\lambda'} - 1\right) a^{\dagger} a\right] a^s : |\beta\rangle \\ &\quad \times \exp\left(|\beta|^2 + \beta^* a - \beta a^{\dagger} + a a^{\dagger}\right) \vdots \\ &= (-1)^s \vdots \int \frac{d^2\beta}{\pi} |\beta|^{2s} \exp\left(-e^{\lambda'} |\beta|^2 + \beta^* a - \beta a^{\dagger} + a a^{\dagger}\right) \vdots \end{aligned} \quad (6.66)$$

Using the integral formula

$$\begin{aligned} &(-1)^n \int \frac{d^2 z}{\pi} z^n z^{*m} e^{-f|z|^2 + \mu z - \nu z^*} \\ &= (-1)^n \frac{1}{\sqrt{f^{m+n}}} \int \frac{d^2 z}{f\pi} z^n z^{*m} e^{-|z|^2 + \mu z/\sqrt{f} - \nu z^*/\sqrt{f}} \\ &= \frac{1}{\sqrt{f^{m+n+2}}} e^{-\mu\nu/f} H_{m,n}\left(\mu/\sqrt{f}, \nu/\sqrt{f}\right), \end{aligned} \quad (6.67)$$

where $H_{m,n}(\cdot, \cdot)$ is a two-variable Hermite polynomial, thus we rewrite Eq. (6.66) as

$$a^{\dagger s} e^{\lambda' a^{\dagger} a} a^s = \frac{1}{e^{\lambda'(s+1)}} \vdots e^{\left(1 - e^{-\lambda'}\right) a^{\dagger} a} H_{s,s}\left(e^{-\lambda'/2} a, e^{-\lambda'/2} a^{\dagger}\right) \vdots \quad (6.68)$$

Substituting Eq. (6.68) into Eq. (6.65) and summing m within $\vdots\ \vdots$, so we have

$$\begin{aligned} \rho(t) &= \left(\frac{1 - e^{\lambda}}{e^{\lambda'}}\right)^{s+1} \frac{1}{s!} \sum_{n=0}^{\infty} \frac{T^n}{n!} e^{2\kappa t (n-s)} \\ &\quad \vdots a^n \exp\left[\left(1 - e^{-\lambda'}\right) a a^{\dagger}\right] H_{s,s}\left(e^{-\lambda'/2} a, e^{-\lambda'/2} a^{\dagger}\right) a^{\dagger n} \vdots \\ &= \left(e^{-\lambda} - 1\right)^{s+1} \frac{e^{2\kappa t}}{s!} \vdots \exp\left[\left(1 - e^{-\lambda'} + T e^{2\kappa t}\right) a a^{\dagger}\right] \\ &\quad \times H_{s,s}\left(e^{-\lambda'/2} a, e^{-\lambda'/2} a^{\dagger}\right) \vdots \\ &= \left(e^{-\lambda} - 1\right)^{s+1} \frac{e^{2\kappa t}}{s!} \vdots \exp\left[e^{2\kappa t}\left(1 - e^{-\lambda}\right) a^{\dagger} a\right] H_{s,s}\left(e^{-\lambda'/2} a, e^{-\lambda'/2} a^{\dagger}\right) \vdots \end{aligned} \quad (6.69)$$

Using the P-represention of $\rho(t)$ in the coherent state representation and the normal ordering product $|z\rangle\langle z| =: \exp(-|z|^2 + z a^{\dagger} + z^* a - a^{\dagger} a):$, we obtain the normal ordering product of $\rho(t)$ as

$$
\begin{aligned}
\rho(t) = &\left(\mathrm{e}^{-\lambda}-1\right)^{s+1}\frac{\mathrm{e}^{2\kappa t}}{s!}\int\frac{\mathrm{d}^2 z}{\pi}\exp\left[\mathrm{e}^{2\kappa t}\left(1-\mathrm{e}^{-\lambda}\right)|z|^2\right] \\
&\times \mathrm{H}_{s,s}\left(\mathrm{e}^{-\lambda'/2}z,\mathrm{e}^{-\lambda'/2}z^*\right)|z\rangle\langle z| \\
= &\left(\mathrm{e}^{-\lambda}-1\right)^{s+1}\frac{\mathrm{e}^{2\kappa t}}{s!}\int\frac{\mathrm{d}^2 z}{\pi}\mathrm{H}_{s,s}\left(\mathrm{e}^{-\lambda'/2}z,\mathrm{e}^{-\lambda'/2}z^*\right) \\
&\times\colon\exp\left\{\left[\mathrm{e}^{2\kappa t}\left(1-\mathrm{e}^{-\lambda}\right)-1\right]|z|^2+za^\dagger+z^*a-a^\dagger a\right\}\colon. \qquad (6.70)
\end{aligned}
$$

Therefore, using the integral formula

$$
\begin{aligned}
&\int\frac{\mathrm{d}^2\xi}{\pi}\mathrm{H}_{m,n}\left(\xi,\xi^*\right)\exp\left(-h|\xi|^2+f\xi+g\xi^*\right) \\
&=\frac{1}{h}\left(\frac{h-1}{h}\right)^{\frac{m+n}{2}}\mathrm{H}_{m,n}\left(\frac{g}{\sqrt{h(h-1)}},\frac{f}{\sqrt{h(h-1)}}\right)\mathrm{e}^{\frac{fg}{h}} \qquad (6.71)
\end{aligned}
$$

and the IWOP method, we get

$$
\begin{aligned}
\rho(t) = &\left(\mathrm{e}^{-\lambda}-1\right)^{s+1}\frac{\mathrm{e}^{2\kappa t}}{s!}\mathrm{e}^{\lambda'} \\
&\times\int\frac{\mathrm{d}^2 z}{\pi}\colon\exp\left\{-\left[\left(1-\mathrm{e}^{\lambda}\right)+\mathrm{e}^{\lambda-2\kappa t}\right]|z|^2\right. \\
&\left.+za^\dagger\mathrm{e}^{\lambda'/2}+z^*a\mathrm{e}^{\lambda'/2}-a^\dagger a\right\}\mathrm{H}_{s,s}\left(z,z^*\right)\colon \\
= &\frac{(-1)^s}{Ts!}\left[\frac{\left(1-\mathrm{e}^{\lambda}\right)T}{1-T\mathrm{e}^{\lambda}}\right]^{s+1}\colon\exp\left[\left(\frac{\mathrm{e}^{\lambda'}}{1-T\mathrm{e}^{\lambda}}-1\right)a^\dagger a\right] \\
&\times\mathrm{H}_{s,s}\left(\frac{a^\dagger\mathrm{e}^{-\kappa t}}{\sqrt{(T\mathrm{e}^{\lambda}-1)T}},\frac{a\mathrm{e}^{-\kappa t}}{\sqrt{(T\mathrm{e}^{\lambda}-1)T}}\right)\colon. \qquad (6.72)
\end{aligned}
$$

Further, using Eq. (2.81) to obtain

$$
\rho(t)=\frac{1}{T}\left[\frac{\left(1-\mathrm{e}^{\lambda}\right)T}{1-T\mathrm{e}^{\lambda}}\right]^{s+1}\colon\mathrm{L}_s\left(\frac{a^\dagger a\mathrm{e}^{-2\kappa t}}{\left(T\mathrm{e}^{\lambda}-1\right)T}\right)\exp\left(\frac{\mathrm{e}^{\lambda}-1}{1-T\mathrm{e}^{\lambda}}a^\dagger a\right)\colon, \qquad (6.73)
$$

which shows that the photon-added chaotic field evolves into a Laguerre-polynomial-weighted chaotic field (also a new light field) in the amplitude decay channel, $T\equiv 1-\mathrm{e}^{-2\kappa t}$. Using the integral formula in Eq. (5.73), we calculate $\mathrm{tr}\rho(t)$ as

$$
\begin{aligned}
\mathtt{tr}\rho(t) &= \mathtt{tr}\left(\int \frac{d^2z}{\pi}|z\rangle\langle z|\rho(t)\right) \\
&= \frac{1}{T}\left[\frac{\left(1-e^{\lambda}\right)T}{1-Te^{\lambda}}\right]^{s+1}\int\frac{d^2z}{\pi}L_s\left(\frac{|z|^2e^{-2\kappa t}}{\left(Te^{\lambda}-1\right)T}\right) \\
&\quad\times\exp\left[\left(\frac{-1+e^{\lambda}}{1-Te^{\lambda}}\right)|z|^2\right] \\
&= \left[\frac{\left(1-e^{\lambda}\right)T}{1-Te^{\lambda}}\right]^{s+1}\left(Te^{\lambda}-1\right)e^{2\kappa t}\int\frac{d^2z}{\pi}L_s\left(|z|^2\right) \\
&\quad\times\exp\left[-\left(e^{\lambda}-1\right)Te^{2\kappa t}|z|^2\right] \\
&= \left[\frac{\left(1-e^{\lambda}\right)T}{1-Te^{\lambda}}\frac{1}{\left(e^{\lambda}-1\right)Te^{2\kappa t}}\right]^{s+1}\left(Te^{\lambda}-1\right) \\
&\quad\times e^{2\kappa t}\left[\left(e^{\lambda}-1\right)Te^{2\kappa t}-1\right]^{s} \\
&= \left[\frac{\left(e^{\lambda}-1\right)T-e^{-2\kappa t}}{\left(Te^{\lambda}-1\right)}\right]^{s}=1.
\end{aligned}
\tag{6.74}
$$

Especially, for $s=0$, no photon addition case, $H_{s,s}(\cdot,\cdot)=1$, it follows from Eq. (6.73) that

$$
\begin{aligned}
\rho_{s=0}(t) &= \frac{1-e^{\lambda}}{1-Te^{\lambda}}:\exp\left[\left(\frac{e^{\lambda'}}{1-Te^{\lambda}}-1\right)a^{\dagger}a\right]: \\
&= \frac{1-e^{\lambda}}{1-Te^{\lambda}}\exp\left[-a^{\dagger}a\ln\frac{1-(1-e^{-2\kappa t})e^{\lambda}}{e^{\lambda}e^{-2\kappa t}}\right] \\
&= \frac{1-e^{\lambda}}{1-\left(1-e^{-2\kappa t}\right)e^{\lambda}}e^{-a^{\dagger}a\ln\left[\left(e^{-\lambda}-1\right)e^{2\kappa t}+1\right]} \\
&= \left(1-e^{f}\right)\exp(fa^{\dagger}a),
\end{aligned}
\tag{6.75}
$$

where

$$
f=-\ln\left[\left(e^{-\lambda}-1\right)e^{2\kappa t}+1\right]. \tag{6.76}
$$

Comparing with the initial density operator $\rho_{s=0}(0)=\rho_c=\left(1-e^{\lambda}\right)e^{\lambda aa^{\dagger}}$, we find that $\rho_{s=0}(t)$ is still a chaotic field with the parameter f instead of λ.

Proof of Eq. (6.71). Using the integral form of two-mode Hermite polynomials, that is

$$
(-1)^n e^{\mu\nu}\int\frac{d^2z}{\pi}z^n z^{*m}e^{-|z|^2+\mu z-\nu z^*}=H_{m,n}(\mu,\nu), \tag{6.77}
$$

and Eq. (6.67), we therefore have

$$
\begin{aligned}
&\int \frac{d^2\xi}{\pi} H_{m,n}(\xi,\xi^*)\exp\left(-h|\xi|^2+f\xi+g\xi^*\right) \\
&= \int \frac{d^2\xi}{\pi}\exp\left(-h|\xi|^2+f\xi+g\xi^*\right)(-1)^n e^{|\xi|^2} \\
&\quad\times \int \frac{d^2z}{\pi} z^n z^{*m} e^{-|z|^2+\xi z-\xi^* z^*} \\
&= (-1)^n \int \frac{d^2z}{\pi} z^n z^{*m} e^{-|z|^2} \int \frac{d^2\xi}{\pi}\exp\left[-(h-1)|\xi|^2\right. \\
&\quad \left.+(f+z)\xi+\left(g-z^*\right)\xi^*\right] \\
&= \frac{(-1)^n}{h-1}\int \frac{d^2z}{\pi} z^n z^{*m} e^{-|z|^2+\frac{1}{h-1}(f+z)(g-z^*)} \\
&= \frac{(-1)^n}{h-1}\int \frac{d^2z}{\pi} z^n z^{*m} e^{-|z|^2\frac{h}{h-1}+\frac{zg-fz^*}{h-1}+\frac{fg}{h-1}} \\
&= \frac{1}{h}\left(\frac{h-1}{h}\right)^{\frac{m+n}{2}} H_{m,n}\left(\frac{g}{\sqrt{h(h-1)}},\frac{f}{\sqrt{h(h-1)}}\right)\exp\left(\frac{fg}{h}\right).
\end{aligned}
\tag{6.78}
$$

So Eq. (6.71) has been proven.

6.6 Preparation of Laguerre-Polynomial-Weighted Coherent States via Beam Splitter and Conditional Measurement

In Sect. 6.4 of this chapter, we examined the evolution of the displaced Fock state in the amplitude decay channel, and the results indicate that the Laguerre-polynomial-weighted coherent states can be generated in the decay channel. So, are there any other theoretical schemes to prepare such polynomial states? This section briefly introduces an effective method, that is, beam splitter and conditional measurement, which can be used not only to prepare Laguerre-polynomial-weighted quantum states, but also to prepare other polynomial-weighted quantum states, such as Hermite-polynomial-weighted quantum states. The theoretical scheme for preparing Laguerre-polynomial-weighted quantum states is as follows: a pure state $|\varphi\rangle_{in}$ and the number state $|m\rangle$ are used as the two inputs of an asymmetric beam splitter, and m photons are conditionally measured on the output port corresponding to the number state, thus the Laguerre-polynomial-weighted quantum state is obtained at the other output port [7].

If a conditional measurement of m photons is performed at the output port corresponding to the state $|m\rangle$, the output quantum state at the other port can be expressed as

$$|\Psi\rangle_{out} = N_{m_a}\langle m|B(\theta)|m\rangle_a \, |\varphi\rangle_{in} \,, \tag{6.79}$$

where N_m is a normalization constant, $B(\theta) = \exp[\theta(a^\dagger b - ab^\dagger)]$ is a optical beam splitter operator, $r = \sin\theta, t = \cos\theta$. The beam splitter operator is actually an entangled operator. When $= \pi/4$, $B(\pi/4)$ is a symmetrical beam splitter, that is, the beam splitter has the same transmission and reflection coefficients. Since the matrix representation of the lossless beam splitter operator satisfies the SU(2) group relationship, the generalized SU(2) coherent state can be used to calculate Eq. (6.79). Here, we use another method, that is the normal ordering representation of the beam splitter operator $B(\theta)$ to calculate the matrix element $_a\langle m|\, B(\theta)\, |m\rangle_a$. The normal ordering product of the beam splitter $B(\theta)$ is

$$B(\theta) =: \exp[(\cos\theta - 1)(a^\dagger a + b^\dagger b) + (a^\dagger b - ab^\dagger)\sin\theta]: . \tag{6.80}$$

Using the coherent state representation of number states, that is

$$|m\rangle = \frac{1}{\sqrt{m}}\frac{\partial^m}{\partial\alpha^m}|\,|\alpha\rangle\,|_{\alpha=0}, \quad |\,|\alpha\rangle = \exp(\alpha a^\dagger)\,|0\rangle\,, \quad \langle\alpha|\,|\alpha\rangle = \mathrm{e}^{|\alpha|^2}, \tag{6.81}$$

we obtain the matrix element $_a\langle m|\, B(\theta)\, |m\rangle_a$ as

$$\begin{aligned} B_m \equiv\ & _a\langle m|\, B(\theta)\, |m\rangle_a \\ =\ & \frac{1}{m!}\frac{\partial^{2m}}{\partial\alpha^m\partial\alpha^{*m}}\langle\alpha|\,|: \exp[(\cos\theta - 1)(a^\dagger a + b^\dagger b) \\ & + (a^\dagger b - ab^\dagger)\sin\theta]\,|\,|\alpha\rangle|_{\alpha,\alpha^*=0} \\ =\ & \frac{1}{m!}\frac{\partial^{2m}}{\partial\alpha^m\partial\alpha^{*m}}\exp[\alpha^*\alpha\cos\theta + \\ & (\alpha^* b - \alpha b^\dagger)\sin\theta]\,\mathrm{e}^{(\cos\theta-1)b^\dagger b}: \Big|_{\alpha,\alpha^*=0} \\ =\ & \frac{1}{m!}\frac{(-1)^m\cos^m\theta\partial^{2m}}{\partial(\alpha^*\cos\theta)^m\partial(-\alpha)^m}: \exp[-(\alpha^*\cos\theta)(-\alpha) \\ & + (\alpha^*\cos\theta)b\tan\theta + (-\alpha)b^\dagger\sin\theta]\mathrm{e}^{(\cos\theta-1)b^\dagger b}: \Big|_{\alpha,\alpha^*=0} . \end{aligned} \tag{6.82}$$

Using the generating function for the two-variable Hermite polynomial and its series expression, the operator B_m can be rewritten as

$$\begin{aligned}B_m &= \frac{(-1)^m \cos^m \theta}{m!} : \mathrm{H}_{m,m}\left(b \tan \theta, b^\dagger \sin \theta\right) \mathrm{e}^{(\cos \theta - 1) b^\dagger b} : \\ &= \frac{(-1)^m \cos^m \theta}{m!} : \sum_{l=0}^{\infty} \frac{(m!)^2 (-1)^l}{l! \left[(m-l)!\right]^2} \\ &\times (b \sin \theta)^{m-l} \left(b^\dagger \sin \theta\right)^{m-l} \mathrm{e}^{(\cos \theta - 1) b^\dagger b} : \\ &= \frac{(-1)^m \cos^m \theta}{m!} \sum_{l=0}^{\infty} \frac{(m!)^2 (-1)^l}{l! \left[(m-l)!\right]^2} \\ &\times (b \sin \theta)^{m-l} : \mathrm{e}^{(\cos \theta - 1) b^\dagger b} : \left(b^\dagger \sin \theta\right)^{m-l}. \end{aligned} \tag{6.83}$$

Further, using the operator identities $: \exp[(\mathrm{e}^\lambda - 1) b^\dagger b] : = \mathrm{e}^{\lambda b^\dagger b}$ and $\mathrm{e}^{\lambda b^\dagger b} b \mathrm{e}^{-\lambda b^\dagger b} = b \mathrm{e}^{-\lambda}$, we have

$$\begin{aligned}B_m &= \frac{(-1)^m \cos^m \theta}{m!} \sum_{l=0}^{\infty} \frac{(m!)^2 (-1)^l}{l! \left[(m-l)!\right]^2} \\ &\times \left(b^\dagger \sin \theta\right)^{m-l} \mathrm{e}^{b^\dagger b \ln \cos \theta} (b \sin \theta)^{m-l} \\ &= \frac{(-1)^m \cos^m \theta}{m!} \sum_{l=0}^{\infty} \frac{(m!)^2 (-1)^l}{l! \left[(m-l)!\right]^2} \\ &\times \left(b^\dagger \sin \theta\right)^{m-l} (b \sin \theta)^{m-l} \mathrm{e}^{b^\dagger b \ln \cos \theta} \\ &= \frac{(-1)^m \cos^m \theta}{m!} : \mathrm{H}_{m,m}\left(b^\dagger \tan \theta, b \tan \theta\right) : \mathrm{e}^{b^\dagger b \ln \cos \theta}. \end{aligned} \tag{6.84}$$

Finally, using the relation between $\mathrm{H}_{m,m}(\cdot, \cdot)$ and $\mathrm{L}_m(\cdot)$ in Eq. (2.81), we get

$$B_m = \cos^m \theta : \mathrm{L}_m\left(b^\dagger b \tan^2 \theta\right) : \mathrm{e}^{b^\dagger b \ln \cos \theta}, \tag{6.85}$$

and

$$|\rangle_{out} = \cos^m \theta : \mathrm{L}_m\left(b^\dagger b \tan^2 \theta\right) : \mathrm{e}^{b^\dagger b \ln \cos \theta} |\varphi\rangle_{in}. \tag{6.86}$$

For any input state $|\varphi\rangle_{in}$, the output state can be expressed as $|\Psi\rangle_{out} \rightarrow B_m |\varphi\rangle_{in}$, so it is very convenient to discuss the features of the output quantum state using Eq. (6.86). According to Eq. (6.86), the process of accompanying m-photon number state input and m-photon measurement can be regarded as a Laguerre polynomial operation of the number state within normal ordering. Specifically, when $m = 0, 1$, there are $B_0 = \mathrm{e}^{b^\dagger b \ln \cos \theta}$ and $B_1 = (1 - b^\dagger b \tanh^2 \theta) : \mathrm{e}^{b^\dagger b \ln \cos \theta} \cos \theta$.

When the input state $|\varphi\rangle_{in}$ is the squeezed state $\operatorname{sech}^{1/2} \lambda \mathrm{e}^{\frac{b^{\dagger 2}}{2} \tan h \lambda} |0\rangle$, where λ is the squeezed parameter, then the output state

$$|\rangle_{out} = \operatorname{sech}^{1/2} \lambda \cos^m \theta : \mathrm{L}_m\left(b^\dagger b \tan^2 \theta\right) : \mathrm{e}^{b^\dagger b \ln \cos \theta} \mathrm{e}^{(b^{\dagger 2} \tan h \lambda)/2} |0\rangle, \tag{6.87}$$

which is a Laguerre-polynomial-weighted non-Gaussian state.

Next, we consider a coherent state as the input state, i.e., $|\varphi\rangle_{in} = |z\rangle$. In this case, it can be seen from Eqs. (6.79) and (6.86) that the output state is

$$\begin{aligned} |\rangle_{out} &= N_m \cos^m\theta \colon \mathrm{L}_m\left(b^\dagger b \tan^2\theta\right) \colon e^{b^\dagger b \ln\cos\theta}\,|z\rangle \\ &= N_m e^{-\frac{1}{2}|z|^2 \sin 2\theta} \cos^m\theta \mathrm{L}_m(\mu b^\dagger)\,|z\cos\theta\rangle \\ &= \bar{N}_m \mathrm{L}_m\left(\mu b\right)|z\cos\theta\rangle\,, \end{aligned} \tag{6.88}$$

where $\mu = z\cos\theta\tan^2\theta$, $\bar{N}_m = N_m\cos^m\theta\exp\left(\frac{1}{2}|z|^2\sin^2\theta\right)$, and we have used the following identity

$$g^{b^\dagger b}\,|\alpha\rangle = \exp\left(\frac{1}{2}|z|^2\sin^2\theta\right)|g\alpha\rangle\,. \tag{6.89}$$

Obviously, Eq. (6.88) is a Laguerre-polynomial-excited coherent state, $|z\cos\theta\rangle$ is a coherent state. When the amplitude of the coherent state $|z\cos\theta\rangle$ decreases compared to the input coherent state $|z\rangle$, it can be understood that the coherent state has undergone a loss process.

Especially, when $\theta = 0$, that is, perfect transmission ($t = 1, r = 0$), there are $\mu = 0$ and $|\rangle_{out} \to |z\rangle$. For $m = 0$, the output state becomes $|\rangle_{out} = |z\cos\theta\rangle$, which is still a coherent state (Gaussian state), but the amplitude decreases. Although the number of photons measured at the auxiliary measurement port is the same as that of the input number state (or in other words, the number of photons remains unchanged), the average photon number at another output port decreases from $|z|^2$ to $|z|^2\cos^2\theta$. This can be further clarified by expanding the coherent state in the number state representation, i.e., $|z\rangle = \sum_{n=0}^{\infty} c_n\,|n\rangle$, where $c_n = \left(z^n e^{-|z|^2/2}\right)/\sqrt{n!}$. Using Eq. (6.88), the output state $|\rangle_{out}$ can be written as

$$|\rangle_{out} = N_m\cos^m\theta\sum_{n=0}^{\infty} c_n\cos^n\theta \colon \mathrm{L}_m\left(b^\dagger b\tan^2\theta\right) \colon \;|n\rangle\,. \tag{6.90}$$

So, when $m = 0$, we have $|\rangle_{out} = \sum_{n=0}^{\infty} c_n\cos^n\theta\,|n\rangle$. Since $\cos^n \le \cos \le 1$, the probability of detecting the number state $|n\rangle$ at the output port decreases with increasing n. This indicates that zero photon assisted detection ($m = 0$) reduces the average number of photons. So, in the above preparation process, the presence of measurements makes the average photon number non-conservative. When $m = 1$, the output state becomes

$$|\rangle_{out} = \bar{N}_1\left(1 - \mu b^\dagger\right)|z\cos\theta\rangle\,, \tag{6.91}$$

which is a superposition of coherent states and excited coherent states. In the number state representation, the output state can be written as

$$|\rangle_{out} = N_1\sum_{n=0}^{\infty} c_n\cos^{n+1}\theta\left(1 - n\tan^2\theta\right)|n\rangle\,. \tag{6.92}$$

It is interesting that if $\tan^2\theta = 1/n$ or $t^2 = n/(n+1)$, the output state will no longer contain the number state $|n\rangle$. When $m = 2$, if $\tan = 1/\sqrt{2}$ is taken, the state $|1\rangle$ will not be detected. If $\tan\theta$ takes other values, $|n \geq 2\rangle$ may not exist in the output state. Based on Eq. (6.88), we can further discuss the non-classical properties of the output state $|\rangle_{out}$, such as photon number distribution, Mandel Q_M-parameter, second-order correlation function, squeezing, and so on. Here, we won't discuss them one by one. Next, let's take the Wigner function as an example for discussion.

In order to obtain the analytical form of the Wigner function for the output state $|\rangle_{out}$, it is necessary to obtain the normalization constant of the state $|\rangle_{out}$. Using the normalization condition ${}_{out}\langle|\rangle_{out} = 1$, and the completeness relation $\pi^{-1}\int d^2\alpha\,|\alpha\rangle\langle\alpha| = 1$, as well as

$$\langle z\cos\theta\,|\alpha\rangle = \exp\left(\frac{1}{2}|z|^2\cos^2\theta - \frac{1}{2}|\alpha|^2 + z^*\alpha\cos\theta\right), \tag{6.93}$$

we obtain

$$\begin{aligned}\bar{N}_m^{-2} &= \langle z\cos\theta|\,\mathrm{L}_m\left(\mu^* b\right)\mathrm{L}_m\left(\mu b^\dagger\right)|z\cos\theta\rangle \\ &= \int\frac{d^2\alpha}{\pi}|\mathrm{L}_m\left(\mu^*\alpha\right)|^2|\langle z\cos\theta|\alpha\rangle|^2 \\ &= \int\frac{d^2\alpha}{\pi}|\mathrm{L}_m\left(\mu^*\alpha\right)|^2 e^{-|z|^2\cos^2\theta-|\alpha|^2+(z^*\alpha+z\alpha^*)\cos\theta}.\end{aligned} \tag{6.94}$$

Using the series expansions of Laguerre polynomials, Eq. (6.94) can be expressed as

$$\begin{aligned}\bar{N}_m^{-2} &= \sum_{l,k=0}^{m}\binom{m}{l}\binom{m}{k}\frac{(-1)^{l+k}}{l!k!}\mu^k\mu^{*l} \\ &\quad\times e^{-|z|^2\cos^2\theta}\int\frac{d^2\alpha}{\pi}\alpha^{*k}\alpha^l e^{-|\alpha|^2+(z^*\alpha+z\alpha^*)\cos\theta} \\ &= \sum_{l,k=0}^{m}\binom{m}{l}\binom{m}{k}\frac{(-1)^k\mu^k\mu^{*l}}{l!k!}\mathrm{H}_{k,l}(z^*\cos\theta, -z\cos\theta),\end{aligned} \tag{6.95}$$

where we have used the integral formula in Eq. (6.77). Thus, Eq. (6.95) is the normalization constant of the output state $|\rangle_{out}$. Especially, when $m = 1$, we have $\bar{N}_1^{-2} = (1 - |z|^2\sin^2\theta)^2 + |z|^2\cos^2\theta\tan^4\theta$.

The negative feature of the Wigner function is a good indicator of non-classicality. For any single-mode quantum system, the Wigner function can be calculated using the definition $W(\gamma) = \mathrm{tr}\,(\rho\Delta(\gamma))$, where $\Delta(\gamma)$ is the single-mode Wigner operator in the coherent state representation represented as [8, 9]

$$\Delta(\gamma) = e^{2|\gamma|^2}\int\frac{d^2\alpha}{\pi}\,|\alpha\rangle\langle-\alpha|\,e^{-2(\alpha\gamma^*-\gamma\alpha)}. \tag{6.96}$$

Substituting Eqs. (6.88) and (6.93) into the definition of $W(\gamma)$, we obtain the Wigner function for the output state $|\rangle_{out}$ as

$$W_m(\gamma) = |\bar{N}_m|^2 e^{2|\gamma|^2-|z|^2\cos^2\theta}\Theta(\mu,\mu^*), \tag{6.97}$$

where

$$\Theta(\mu,\mu^*) = \int\frac{d^2\alpha}{\pi^2}\mathrm{L}_m\left(-\mu\alpha^*\right)\mathrm{L}_m\left(\mu^*\alpha\right)e^{-|\alpha|^2+\alpha(z^*\cos\theta-2\gamma^*)-\alpha^*(z\cos\theta-2\gamma)}. \tag{6.98}$$

It is not difficult to see that the Wigner function $W_m(\gamma)$ in the phase space is a real number because $\Theta^*(\mu,\mu^*) = \Theta(\mu,\mu^*)$.

Further, by integrating Eq. (6.98), we obtain

$$W_m(\gamma) = W_0(\gamma)F_m(\gamma), \tag{6.99}$$

where $W_0(\gamma) = \pi^{-1}\exp\left(-2|\gamma - z\cos\theta|^2\right)$ is the Wigner function for the coherent state $|z\cos\theta\rangle$, and $F_m(\gamma)$ refers to its non-Gaussian term, i.e.,

$$F_m(\gamma) = |\bar{N}_m|^2\sum_{j,l=0}^{m}\frac{\mu^l\mu^{*j}}{l!j!}\binom{m}{l}\binom{m}{j}\mathrm{H}_{l,j}(z^*\cos\theta-2\gamma^*, z\cos\theta-2\gamma), \tag{6.100}$$

which is caused by the conditional measurement. Especially, when $m = 1$, thus Eq. (6.100) becomes

$$\begin{aligned}F_1(\gamma) = &|\bar{N}_1|^2\{1+|\mu|^2(1-|z\cos\theta-2\gamma|^2)\\&+[\mu(z^*\cos\theta-2\gamma^*)+c.c]\}.\end{aligned} \tag{6.101}$$

Since $W_0(\gamma)$ is always positive, the negative region of the Wigner function is determined by $F_1(\gamma) < 0$.

References

1. Yu ZS, Ren GH, Yu ZY, Wei CHN, Fan HY (2018) Time evolution of the wigner operator as a quasi-density operator in amplitude dessipative channel. Int J Theor Phys 57(6):1888–1893
2. Fan HY, Hu LY (2008) Operator-sum representation of density operators as solutions to master equations obtained via the entangled state approach. Mod Phys Lett B 22(25):2435–2468
3. Meng XG, Wang JS, Liang BL (2023) Entangled state representations in quantum optics. Springer, Singapore
4. Hu LY, Chen F, Wang ZS, Fan HY (2011) Time evolution of distribution functions in dissipative environments. Chin Phys B 20(7):074204
5. Xu XF, Fan HY (2015) A new optical field generated as an output of the displaced Fock state in an amplitude dissipative channel. Chin Phys B 24(1):149–152

6. Fan HY, He R (2015) Photon-added chaotic field and its damping in a thermo enviroment. Can J Phys 93(4):456–459
7. Ye W, Zhou WD, Zhang HL, Liu CJ, Huang JH, Hu LY (2017) Laguerre polynomial excited coherent state: generation and nonclassical properties. Laser Phys Lett 14(11):115201
8. Meng XG, Wang Z, Fan HY, Wang JS (2012) Nonclassicality and decoherence of photon-subtracted squeezed vacuum states. J Opt Soc Am B 29(11):3141–3149
9. Fan HY, Ruan TN (1983) Coherent state formulation of the Weyl correspondence and the Wigner function. Commun Theor Phys 2(6):1563–1574

Chapter 7
Light Fields Represented by Binomial and Negative Binomial

This chapter points out that the Fock space can be classified as the binomial states and the negative binomial states of light fields, and we also discuss the evolutions of the two types of light fields.

7.1 Fock Space Completeness Characterized by Binomial States

The binomial distribution is $\binom{n}{l}\sigma^l (1-\sigma)^{n-l}$, thus the density operator of the binomial state describing a light field is [1]

$$\sum_{l=0}^{n}\binom{n}{l}\sigma^l (1-\sigma)^{n-l} |l\rangle\langle l| \equiv \rho_n(\sigma), \tag{7.1}$$

where $|l\rangle$ is a number state. According to the binomial theorem, we get $\mathtt{tr}\rho_n(\sigma)=1$. Using the generating function for the Laguerre polynomial in Eq. (2.34), we have

$$\begin{aligned}1 &= : \mathrm{e}^{a^\dagger a}\mathrm{e}^{-a^\dagger a} : \\ &= \sigma\frac{1}{1-(1-\sigma)} : \mathrm{e}^{\frac{1-\sigma}{-\sigma}\left(\frac{\sigma}{\sigma-1}a^\dagger a\right)}\mathrm{e}^{-a^\dagger a} : \\ &= \sigma\sum_{n=0}^{\infty}(1-\sigma)^n : \mathrm{L}_n\left(\frac{\sigma}{\sigma-1}a^\dagger a\right)\mathrm{e}^{-a^\dagger a} : .\end{aligned} \tag{7.2}$$

Further, using the series expansions of Laguerre polynomials $\mathrm{L}_n(x)$ in Eq. (3.48), thus Eq. (7.2) can be expanded as

H. Fan and L. Hu, *Dissipation and Diffusion of Quantum Light Field*,
https://doi.org/10.1007/978-981-95-1528-8_7

$$
\begin{aligned}
1 &= \sigma \sum_{n=0}^{\infty} (1-\sigma)^n \sum_{l=0}^{n} \binom{n}{l} \frac{(-1)^l}{l!} :\left(\frac{\sigma}{\sigma-1} a^\dagger a\right)^l \mathrm{e}^{-a^\dagger a}: \\
&= \sigma \sum_{n=0}^{\infty} (1-\sigma)^n \sum_{l=0}^{n} \binom{n}{l} \frac{(-1)^l}{l!} \left(\frac{\sigma}{\sigma-1}\right)^l a^{\dagger l}|0\rangle\langle 0|a^l \\
&= \sigma \sum_{n=0}^{\infty} \sum_{l=0}^{n} \binom{n}{l} \sigma^l (1-\sigma)^{n-l} |l\rangle\langle l|,
\end{aligned} \tag{7.3}
$$

that is, 1 can be decomposed into

$$
1 = \sigma \sum_{n=0}^{\infty} \rho_n. \tag{7.4}
$$

This indicates that the completeness of Fock space can be characterized and partitioned using binomial states. Or in other words, the binomial state can be obtained by decomposing 1. In fact, it can be proven that when a number state undergoes the amplitude decay channel, it evolves into a binomial state.

If we decompose 1 into

$$
1 = \frac{1}{2}\left[1 + (1-2\sigma)^n\right] + \frac{1}{2}\left[1 - (1-2\sigma)^n\right], \tag{7.5}
$$

where the first part on the right side of Eq. (7.5) can be expressed as

$$
\begin{aligned}
&\frac{1}{2}\left[1 + (1-2\sigma)^n\right] \\
&= \frac{1}{2}\left[(1-\sigma+\sigma)^n + (1-\sigma-\sigma)^n\right] \\
&= \frac{1}{2}\sum_{m=0}^{n} \binom{n}{m} \sigma^m (1-\sigma)^{n-m} + \frac{1}{2}\sum_{m=0}^{n} \binom{n}{m} (-\sigma)^m (1-\sigma)^{n-m} \\
&= \sum_{m=0}^{[n/2]} \binom{n}{2m} \sigma^{2m} (1-\sigma)^{n-2m},
\end{aligned} \tag{7.6}
$$

while the second part on the right side of Eq. (7.5) can be rewritten as

$$\begin{aligned}
&\frac{1}{2}\left[1-(1-2\sigma)^n\right] \\
&=\frac{1}{2}\left[(1-\sigma+\sigma)^n-(1-\sigma-\sigma)^n\right] \\
&=\frac{1}{2}\sum_{m=0}^{n}\binom{n}{m}\sigma^m(1-\sigma)^{n-m}-\frac{1}{2}\sum_{m=0}^{n}\binom{n}{m}(-\sigma)^m(1-\sigma)^{n-m} \\
&=\sum_{m=0}^{[(n-1)/2]}\binom{n}{2m+1}\sigma^{2m+1}(1-\sigma)^{n-2m-1}. \qquad (7.7)
\end{aligned}$$

Correspondingly, we can introduce the even binomial states

$$\sum_{m=0}^{[n/2]}\binom{n}{2m}\sigma^{2m}(1-\sigma)^{n-2m}|2m\rangle\langle 2m| \equiv \rho_{even}(\sigma), \qquad (7.8)$$

where its trace is

$$\mathtt{tr}\rho_{even}(\sigma)=\frac{1}{2}\left[1+(1-2\sigma)^n\right], \qquad (7.9)$$

and the odd binomial states

$$\sum_{m=0}^{[(n-1)/2]}\binom{n}{2m+1}\sigma^{2m+1}(1-\sigma)^{n-2m-1}|2m+1\rangle\langle 2m+1| \equiv \rho_{odd}(\sigma), \qquad (7.10)$$

where its trace is

$$\mathtt{tr}\rho_{odd}(\sigma)=\frac{1}{2}\left[1-(1-2\sigma)^n\right]. \qquad (7.11)$$

7.2 Fock Space Completeness Characterized by Negative Binomial States

We can further use Eq. (7.4) to find that the completeness of Fock space can be characterized using the negative binomial states [2]. Using the reallocate formula for summation

$$\sum_{n=0}^{\infty}\sum_{l=0}^{n}A_{n-l}B_l=\sum_{s=0}^{\infty}\sum_{m=0}^{\infty}A_sB_m \qquad (7.12)$$

to rewritten Eq. (7.4) as

$$\begin{aligned}1 &= \sigma\sum_{n=0}^{\infty}\rho_n \\ &= \sigma\sum_{n=0}^{\infty}\sum_{l=0}^{n}\binom{n}{n-l}(1-\sigma)^{n-l}\sigma^l|l\rangle\langle l| \\ &= \sum_{m=0}^{\infty}\sum_{s=0}^{\infty}\binom{m+s}{s}(1-\sigma)^s\sigma^{m+1}|m\rangle\langle m|,\end{aligned} \tag{7.13}$$

where $\binom{m+s}{s} = \frac{(m+s)!}{s!m!} = \binom{m+s}{m}$, and letting $1-\sigma=\gamma$, thus Eq. (7.13) becomes

$$1 = \sum_{s=0}^{\infty}\sum_{m=0}^{\infty}\binom{m+s}{m}\gamma^s(1-\gamma)^{m+1}|m\rangle\langle m|. \tag{7.14}$$

Denoting

$$\sum_{m=0}^{\infty}\binom{m+s}{m}\gamma^{s+1}(1-\gamma)^m|m\rangle\langle m| \equiv \rho_s, \tag{7.15}$$

which refers to a negative binomial state, thus Eq. (7.13) becomes

$$1 = \frac{1-\gamma}{\gamma}\sum_{s=0}^{\infty}\rho_s, \tag{7.16}$$

that is, the Fock space can also be divided according to the negative binomial states. Further, using $a|m\rangle = \sqrt{m}|m-1\rangle$, it can also be proven that

$$\begin{aligned}\rho_s &= \frac{\gamma^{s+1}}{s!(1-\gamma)^s}a^s\sum_{m=0}^{\infty}(1-\gamma)^m|m\rangle\langle m|a^{\dagger s} \\ &= \frac{1}{s!(n_c)^s}a^s\rho_c a^{\dagger s},\end{aligned} \tag{7.17}$$

where

$$n_c = \frac{1-\gamma}{\gamma},\quad \rho_c = \sum_{m=0}^{\infty}\gamma(1-\gamma)^m m\rangle\langle m|. \tag{7.18}$$

If we rewrite the chaotic light field $\rho_c = \gamma e^{a^\dagger a\ln(1-\gamma)}$ as

$$\rho_c = \frac{\gamma}{1-\gamma}{:}e^{\frac{\gamma}{\gamma-1}aa^\dagger}{:}, \tag{7.19}$$

thus Eq. (7.17) can be formed as

$$\rho_s = \frac{1}{s!\,(n_c)^{s+1}} \vdots a^s e^{\frac{\gamma}{\gamma-1} a a^\dagger} a^{\dagger s} \vdots . \tag{7.20}$$

7.3 Binomial States Serving as the Evolutions of Number States for Amplitude Decay

Next, we give new physical meanings to binomial states [3]. Firstly, we rewrite the density operator $\rho_n(\sigma)$ as

$$\begin{aligned}
\rho_n(\sigma) &= \sum_{m=0}^{n} \sigma^{n-m} \binom{n}{m} (1-\sigma)^m |n-m\rangle \langle n-m| \\
&= \sum_{m=0}^{n} \binom{n}{m} \frac{(1-\sigma)^m \sigma^{n-m}}{(n-m)!} a^{\dagger n-m} |0\rangle \langle 0| a^{n-m} \\
&= \sum_{m=0}^{n} : \frac{(-1)^n n!\,(\sigma-1)^{2n-m}}{m!\,[(n-m)!]^2} \left(\frac{a^\dagger \sqrt{\sigma}}{1-\sigma}\right)^{n-m} \left(\frac{-a\sqrt{\sigma}}{1-\sigma}\right)^{n-m} \exp\left(-a^\dagger a\right) : \\
&= \frac{(-1)^{n+1}}{(1-\sigma)\,n!} \int \frac{d^2\beta}{\pi} |\beta|^{2n} : \exp\left\{\frac{1}{1-\sigma}[|\beta|^2\right. \\
&\quad \left. + \sqrt{\sigma}\left(\beta a^\dagger - \beta^* a\right) - a^\dagger a]\right\} : ,
\end{aligned} \tag{7.21}$$

where in the final step, we used the following integral formula [4]

$$\begin{aligned}
&\int \frac{d^2\beta}{\pi} \beta^n \beta^{*n} \exp\left(\zeta|\beta|^2 + \xi\beta + \eta\beta^*\right) \\
&= e^{-\xi\eta} \sum_{m=0}^{n} : \frac{n!n!}{(n-m)!m!\,(n-m)!\,(-\zeta)^{2n-m+1}} \xi^{n-m} \eta^{n-m} .
\end{aligned} \tag{7.22}$$

Since

$$\frac{(-1)^n |\beta|^{2n}}{n!} e^{-|\beta|^2} = -\langle\beta|n\rangle\langle n|\beta\rangle , \tag{7.23}$$

where $|\beta\rangle$ is a coherent state, thus we have

$$\begin{aligned}
\rho_n(\sigma) &= \frac{-1}{1-\sigma} \int \frac{d^2\beta}{\pi} \langle\beta|n\rangle\langle n|\beta\rangle e^{|\beta|^2} \\
&\quad \times : \exp\left[\frac{1}{\sigma-1}\left(a^\dagger\sqrt{\sigma} + \beta^*\right)\left(a\sqrt{\sigma} - \beta\right) - a^\dagger a\right] : ,
\end{aligned} \tag{7.24}$$

where

$$\frac{1}{\sigma-1}:\exp\left[\frac{1}{\sigma-1}\left(a^{\dagger}\sqrt{\sigma}+\beta^{*}\right)\left(a\sqrt{\sigma}-\beta\right)-a^{\dagger}a\right]:$$

$$=\int\frac{d^2\alpha}{\pi}e^{-|\alpha|^2(\sigma-1)}:e^{\beta^{*}\alpha-\beta\alpha^{*}+a^{\dagger}\alpha\sqrt{\sigma}+a\alpha^{*}\sqrt{\sigma}-a^{\dagger}a}:. \tag{7.25}$$

Letting

$$P(\alpha,0)=e^{|\alpha|^2}\int\frac{d^2\beta}{\pi}\langle-\beta|n\rangle\langle n|\beta\rangle e^{|\beta|^2+\beta^{*}\alpha-\beta\alpha^{*}}, \tag{7.26}$$

which is the P-representation of the pure number state $|n\rangle\langle n|$ in the coherent state representation, thus Eq. (7.24) can be also written as

$$\begin{aligned}\rho_n(\sigma)&=\int\frac{d^2\beta}{\pi}\langle\beta|n\rangle\langle n|\beta\rangle e^{|\beta|^2}\int\frac{d^2\alpha}{\pi}e^{-|\alpha|^2(\sigma-1)}\\&\quad\times:e^{\beta^{*}\alpha-\beta\alpha^{*}+a^{\dagger}\alpha\sqrt{\sigma}+a\alpha^{*}\sqrt{\sigma}-a^{\dagger}a}:\\&=\int\frac{d^2\alpha}{\pi}P(\alpha,0)e^{-|\alpha|^2\sigma}:e^{a^{\dagger}\alpha\sqrt{\sigma}+a\alpha^{*}\sqrt{\sigma}-a^{\dagger}a}:.\end{aligned} \tag{7.27}$$

Noting that the coherent state $|\alpha\rangle=\exp\left(-|\alpha|^2/2+\alpha a^{\dagger}\right)|0\rangle$ possesses the completeness relation

$$\int\frac{d^2\alpha}{\pi}|\alpha\rangle\langle\alpha|=\int\frac{d^2\alpha}{\pi}:e^{-|\alpha|^2+a^{\dagger}\alpha+a\alpha^{*}-a^{\dagger}a}:=1 \tag{7.28}$$

and the identity

$$e^{a^{\dagger}a\ln\sqrt{\sigma}}|\alpha\rangle=e^{-|\alpha|^2/2+\alpha a^{\dagger}\sqrt{\sigma}}|0\rangle, \tag{7.29}$$

thus Eq. (7.27) can be represented as

$$\begin{aligned}\rho_n(\sigma)&=\int\frac{d^2\alpha}{\pi}e^{(1-\sigma)|\alpha|^2}P(\alpha,0)e^{a^{\dagger}a\ln\sqrt{\sigma}}|\alpha\rangle\langle\alpha|e^{a^{\dagger}a\ln\sqrt{\sigma}}\\&=\sum_{m=0}^{\infty}\frac{(1-\sigma)^n}{n!}e^{\ln\sqrt{\sigma}a^{\dagger}a}\int\frac{d^2\alpha}{\pi}P(\alpha,0)|\alpha|^{2m}|\alpha\rangle\langle\alpha|e^{\ln\sqrt{\sigma}a^{\dagger}a}\\&=\sum_{m=0}^{\infty}\frac{(1-\sigma)^m}{m!}e^{\ln\sqrt{\sigma}a^{\dagger}a}a^m\int\frac{d^2\alpha}{\pi}P(\alpha,0)|\alpha\rangle\langle\alpha|a^{\dagger m}e^{\ln\sqrt{\sigma}a^{\dagger}a}.\end{aligned} \tag{7.30}$$

Since the inverse of Eq. (7.26) is

$$\int\frac{d^2\alpha}{\pi}P(\alpha,0)|\alpha\rangle\langle\alpha|=|n\rangle\langle n|, \tag{7.31}$$

we have

$$\rho_n(\sigma) = \sum_{m=0}^{n} \frac{(1-\sigma)^m}{m!} e^{\ln\sqrt{\sigma} a^\dagger a} a^m |n\rangle\langle n| a^{\dagger m} e^{\ln\sqrt{\sigma} a^\dagger a}. \quad (7.32)$$

By comparing the solution form of the amplitude decay channel in Eq. (6.20), we see that if letting $\sigma = e^{-2\kappa t}$ (κ is the decay rate) in Eq. (7.32), thus Eq. (7.32) can be recognized as the evolution of the initial number state $|n\rangle\langle n| \equiv \rho(0)$ in the amplitude decay channel, that is

$$\rho_n(\sigma) \longrightarrow \rho_n(t) = \sum_{m=0}^{n} \frac{\left(1-e^{-2\kappa t}\right)^m}{m!} e^{-\kappa t a^\dagger a} a^m |n\rangle\langle n| a^{\dagger m} e^{-\kappa t a^\dagger a}. \quad (7.33)$$

In other words, the binomial states can be served as the evolutions of number states in the amplitude decay channel. So, it is easy to obtain the evolved average photon number, that is

$$\mathtt{tr}\left(\rho_n(t) a^\dagger a\right) = \sum_{m=0}^{n} \binom{n}{m} e^{-2m\kappa t} \left(1-e^{-2\kappa t}\right)^{n-m} \langle n| a^\dagger a |n\rangle = n e^{-2\kappa t}, \quad (7.34)$$

which precisely reflects the influence of the amplitude decay on number states, that is, the decrease of photon numbers. Similarly, we have

$$\mathtt{tr}\left(\rho(t)\left(a^\dagger a\right)^2\right) = \left(n e^{-2\kappa t}\right)^2 + n e^{-2\kappa t}\left(1-e^{-2\kappa t}\right), \quad (7.35)$$

thus the Mandel-Q_M parameter of the number state in the amplitude decay channel is

$$Q_M = \frac{\mathtt{tr}\left(\rho(t)\left(a^\dagger a\right)^2\right) - \left(\mathtt{tr}\left(\rho(t) a^\dagger a\right)\right)^2}{\mathtt{tr}\left(\rho(t) a^\dagger a\right)} - 1 = -e^{-2\kappa t} \overset{t\to\infty}{\to} 0. \quad (7.36)$$

It can be seen that at time t, $Q_M < 0$ means that quantum state follows a sub-Poisson distribution. In the case of infinite time t, $Q_M = 0$ follows Poisson statistics distribution.

7.4 Negative Binomial States: Generation and Decay

7.4.1 Negative Binomial States Caused by Partial Tracing of Photon-Added Two-Mode Squeezed States

The research shows that the negative binomial state of the light field can be obtained by a partial trace of a photon-added two-mode squeezed state ($C_\iota b^{\dagger\iota} S_2|00\rangle$, a non-Gaussian quantum state), where C_ι is the normalization coefficient, $S_2 = \exp[\lambda(a^\dagger b^\dagger - ab)]$ is the two-mode squeezed operator, and the partial trace is for the b-mode. The concise form of the two-mode squeezed vacuum state $S_2|00\rangle$ is [5, 6]

$$S_2|00\rangle = \mathrm{sech}\lambda \mathrm{e}^{a^\dagger b^\dagger \tanh\lambda}|00\rangle, \quad |00\rangle = |0\rangle_a |0\rangle_b. \tag{7.37}$$

Firstly, we calculate the normalization constant C_ι. For this reason, introducing a two-mode coherent state

$$|z, \tilde{z}'\rangle = \exp\left(-\frac{|z|^2}{2} - \frac{|\tilde{z}'|^2}{2} + za^\dagger + z'b^\dagger\right)|00\rangle, \tag{7.38}$$

which obeys the completeness relation

$$\int \frac{\mathrm{d}^2 z \mathrm{d}^2 z'}{\pi^2} |z, \tilde{z}'\rangle\langle z, \tilde{z}'| = 1, \tag{7.39}$$

thus we have

$$\begin{aligned}
&\langle 00| S_2^\dagger b^l b^{\dagger l} S_2|00\rangle \\
&= \mathrm{sech}^2\lambda \langle 00| \mathrm{e}^{ab\tanh\lambda} b^l b^{\dagger l} \mathrm{e}^{a^\dagger b^\dagger \tanh\lambda}|00\rangle \\
&= \mathrm{sech}^2\lambda \langle 00| \mathrm{e}^{ab\tanh\lambda} b^l \int \frac{\mathrm{d}^2 z \mathrm{d}^2 z'}{\pi^2} |z, \tilde{z}'\rangle\langle z, \tilde{z}'| b^{\dagger l} \mathrm{e}^{a^\dagger b^\dagger \tanh\lambda}|00\rangle \\
&= \mathrm{sech}^2\lambda \int \frac{\mathrm{d}^2 z'}{\pi} |\tilde{z}'|^{2l} \mathrm{e}^{-|z'|^2 \mathrm{sech}^2\lambda} = l! \cosh^{2l}\lambda.
\end{aligned} \tag{7.40}$$

Thus, $C_l = \left(\mathrm{sech}^l\lambda\right)/\sqrt{l!}$, and the corresponding density operator reads

$$\rho(0) = \frac{\mathrm{sech}^{2l}\lambda}{l!} b^{\dagger l} S_2|00\rangle\langle 00| S_2^\dagger b^l \equiv |\psi\rangle_u \langle\psi|, \tag{7.41}$$

where

$$
\begin{aligned}
|\psi\rangle_l &= \frac{\operatorname{sech}^l \lambda}{\sqrt{l!}} b^{\dagger l} S_2 |00\rangle \\
&= \frac{\operatorname{sech}^{l+1} \lambda}{\sqrt{l!}} b^{\dagger l} \mathrm{e}^{a^\dagger b^\dagger \tanh \lambda} |00\rangle \\
&= \frac{\operatorname{sech}^{l+1} \lambda}{\sqrt{l!}} b^{\dagger l} \sum_{n=0}^{\infty} \frac{\left(a^\dagger b^\dagger \tanh \lambda\right)^n}{n!} |00\rangle \\
&= \sum_{n=0}^{\infty} \sqrt{\binom{n+l}{l} \operatorname{sech}^{l+1} \lambda (1 - \operatorname{sech}^2 \lambda)^n} \, |n, l+n\rangle . \qquad (7.42)
\end{aligned}
$$

Clearly, $\sqrt{\binom{n+l}{l} \operatorname{sech}^{l+1} \lambda (1 - \operatorname{sech}^2 \lambda)^n}$ is just a negative binomial distribution, and

$$
\begin{aligned}
b \, |\psi(\beta)\rangle_s &= \sum_{n=0}^{\infty} \sqrt{\binom{n+s}{s} \gamma^{s+1} (1-\gamma)^n n} \, |n-1, \tilde{s}+1+\tilde{n}+1\rangle \\
&= \sum_{n=0}^{\infty} \sqrt{\binom{n+s+1}{s} \gamma^{s+1} (1-\gamma)^{n+1}} \, |n, \tilde{s}+1+\tilde{n}\rangle \\
&= \sqrt{\frac{(s+1)(1-\gamma)}{\gamma}} \, |\psi(\beta)\rangle_{s+1} . \qquad (7.43)
\end{aligned}
$$

If partially tracing the b-mode, thus we have

$$
\begin{aligned}
\mathrm{tr}_b \rho(0) &= \frac{\operatorname{sech}^{2l} \lambda}{l!} \mathrm{tr}_b \left(b^{\dagger l} S_2 |00\rangle \langle 00| S_2^\dagger b^l \right) \\
&= \frac{\operatorname{sech}^{2(l+1)} \lambda}{l!} \mathrm{tr}_b \left(\int \frac{\mathrm{d}^2 z'}{\pi} b^{\dagger l} \mathrm{e}^{b^\dagger a^\dagger \tanh \lambda} |00\rangle \langle 00| \mathrm{e}^{ba \tanh \lambda} b^l \left|z'\right\rangle \left\langle z'\right| \right) \\
&= \frac{\operatorname{sech}^{2(l+1)} \lambda}{l!} \int \frac{\mathrm{d}^2 z'}{\pi} \left\langle z'\right| b^{\dagger l} \mathrm{e}^{b^\dagger a^\dagger \tanh \lambda} |00\rangle \langle 00| \mathrm{e}^{ba \tanh \lambda} b^l \left|z'\right\rangle \\
&= \frac{\operatorname{sech}^{2l} \lambda}{l!} \int \frac{\mathrm{d}^2 z'}{\pi} \left\langle z'\right| z'^l z'^{*l} \mathrm{e}^{z'^* a^\dagger \tanh \lambda} |00\rangle \langle 00| \mathrm{e}^{z' a \tanh \lambda} \left|z'\right\rangle \\
&= \frac{\operatorname{sech}^{2l} \lambda}{l!} \int \frac{\mathrm{d}^2 z'}{\pi} \left\langle z'\right| z'^l z'^{*l} \mathrm{e}^{z'^* a^\dagger \tanh \lambda} \\
&\quad |0\rangle_{aa} \langle 0| \mathrm{e}^{z' a \tanh \lambda} \left\langle z' \right| 0\rangle_{bb} \langle 0| z'\rangle . \qquad (7.44)
\end{aligned}
$$

Noting $\langle 0 \left| z' \right\rangle = \mathrm{e}^{-|z'|^2/2}$ and

$$
\mathrm{e}^{z'^* a^\dagger \tanh \lambda} |0\rangle_a = \left| z'^* \tanh \lambda \right\rangle, \qquad {}_a\langle 0| \mathrm{e}^{z' a \tanh \lambda} = \left\langle z'^* \tanh \lambda \right| \qquad (7.45)
$$

is an unnormalized coherent state of a-mode, so we have

$$z^{'*l}e^{z^{'*}a^{\dagger}\tanh\lambda}|0\rangle = z^{'*l}\left|z^{'*}\tanh\lambda\right\rangle = \left(\frac{a^{l}}{\tanh\lambda}\right)^{l}\left|z^{'*}\tanh\lambda\right\rangle$$
$$\langle z^{'*}\tanh\lambda|z^{'l} = \left\langle z^{'*}\tanh\lambda\right|\left(\frac{a^{\dagger l}}{\tanh\lambda}\right)^{l}. \tag{7.46}$$

Substituting Eq. (7.46) into Eq. (7.44) leads to

$$\mathtt{tr}_{b}\rho\,(0) = \frac{\mathrm{sech}^{2l}\lambda}{l!}\frac{1}{\tanh^{2l}\lambda}a^{l}\int\frac{d^{2}z'}{\pi}\left|z^{'*}\tanh\lambda\right\rangle\langle z^{'*}\tanh\lambda|e^{-|z'|^{2}}a^{\dagger l}. \tag{7.47}$$

Using the IWOP method to obtain

$$\begin{aligned}\mathtt{tr}_{b}\rho\,(0) &= \frac{\mathrm{sech}^{2l}\lambda}{l!}\frac{1}{\tanh^{2l}\lambda}a^{l}\int\frac{d^{2}z'}{\pi}:e^{-|z'|^{2}+z^{'*}a^{\dagger}\tanh\lambda+z'a\tanh\lambda-a^{\dagger}a}:a^{\dagger l}\\ &= \frac{\mathrm{sech}^{2(l+1)}\lambda}{l!}\frac{1}{\tanh^{2l}\lambda}a^{l}:e^{a^{\dagger}a(\tanh^{2}\lambda-1)}:a^{\dagger l}\\ &= \frac{\mathrm{sech}^{2(l+1)}\lambda}{l!}\frac{1}{(1-\mathrm{sech}^{2}\lambda)^{l}}a^{l}e^{a^{\dagger}a\ln\tanh^{2}\lambda}a^{\dagger l},\end{aligned} \tag{7.48}$$

letting $\mathrm{sech}^{2}\lambda = \gamma$, thus Eq. (7.48) becomes

$$\mathtt{tr}_{b}\rho\,(0) = \frac{\gamma^{l+1}}{l!\,(1-\gamma)^{l}}a^{l}e^{a^{\dagger}a\ln(1-\gamma)}a^{\dagger l} = \frac{\gamma^{l+1}}{l!\,(1-\gamma)^{l}}a^{l}\,(1-\gamma)^{a^{\dagger}a}\,a^{\dagger l}, \tag{7.49}$$

which is just the negative binomial state of light field because it is equivalent to

$$\begin{aligned}\mathtt{tr}_{b}\rho\,(0) &= \frac{\gamma^{l+1}}{l!}\frac{1}{(1-\gamma)^{l}}a^{l}\sum_{n=0}^{\infty}(1-\gamma)^{n}\,|n\rangle\,\langle n|\,a^{\dagger l}\\ &= \sum_{n=0}^{\infty}\binom{n+l}{n}\gamma^{l+1}(1-\gamma)^{n}\,|n\rangle\,\langle n|\,.\end{aligned} \tag{7.50}$$

According to the negative binomial theorem

$$(1+x)^{-(s+1)} = \sum_{n=0}^{\infty}\frac{(n+s)!}{n!s!}(-x)^{n}, \tag{7.51}$$

we get

$$\mathtt{tr}_{a}(\mathtt{tr}_{b}\rho\,(0)) = \gamma^{s+1}\sum_{n=0}^{\infty}\frac{(n+s)!}{n!s!}(1-\gamma)^{n} = 1. \tag{7.52}$$

From this, it can be seen that the expected values of observables of a-mode are exactly equivalent to calculating their expected values in the corresponding negative binomial state light field.

7.4.2 Decay of Negative Binomial States

Substituting Eq. (7.17) as the initial density operator into the solution of the maser equation describing the amplitude decay channel [see also Eq. (6.20)], and using the operator identity

$$e^{fa^\dagger a} a e^{-fa^\dagger a} = a e^{-f}, \tag{7.53}$$

we get

$$\begin{aligned} s!\,(n_c)^s\,\rho(t) &= \sum_{m=0}^{\infty} \frac{T^m}{m!} e^{-\kappa t a^\dagger a} a^m a^s \rho_c a^{\dagger s} a^{\dagger m} e^{-\kappa t a^\dagger a} \\ &= \sum_{m=0}^{\infty} \frac{T^m}{m!} e^{2\kappa t(m+s)} a^{m+s} e^{-\kappa t a^\dagger a} \rho_c e^{-\kappa t a^\dagger a} a^{\dagger m+s} \\ &= \gamma \sum_{m=0}^{\infty} \frac{T^m}{m!} e^{2\kappa t(m+s)} a^{m+s} e^{(\lambda - 2\kappa t) a^\dagger a} a^{\dagger m+s}, \end{aligned} \tag{7.54}$$

where $\lambda = \ln(1-\gamma)$. We encounter obstacles in summing m here because there is the term $e^{(\lambda-2\kappa t)a^\dagger a}$ is sandwiched between a^{m+s} and $a^{\dagger m+s}$. Therefore, using the operator identity [7–9]

$$\begin{aligned} e^{\lambda a^\dagger a} &= e^{-\lambda} \vdots \exp\left[\left(1 - e^{-\lambda}\right) a a^\dagger\right] \vdots, \\ \text{or } \vdots e^{\lambda a a^\dagger} \vdots &= \frac{1}{1-\lambda} e^{a^\dagger a \ln \frac{1}{1-\lambda}} \end{aligned} \tag{7.55}$$

to rewrite Eq. (7.54) as

$$\begin{aligned} s!\,(n_c)^s\,\rho(t) = \gamma e^{-(\lambda - 2\kappa t)} \sum_{m=0}^{\infty} \frac{T^m}{m!} e^{2\kappa t(m+s)} \\ \times \vdots a^{m+s} \exp\left[\left(1 - e^{2\kappa t - \lambda}\right) a a^\dagger\right] a^{\dagger m+s} \vdots. \end{aligned} \tag{7.56}$$

Noting the the operators a and $a^\dagger$ are commutative in the symbol $\vdots\ \vdots$, so we can sum m in the right side of Eq. (7.56), the result is

$$\begin{aligned} s!\,(n_c)^s\,\rho\,(t) &= \gamma \mathrm{e}^{-(\lambda-2\kappa t)} \mathrm{e}^{2\kappa t s} a^s \vdots \exp\left\{\left[T\mathrm{e}^{2\kappa t} + \left(1-\mathrm{e}^{2\kappa t-\lambda}\right)\right] aa^\dagger\right\} \vdots a^{\dagger s} \\ &= \gamma \mathrm{e}^{-(\lambda-2\kappa t)} \mathrm{e}^{2\kappa t s} a^s \vdots \exp\left\{\mathrm{e}^{2\kappa t}\left(1-\mathrm{e}^{-\lambda}\right) aa^\dagger\right\} \vdots a^{\dagger s} \\ &= \frac{\gamma \mathrm{e}^{2\kappa t}}{1-\gamma} \mathrm{e}^{2\kappa t s} a^s \vdots \exp\left(\frac{\gamma \mathrm{e}^{2\kappa t}}{\gamma-1} aa^\dagger\right) \vdots a^{\dagger s}, \end{aligned} \tag{7.57}$$

where

$$\begin{aligned} &\frac{\gamma \mathrm{e}^{2\kappa t}}{1-\gamma} \vdots \exp\left(\frac{\gamma \mathrm{e}^{2\kappa t}}{\gamma-1} aa^\dagger\right) \vdots \\ &= \frac{\gamma \mathrm{e}^{2\kappa t}}{1-\gamma(1-\mathrm{e}^{2\kappa t})} \exp\left[aa^\dagger \ln \frac{\gamma-1}{\gamma(1-\mathrm{e}^{2\kappa t})-1}\right] \\ &= \frac{\gamma \mathrm{e}^{2\kappa t}}{1-\gamma(1-\mathrm{e}^{2\kappa t})} \exp\left\{aa^\dagger \ln\left[1-\frac{\gamma \mathrm{e}^{2\kappa t}}{1-\gamma(1-\mathrm{e}^{2\kappa t})}\right]\right\} \\ &= \gamma' \mathrm{e}^{a^\dagger a \ln(1-\gamma')}. \end{aligned} \tag{7.58}$$

Since

$$\gamma' = \frac{\gamma \mathrm{e}^{2\kappa t}}{1-\gamma(1-\mathrm{e}^{2\kappa t})} = \frac{1}{\mathrm{e}^{-2\kappa t}(1-\gamma)+\gamma}\gamma < \gamma, \tag{7.59}$$

thus Eq. (7.56) becomes

$$\begin{aligned} \rho\,(t) &= \frac{1}{s!\left(n_c \mathrm{e}^{-2\kappa t}\right)^s} a^s \gamma' \mathrm{e}^{a^\dagger a \ln(1-\gamma')} a^{\dagger s} = \frac{1}{s!\left(n_c \mathrm{e}^{-2\kappa t}\right)^s} a^s \rho_c' a^{\dagger s} \\ &= \frac{1}{s!\left(n_c \mathrm{e}^{-2\kappa t}\right)^s} \sum_{n=0}^{\infty} \gamma' \left(1-\gamma'\right)^n a^s \left|n\right\rangle \left\langle n\right| a^{\dagger s} \\ &= \frac{1}{s!\left(n_c \mathrm{e}^{-2\kappa t}\right)^s} \sum_{n=0}^{\infty} \gamma' \left(1-\gamma'\right)^n \frac{n!}{(n-s)!} \left|n-s\right\rangle \left\langle n-s\right| \\ &= \frac{\left(1-\gamma'\right)^s}{\gamma'^s} \frac{1}{\left(\frac{1-\gamma}{\gamma}\mathrm{e}^{-2\kappa t}\right)^s} \sum_{n'=0}^{\infty} \gamma'^{s+1} \left(1-\gamma'\right)^{n'} \binom{n'+s}{n'} \left|n'\right\rangle\left\langle n'\right| \\ &= \sum_{n=0}^{\infty} \gamma'^{s+1} \left(1-\gamma'\right)^n \binom{n+s}{n} \left|n\right\rangle \left\langle n\right|, \end{aligned} \tag{7.60}$$

where it can be easily proved that $\mathtt{tr}\rho\,(t) = 1$, and in the last step we have used the following relation:

$$\frac{1-\gamma'}{1-\gamma} \frac{\gamma}{\gamma'} = \mathrm{e}^{-2\kappa t}. \tag{7.61}$$

From above results, it can be concluded that the negative binomial state still evolves into a new negative binomial state after passing through the amplitude decay channel, just the parameter changed from γ to γ', $\gamma' < 1$ [10, 11]. And the average number of photons at time t reads

$$\mathrm{tr}\left(\rho\left(t\right)a^{\dagger}a\right) = \frac{1-\gamma'}{\gamma'} = \frac{1-\gamma}{\gamma}\mathrm{e}^{-2\kappa t} = n_c \mathrm{e}^{-2\kappa t}, \tag{7.62}$$

that is, the average photon number n_c exhibits the exponential decay during the amplitude decay channel.

References

1. Fan HY, Jing SC (1995) Even and odd binomial states. Commun Theor Phys 24(1):125–128
2. Fan HY, Li S (2005) Constructing even-and odd-negative binomial states. Commun Theor Phys 43(3):519–522
3. Fan HY, Ren G (2010) Evolution of number state to density operator of binomial distribution in the amplitude dissipative channel. Chin Phys Lett 27(5):050302
4. Fan HY, Lu HL, Fan Y (2006) Newton-Leibniz integration for ket-bra operators in quantum mechanics and derivation of entangled state representations. Ann Phys 321(2):480–494
5. Meng XG, Wang JS, Liang BL, Han CX (2018) Evolution of a two-mode squeezed vacuum for amplitude decay via continuous-variable entangled state approach. Front Phys 13(5):130322
6. Meng XG, Li KC, Wang JS, Yang ZS, Zhang XY, Zhang ZT, Liang BL (2020) Multi-variable special polynomials using an operator ordering method. Front Phys 15(5):52501
7. Fan HY (2004) Entangled states, squeezed states gained via the route of developing Dirac's symbolic method and their applications. Int J Mod Phys B 18(10–11):1387–1455
8. Xu SM, Zhang YH, Xu XL, Li HQ, Wang JS (2016) New approach for anti-normally and normally ordering bosonic-operator functions in quantum optics. Chin Phys B 25(12):120301
9. Meng XG, Wang JS, Liang BL (2023) Entangled state representations in quantum optics. Springer, Singapore
10. Chen F, Fan HY (2014) Evolution law of a negative binomial state in an amplitude dissipative channel. Chin Phys B 23(3):030304
11. Da C, Chen QF, Fan HY (2014) Evolution law of the negative binomial state in laser channel and its photon-number decay formula. Int J Theor Phys 53(12):4372–4780

Chapter 8
Quantum Decay for Common Quantum States

Squeezed light fields are a very important type of light sources, which have wide applications in optical communication and weak signal detection. When the quantum fluctuations of a certain observable orthogonal component of light fields are squeezed below the corresponding standard quantum limit, so this type of light fields are called squeezed state light fields. Due to their quantum noises being lower than those of all classical optical fields, including lasers, and the fact that two-mode orthogonal squeezed state light fields can directly provide the necessary nonlocal quantum entanglement for quantum information processing, the generations and applications of squeezed state light fields are important research topics in contemporary scientific and technological fields, such as quantum optics, quantum measurement, and quantum communication. Of course, reducing the noise of one orthogonal component of a squeezed light field comes at the cost of increasing the noise of the other orthogonal component of the light field.

8.1 Evolution Operator Maintaining Squeezing

In theory, the most direct, concise and physically clear method to describe a squeezed operator was first proposed by Fan in his doctoral dissertation paper, which is to use the IWOP method to perform the following integral [1, 2]

$$\int_{-\infty}^{\infty} \frac{\mathrm{d}x}{\sqrt{\mu}} |x/\mu\rangle\langle x| \equiv S, \ \mu = \mathrm{e}^{\lambda}, \tag{8.1}$$

where $|x\rangle$ is the eigen-state of coordinate operator, and Eq. (8.1) reflects the quantum mapping caused by the classical scale transformation $x \longrightarrow x/\mu$. The integral result of Eq. (8.1) is

H. Fan and L. Hu, *Dissipation and Diffusion of Quantum Light Field*,
https://doi.org/10.1007/978-981-95-1528-8_8

$$\begin{aligned} S &= \operatorname{sech}^{1/2}\lambda\colon \exp\left[\frac{1}{2}\left(a^2 - a^{\dagger 2}\right)\tanh\lambda + \left(\operatorname{sech}\lambda - 1\right)a^{\dagger}a\right]\colon \\ &= \exp\left[\frac{\lambda}{2}\left(a^2 - a^{\dagger 2}\right)\right], \end{aligned} \tag{8.2}$$

where : : means normal ordering, $a^{\dagger}$, a are, respectively photon creation and annihilation operators, which obey the commutative relation $\left[a, a^{\dagger}\right] = 1$, and we have used the operator identity

$$e^{\lambda a^{\dagger}a} =\colon \exp\left[\left(e^{\lambda} - 1\right)a^{\dagger}a\right]\colon. \tag{8.3}$$

Equations (8.1) and (8.2) explain that the scale transformation $x \to x/\mu$ of classical coordinate maps to the squeezed operator in the Hilbert space. For single-mode case, there are two kinds of squeezed states. One is the number-phase squeezed state, its time-evolution is determined by the following Hamiltonian

$$H = \omega a^{\dagger}a + f(t)\, a^{\dagger}\sqrt{N+1} + f^{*}(t)\sqrt{N+1}a, \quad N = a^{\dagger}a, \tag{8.4}$$

and its corresponding squeezed operator has the following form: $\exp\left[\lambda\left(a^{\dagger}\sqrt{N+1} - \sqrt{N+1}a\right)\right]$; Another type of squeezing is generated by the degenerate parametric amplification process, and its evolution is determined by the following Hamiltonian

$$H = \omega a^{\dagger}a + f(t)\, a^{\dagger 2} + f^{*}(t)\, a^2, \tag{8.5}$$

and its corresponding squeezed operator is $\exp\left[\lambda\left(a^2 - a^{\dagger 2}\right)/2\right]$. Note that the two Hamiltonians above are clearly time-dependent, and the time-evolution operator cannot be written in the form of $\exp(-\mathrm{i}Ht)$, so it is necessary to consider the following form of the chronological operator, i.e.,

$$U(t, t_0) = T\exp\left[-\mathrm{i}\int_{t_0}^{t} H\left(t'\right)\mathrm{d}t'\right]. \tag{8.6}$$

Since

$$\begin{aligned} \left[a^{\dagger}\sqrt{N+1}, \sqrt{N+1}a\right] &= -2\left(N + \frac{1}{2}\right), \\ \left[\sqrt{N+1}a, N + \frac{1}{2}\right] &= \sqrt{N+1}a, \\ \left[a^{\dagger}\sqrt{N+1}, N + \frac{1}{2}\right] &= -a^{\dagger}\sqrt{N+1} \end{aligned} \tag{8.7}$$

have the same Lie algebraic structure as the operator $(a^{\dagger}a, a^{\dagger 2}, a^2)$, we only consider the case of Eq. (8.5), and derive the evolution operator $U(t, t_0)$. To our knowledge, although the issue of how coherent states maintain coherence in the time-evolution has been studied, there is still no relevant research on how the squeezing can be maintained during the process of evolution. Next, we will use the coherent state representation and Ricardif equation to derive the operator $U(t, t_0)$, this is a new method, and then examine whether the squeezing can be maintained during the time-evolution process [3].

Letting $H = H_0 + H_I$, $H_0 = \omega a^\dagger a$ and $H_I = f(t) a^{\dagger 2} + f^*(t) a^2$ in the interaction representation, the operator $U_I(t, t_0)$ obeys the equation

$$\mathrm{i}\frac{\partial U_I(t, t_0)}{\partial t} = H_I(t) U_I(t, t_0), \tag{8.8}$$

which meets the initial conditions

$$U_I(t_0, t_0) = 1, \tag{8.9}$$

and where

$$\begin{aligned} H_I(t) &= \mathrm{e}^{\mathrm{i}H_0 t}\left[f(t) a^{\dagger 2} + f^*(t) a^2\right]\mathrm{e}^{-\mathrm{i}H_0 t} \\ &= \mathrm{e}^{\mathrm{i}2\omega t} f(t) a^{\dagger 2} + f^*(t)\mathrm{e}^{-\mathrm{i}2\omega t} a^2. \end{aligned} \tag{8.10}$$

In order to solve Eq. (8.8), we introduce the coherent state representation

$$\int \frac{\mathrm{d}^2 z}{\pi} |z\rangle\langle z| = 1, \quad |z\rangle = \mathrm{e}^{z a^\dagger - \frac{1}{2}|z|^2}|0\rangle, \tag{8.11}$$

where $|z\rangle\langle z|$ can be represented as the following normal ordering form:

$$|z\rangle\langle z| =: \mathrm{e}^{-|z|^2 + z a^\dagger + z^* a - a^\dagger a}: \tag{8.12}$$

since the normal ordering product of vacuum projector is $|0\rangle\langle 0| =: \mathrm{e}^{-a^\dagger a}:$. In the coherent state representation, we have

$$\begin{aligned} &|z\rangle\langle z| a^\dagger \to z^* |z\rangle\langle z|, \\ &|z\rangle\langle z| a =: \mathrm{e}^{-|z|^2 + z a^\dagger + z^* a - a^\dagger a}: a = \left(z + \frac{\partial}{\partial z^*}\right)|z\rangle\langle z|, \end{aligned} \tag{8.13}$$

which means $a^\dagger \to z^*$ and $a \to z + \frac{\partial}{\partial z^*}$. Using Eq. (8.13), we obtain

$$U_I(z, z^*) \equiv \langle z| U_I(t, t_0) |z\rangle = \mathtt{tr}\left(|z\rangle\langle z| U_I(t, t_0)\right), \tag{8.14}$$

which yields

$$\mathrm{i}\frac{\partial U_I(z, z^*)}{\partial t} = \left[F(t) z^{*2} + F^*(t)\left(z + \frac{\partial}{\partial z^*}\right)^2\right] U_I(z, z^*), \tag{8.15}$$

where

$$F(t) \equiv f(t)\,\mathrm{e}^{\mathrm{i}2\omega t}, \quad U_I\left(z, z^*\right)|_{t=t_0} = 1. \tag{8.16}$$

Without loss of generality, the formal solution of Eq. (8.15) reads

$$\begin{aligned} U_I\left(z, z^*\right) = \exp\{&-\mathrm{i}[z^{*2}(E(t) + X(t)) + z^* z J(t) \\ &+ z^2\left(E^*(t) + Y(t)\right) + R(t)]\}, \end{aligned} \tag{8.17}$$

where

$$E(t) \equiv \int_{t_0}^{t} F(\lambda)\,\mathrm{d}\lambda, \tag{8.18}$$

and $E(t)$, $X(t)$, $Y(t)$, $J(t)$, $R(t)$ are undetermined, the corresponding initial conditions are

$$X(t)|_{t=t_0} = 0, \ J(t)|_{t=t_0} = 0, \ Y(t)|_{t=t_0} = 0, \ R(t)|_{t=t_0} = 0. \tag{8.19}$$

Using Eq. (8.15), we see

$$\begin{aligned} \mathrm{i}\frac{\partial}{\partial t} U_I\left(z, z^*\right) = \{&z^{*2}\left[F(t) + \dot{X}(t)\right] + z^2\left[F^*(t) + \dot{Y}(t)\right] \\ &+ z^* z \dot{J}(t) + \dot{R}(t)\} U_I\left(z, z^*\right). \end{aligned} \tag{8.20}$$

For convenience, omitting the variable t below, we have

$$\begin{aligned} \frac{\partial^2}{\partial z^{*2}} U_I\left(z, z^*\right) = [&-\mathrm{i}2(E + X) - 4z^{*2}(E + X)^2 \\ &-z^2 J^2 - 4|z|^2 J(X + E)] U_I\left(z, z^*\right). \end{aligned} \tag{8.21}$$

Substituting Eqs. (8.20) and (8.21) into Eq. (8.15) leads to

$$\begin{aligned} \mathrm{i}\frac{\partial}{\partial t} U_I\left(z, z^*\right) = [&z^{*2}\left(F - 4F^* E^2 - 4F^* X^2 - 8F^* E X\right) \\ &+z^2 F^*(1 - \mathrm{i}J)^2 - \mathrm{i}2F^*(E + X) \\ &-4|z|^2\left(\mathrm{i}FE + \mathrm{i}F^* X + F^* J X + F^* E J\right)] U_I\left(z, z^*\right). \end{aligned} \tag{8.22}$$

By comparing the coefficients of z^{*2}, z^2 and $|z|^2$ on both sides of Eqs. (8.20) and (8.22), a set of differential equations can be obtained as

$$\dot{X} = -4F^*(E + X)^2, \tag{8.23}$$

$$\dot{Y} = -\mathrm{i}2F^* J - F^* J^2, \tag{8.24}$$

$$\dot{J} = -4F^*(\mathrm{i}E + \mathrm{i}X + JX + EJ), \tag{8.25}$$

$$\dot{R} = -\mathrm{i}2F^*(E + X). \tag{8.26}$$

Letting $G = E + X$, thus Eq. (8.26) becomes

$$\dot{G} = -4F^*G^2 + F, \tag{8.27}$$

which is Riccati equation, the most interesting first-order nonlinear differential equation. Taking $G = (\ln \psi)_t/(4F^*)$, and differentiating time t, we obtain

$$\begin{aligned} \dot{G} &= \frac{\mathrm{d}}{\mathrm{d}t}\left(\frac{1}{4F^*\psi}\frac{\mathrm{d}\psi}{\mathrm{d}t}\right) \\ &= -\frac{1}{4F^*\psi^2}\frac{\mathrm{d}^2\psi}{\mathrm{d}t^2} + \frac{1}{\psi}\frac{\mathrm{d}}{\mathrm{d}t}\left(\frac{1}{4F^*}\frac{\mathrm{d}\psi}{\mathrm{d}t}\right). \end{aligned} \tag{8.28}$$

So, Eq. (8.27) reduces to

$$\frac{\mathrm{d}}{\mathrm{d}t}\left(\frac{1}{4F^*}\frac{\mathrm{d}\psi}{\mathrm{d}t}\right) - F\psi = 0. \tag{8.29}$$

Without loss of generality, assuming $F(t) = f(t)\mathrm{e}^{\mathrm{i}2\omega t} = F^*(t)$ is a real number, thus the solution of Eq. (8.29) has the following form:

$$\psi = \alpha \mathrm{e}^{2E(t)} + \beta \mathrm{e}^{-2E(t)}, \tag{8.30}$$

where α, β are the integral constants. Therefore, we have

$$\begin{aligned} X = G - E &= \frac{1}{4F}(\ln \psi)_t - E \\ &= \frac{1}{4F}\left[\ln\left(\alpha \mathrm{e}^{2E(t)} + \beta \mathrm{e}^{-2E(t)}\right)\right]_t - E \\ &= \frac{1}{2}\frac{\alpha \mathrm{e}^{2E} - \beta \mathrm{e}^{-2E}}{\alpha \mathrm{e}^{2E} + \beta \mathrm{e}^{-2E}} - E. \end{aligned} \tag{8.31}$$

Since $E|_{t=t_0} = 0$, $X|_{t=t_0} = 0$, we have $\alpha = \beta$ that leads to

$$X = \frac{1}{2}\tanh 2E - E. \tag{8.32}$$

Since $F(t)$ is a real number, thus using Eq. (8.32), we can obtain the solutions of Eqs. (8.24)–(8.26), i.e.,

$$Y = \frac{1}{2}\tanh 2E - E, \tag{8.33}$$

$$J = \mathrm{i}\left(\mathrm{sech}2E - 1\right), \tag{8.34}$$

$$R = -\frac{\mathrm{i}}{2}\ln\cosh 2E. \tag{8.35}$$

Substituting Eqs. (8.32) and (8.33)–(8.35) into Eq. (8.17), we obtain

$$U_I\left(z,z^*\right)=\exp\left\{-\mathrm{i}\left[\frac{1}{2}\left(z^{*2}+z^2\right)\tanh 2E\right.\right.$$
$$\left.\left.+\mathrm{i}\left|z\right|^2\left(\mathrm{sech}2E-1\right)-\frac{\mathrm{i}}{2}\ln\cosh 2E\right]\right\}. \tag{8.36}$$

According to the fact that the operator's expected value in the coherent state representation can uniquely determine the operator characteristics itself, and using Eq. (8.36), we have

$$\begin{aligned}U_I\left(t,t_0\right)&=\exp\left(-\frac{\mathrm{i}}{2}a^{\dagger 2}\tanh 2E\right):\exp\left[a^{\dagger}a\left(\mathrm{sech}2E-1\right)\right]:\\&\quad\times\exp\left(-\frac{\mathrm{i}}{2}a^{2}\tanh 2E\right)\exp\left(-\frac{1}{2}\ln\cosh 2E\right)\\&=\exp\left(-\frac{\mathrm{i}}{2}a^{\dagger 2}\tanh 2E\right)\exp\left[\left(a^{\dagger}a+\frac{1}{2}\right)\ln\mathrm{sech}2E\right]\\&\quad\times\exp\left(-\frac{\mathrm{i}}{2}a^{2}\tanh 2E\right).\end{aligned}\tag{8.37}$$

Using the identity

$$U_s\left(t,t_0\right)=\mathrm{e}^{-\mathrm{i}H_0t}U_I\left(t,t_0\right)\mathrm{e}^{\mathrm{i}H_0t_0}, \tag{8.38}$$

we can obtain the time-evolution operator $U_s\left(t,t_0\right)$ in the Schrödinger representation, i.e.,

$$\begin{aligned}U_s\left(t,t_0\right)&=\exp\left(-\frac{\mathrm{i}}{2}a^{\dagger 2}\mathrm{e}^{-\mathrm{i}2\omega t}\tanh 2E\right)\\&\quad\times\exp\left[-\mathrm{i}\omega\left(t-t_0\right)a^{\dagger}a+\left(a^{\dagger}a+\frac{1}{2}\right)\ln\mathrm{sech}2E\right]\\&\quad\times\exp\left(-\frac{\mathrm{i}}{2}a^{2}\mathrm{e}^{\mathrm{i}2\omega t_0}\tanh 2E\right),\end{aligned}\tag{8.39}$$

thus the Feynman matrix element is

$$\begin{aligned}\left\langle z\right|U_s\left(t,t_0\right)\left|z_0\right\rangle&=\left\langle z\right|\mathrm{e}^{-\frac{\mathrm{i}}{2}a^{\dagger 2}\mathrm{e}^{-\mathrm{i}2\omega t}\tanh 2E}\mathrm{e}^{-\mathrm{i}\omega(t-t_0)a^{\dagger}a+\left(a^{\dagger}a+\frac{1}{2}\right)\ln\mathrm{sech}2E}\\&\quad\times\mathrm{e}^{-\frac{\mathrm{i}}{2}a^{2}\mathrm{e}^{\mathrm{i}2\omega t_0}\tanh 2E}\left|z_0\right\rangle\\&=\left\langle z\right|\mathrm{e}^{-\frac{\mathrm{i}}{2}a^{\dagger 2}\mathrm{e}^{-\mathrm{i}2\omega t}\tanh 2E}:\exp\left[\left(\mathrm{e}^{-\mathrm{i}\omega(t-t_0)+\ln\mathrm{sech}2E}-1\right)a^{\dagger}a\right]:\\&\quad\times\mathrm{e}^{-\frac{\mathrm{i}}{2}a^{2}\mathrm{e}^{\mathrm{i}2\omega t_0}\tanh 2E}\mathrm{e}^{-\frac{1}{2}\ln\cosh 2E}\left|z_0\right\rangle\\&=\mathrm{e}^{-\frac{\mathrm{i}}{2}z^{*2}\mathrm{e}^{-\mathrm{i}2\omega t}\tanh 2E}\mathrm{sech}^{1/2}2E\\&\quad\times\exp\left[\left(\mathrm{e}^{-\mathrm{i}\omega(t-t_0)}\mathrm{sech}2E-1\right)z^{*}z_0\right]\mathrm{e}^{-\frac{\mathrm{i}}{2}z_0^{2}\mathrm{e}^{\mathrm{i}2\omega t_0}\tanh 2E}.\end{aligned}\tag{8.40}$$

Similarly, for the number-phase squeezing, the corresponding time-evolution operator can be obtained from Eq. (8.4) as follows:

$$
\begin{aligned}
& \mathrm{e}^{-\mathrm{i}a^{\dagger}\sqrt{N+1}\mathrm{e}^{-\mathrm{i}2\omega t}\tanh 2E}\mathrm{e}^{-\mathrm{i}\omega(t-t_0)\left(N+\frac{1}{2}\right)+(N+1)\ln \operatorname{sech}2E} \\
& \times \mathrm{e}^{-\mathrm{i}\sqrt{N+1}a\mathrm{e}^{\mathrm{i}2\omega t_0}\tanh 2E}.
\end{aligned} \tag{8.41}
$$

Next, let's see if the operator $U_s(t, t_0)$ can maintain squeezing during the evolution process. Assuming the initial state is a squeezed vacuum state

$$
S|0\rangle = \operatorname{sech}^{1/2}\lambda \mathrm{e}^{\frac{\tanh\lambda}{2}a^{\dagger 2}}|0\rangle, \tag{8.42}
$$

thus acting the operator $U_s(t, t_0)$ on the squeezed vacuum state, we have

$$
\begin{aligned}
U_s(t, t_0) S|0\rangle = & \operatorname{sech}^{1/2}\lambda \mathrm{e}^{-\frac{\mathrm{i}}{2}a^{\dagger 2}\mathrm{e}^{-\mathrm{i}2\omega t}\tanh 2E} \\
& \times \mathrm{e}^{-\mathrm{i}\omega(t-t_0)a^{\dagger}a+\left(a^{\dagger}a+\frac{1}{2}\right)\ln \operatorname{sech}2E} \\
& \times \mathrm{e}^{-\frac{\mathrm{i}}{2}a^{2}\mathrm{e}^{\mathrm{i}2\omega t_0}\tanh 2E}\mathrm{e}^{-\frac{\tanh\lambda}{2}a^{\dagger 2}}|0\rangle.
\end{aligned} \tag{8.43}
$$

Using the integral formula in Eq. (1.57) and the IWOP method, we obtain

$$
\begin{aligned}
\mathrm{e}^{fa^2}\mathrm{e}^{ga^{\dagger 2}} & = \int \frac{\mathrm{d}^2 z}{\pi}\mathrm{e}^{fz^2}|z\rangle\langle z|\mathrm{e}^{gz^{*2}} \\
& = \int \frac{\mathrm{d}^2 z}{\pi} : \mathrm{e}^{-|z|^2+za^{\dagger}+z^{*}a+fz^2+gz^{*2}-a^{\dagger}a} : \\
& = \frac{1}{\sqrt{1-4fg}}\exp\left(\frac{ga^{\dagger 2}}{1-4fg}\right)\exp\left(\frac{-a^{\dagger}a}{1-4fg}\right) \\
& \quad \times \exp\left(\frac{fa^2}{1-4fg}\right).
\end{aligned} \tag{8.44}
$$

Thus, Eq. (8.43) becomes

$$
\begin{aligned}
U_s(t, t_0) S|0\rangle = & \sqrt{\frac{\operatorname{sech}\lambda \operatorname{sech}2E}{1-\mathrm{i}\tanh 2E\tanh\lambda \mathrm{e}^{2\mathrm{i}\omega t_0}}}\mathrm{e}^{-\frac{\mathrm{i}}{2}a^{\dagger 2}\mathrm{e}^{-\mathrm{i}2\omega t}\tanh 2E} \\
& \times \mathrm{e}^{[-\mathrm{i}\omega(t-t_0)+\ln \operatorname{sech}2E]a^{\dagger}a}\exp\left[-\frac{a^{\dagger 2}\tanh\lambda}{2\left(1-\mathrm{i}\tanh 2E\tanh\lambda \mathrm{e}^{2\mathrm{i}\omega t_0}\right)}\right]|0\rangle \\
= & \sqrt{\frac{\operatorname{sech}\lambda \operatorname{sech}2E}{1-\mathrm{i}\tanh 2E\tanh\lambda \mathrm{e}^{2\mathrm{i}\omega t_0}}}\mathrm{e}^{-\frac{\mathrm{i}}{2}a^{\dagger 2}\mathrm{e}^{-\mathrm{i}2\omega t}\tanh 2E} \\
& \times \exp\left[-\frac{\mathrm{e}^{-\mathrm{i}2\omega(t-t_0)+2\ln \operatorname{sech}2E}\tanh\lambda}{2\left(1-\mathrm{i}\tanh 2E\tanh\lambda \mathrm{e}^{\mathrm{i}2\omega t_0}\right)}a^{\dagger 2}\right]|0\rangle
\end{aligned}
$$

$$= \sqrt{\frac{\operatorname{sech}\lambda \operatorname{sech} 2E}{1 - \mathrm{i} \tanh 2E \tanh \lambda \mathrm{e}^{\mathrm{i}2\omega t_0}}} \exp\left(-\frac{\mathrm{i}}{2}\mathrm{e}^{-\mathrm{i}2\omega t} \tanh 2E\right)$$
$$\times \exp\left[-\frac{\operatorname{sech}^2 2E \tanh \lambda \mathrm{e}^{-\mathrm{i}2\omega(t-t_0)}}{2\left(1 - \mathrm{i} \tanh 2E \tanh \lambda \mathrm{e}^{\mathrm{i}2\omega t_0}\right)} a^{\dagger 2}\right] |0\rangle . \tag{8.45}$$

From this, it can be seen that when the initial state is in a squeezed state, the final state remains in a squeezed state.

In this section, using the coherent state representation to solve the Riccati equation analytically, we derived the dynamic evolution operator that maintains the squeezed characteristics.

8.2 Evolution of Squeezed Chaotic Light Field for Amplitude Decay

8.2.1 Squeezed Chaotic Light Field

In the Fock representation, squeezed chaotic light field serves as a generating field for the squeezed number states [4, 5], i.e.,

$$\rho(0) = \left(1 - \mathrm{e}^k\right) \sum_{n=0}^{\infty} \mathrm{e}^{kn} S^{-1} |n\rangle \langle n| S, \tag{8.46}$$

where $|n\rangle = \frac{a^{\dagger n}}{\sqrt{n!}} |0\rangle$ is a number state, $S^{-1}|n\rangle$ is a squeezed number state that can be represented as the Hermite polynomial excited squeezed vacuum, that is

$$S^{-1}|n\rangle = \frac{\operatorname{sech}^{\frac{1}{2}}\lambda}{\sqrt{2^n n!}} \tanh^{\frac{n}{2}} \lambda \mathrm{H}_n\left(\frac{a^\dagger}{\sqrt{\sinh 2\lambda}}\right) \exp\left(\frac{a^{\dagger 2}}{2} \tanh \lambda\right) |0\rangle . \tag{8.47}$$

Thus, the normal ordering product of the corresponding density operator $\rho(0)$ can be found in the following form:

$$\rho(0) = \frac{1 - \mathrm{e}^k}{\sqrt{\cosh^2 \lambda - \mathrm{e}^{2k} \sinh^2 \lambda}} \exp\left[-\frac{(\mathrm{e}^{2k} - 1) \tanh \lambda}{2\left(1 - \mathrm{e}^{2k} \tanh^2 \lambda\right)} a^{\dagger 2}\right]$$
$$\times : \exp\left[a^\dagger a \left(\frac{\mathrm{e}^{2k}}{\cosh^2 \lambda - \mathrm{e}^{2k} \sinh^2 \lambda} - 1\right)\right] :$$
$$\exp\left[-\frac{(\mathrm{e}^{2k} - 1) \tanh \lambda}{2\left(1 - \mathrm{e}^{2k} \tanh^2 \lambda\right)} a^2\right] . \tag{8.48}$$

Taking $k = -\omega\hbar/(k_B T)$, we thus have

$$\left(\mathrm{e}^{-k}-1\right)^{-1}=\left(\mathrm{e}^{\frac{\omega\hbar}{k_B T}}-1\right)^{-1}\equiv\bar{n}, \tag{8.49}$$

and

$$\frac{\mathrm{e}^{k}-1}{\mathrm{e}^{k}+1}=-\frac{1}{2\bar{n}+1}. \tag{8.50}$$

Letting

$$(2\bar{n}+1)\,\mathrm{e}^{2\lambda}+1=2\tau_1^2,\quad (2\bar{n}+1)\,\mathrm{e}^{-2\lambda}+1=2\tau_2^2, \tag{8.51}$$

we therefore have

$$\begin{aligned}\tau_1^2\tau_2^2&=\frac{\cosh^2\lambda-\mathrm{e}^{2k}\sinh^2\lambda}{\left(1-\mathrm{e}^{k}\right)^2},\\ \tau_1^2-\tau_2^2&=(2\bar{n}+1)\sinh 2\lambda,\\ \tau_1^2+\tau_2^2&=(2\bar{n}+1)\cosh 2\lambda+1\end{aligned} \tag{8.52}$$

and

$$\begin{aligned}-\frac{1}{2\tau_1^2}-\frac{1}{2\tau_2^2}&=-1+\frac{\mathrm{e}^{k}}{\cosh^2\lambda-\mathrm{e}^{2k}\sinh^2\lambda},\\ \frac{\tau_1^2-\tau_2^2}{4\tau_1^2\tau_2^2}&=\frac{\left(1-\mathrm{e}^{2k}\right)\tanh\lambda}{2\left(1-\mathrm{e}^{2k}\tanh^2\lambda\right)}.\end{aligned} \tag{8.53}$$

Therefore, the squeezed chaotic light field in Eq. (8.46) can be rewritten as

$$\rho(0)=\left(1-\mathrm{e}^{k}\right)\sum_{n=0}^{\infty}\mathrm{e}^{kn}S^{-1}\left|n\right\rangle\left\langle n\right|S=\frac{1}{\tau_1\tau_2}\colon\exp\left(-\frac{Q^2}{2\tau_1^2}-\frac{P^2}{2\tau_2^2}\right)\colon, \tag{8.54}$$

where $Q=\left(a+a^{\dagger}\right)/\sqrt{2}$, $P=\left(a-a^{\dagger}\right)/(\sqrt{2}\mathrm{i})$. It can be seen from Eq. (8.54) that the density operator $\rho(0)$ of the squeezed chaotic light field is a two-dimensional normal distribution within the normal ordering. We can verify the correctness of the density operator $\rho(0)$ in Eq. (8.54) by calculating its normalization, that is

$$\begin{aligned}\mathtt{tr}\rho(0)&=\mathtt{tr}\left(\int\frac{\mathrm{d}^2z}{\pi}\left|z\right\rangle\left\langle z\right|\rho(0)\right)\\ &=\frac{1}{\tau_1\tau_2}\int\frac{\mathrm{d}\tau_1\mathrm{d}\tau_2}{2\pi}\left\langle z\right|\colon\exp\left(-\frac{P^2}{2\tau_1^2}-\frac{Q^2}{2\tau_2^2}\right)\colon\left|z\right\rangle=1.\end{aligned} \tag{8.55}$$

Next, we will discuss the photon number distribution of the state $\rho(0)$. Using the unnormalized coherent state $\left|\alpha\right\rangle=\exp\left(\alpha a^{\dagger}\right)\left|0\right\rangle$, and the relations $\left|n\right\rangle=\frac{1}{\sqrt{n!}}\frac{\mathrm{d}^n}{\mathrm{d}\alpha^n}\left|\alpha\right\rangle\big|_{\alpha=0}$ and $\left\langle\beta\right|\alpha\rangle=\mathrm{e}^{\alpha\beta^*}$, as well as the normal ordering of $\rho(0)$ in Eq. (8.48), the probability of finding m photons in this field is

$$\begin{aligned}
\mathcal{P}(m) &= \langle m|\, \rho\,(0)\, |m\rangle \\
&= \frac{2\sqrt{fg}}{m!} \frac{\mathrm{d}^{2m}}{\mathrm{d}\beta^{*m}\mathrm{d}\alpha^m} \langle\beta|:\exp\left[\frac{g\left(a-a^\dagger\right)^2 - f\left(a+a^\dagger\right)^2}{2}\right]:|\alpha\rangle|_{\alpha,\beta^*=0} \\
&= \frac{2\sqrt{fg}}{m!} \frac{\mathrm{d}^{2m}}{\mathrm{d}\beta^{*m}\mathrm{d}\alpha^m} \exp\left[A_1\alpha\beta^* + A_2\left(\alpha^2+\beta^{*2}\right)\right]\Big|_{\alpha,\beta^*=0}, \qquad (8.56)
\end{aligned}$$

where A_1 and A_2 are, respectively,

$$A_1 = \frac{\bar{n}\,(\bar{n}+1)}{\bar{n}^2 + (2\bar{n}+1)\cosh^2 r}, \quad A_2 = \frac{(2\bar{n}+1)\sinh 2r}{4\left(\bar{n}^2+(2\bar{n}+1)\cosh^2 r\right)}. \qquad (8.57)$$

Noting that

$$\begin{aligned}
&\frac{\partial^{2m}}{\partial t^m \partial\tau^m} \exp\left(-t^2-\tau^2+2x\tau t\right)\Big|_{t,\tau=0} \\
&= \sum_{n,l,k=0}^{\infty} \frac{(-1)^{n+l}}{n!l!k!} (2x)^k \frac{\partial^{2m}}{\partial t^m\partial\tau^m} \tau^{2n+k} t^{2l+k}\Bigg|_{t,\tau=0} \\
&= 2^m m! \sum_{n=0}^{[m/2]} \frac{m!}{2^{2n}\,(n!)^2\,(m-2n)!} x^{m-2n}, \qquad (8.58)
\end{aligned}$$

and the new form of the Legendre polynomial $\mathrm{P}_m\,(x)$ (equivalent to the form of the traditional Legendre polynomial)

$$x^m \sum_{l=0}^{[m/2]} \frac{m!}{2^{2l}\,(l!)^2\,(m-2l)!} \left(1-\frac{1}{x^2}\right)^l = \mathrm{P}_m\,(x)\,, \qquad (8.59)$$

thus Eq. (8.56) becomes

$$\begin{aligned}
\mathcal{P}(m) &= A_2^m \frac{2\sqrt{fg}}{m!} \frac{\mathrm{d}^{2m}}{\mathrm{d}\beta^{*m}\mathrm{d}\alpha^m} \exp\left(-\alpha^2-\beta^{*2}+\frac{A_1}{A_2}\alpha\beta^*\right)\Bigg|_{\alpha=\beta^*=0} \\
&= 2\sqrt{fg}E^{m/2}\mathrm{P}_m\left(A_1/\sqrt{E}\right), \qquad (8.60)
\end{aligned}$$

which is the photon number distribution of the squeezed chaotic light field, and

$$E = \frac{\bar{n}^2-(2\bar{n}+1)\sinh^2 r}{\bar{n}^2+(2\bar{n}+1)\cosh^2 r}. \qquad (8.61)$$

8.2.2 Decay of Squeezed Chaotic Field

Here, we want to investigate the evolution of the squeezed chaotic field in the amplitude decay channel [5]. For this, we first derive the anti-normal ordering product of the density operator $\rho(0)$. Inserting Eqs. (8.48) or (8.54) into the formula that converts the density operator into its anti-normal ordering in Eq. (2.76) and using the integral formula in Eq. (1.57), we can get the anti-normal ordering product of $\rho(0)$, that is

$$\begin{aligned}\rho(0) &= 2\sqrt{fg}\int\frac{d^2\beta}{\pi}\vdots\exp\Big[-|\beta|^2-\frac{f}{2}(\beta-\beta^*)^2\\&\quad+\frac{g}{2}(\beta+\beta^*)^2+\beta^*a-\beta a^\dagger+a^\dagger a\Big]\vdots\\&=2\sqrt{fg}\int\frac{d^2\beta}{\pi}\vdots\exp\Big[(f+g-1)|\beta|^2-a^\dagger\beta\\&\quad+a\beta^*+\frac{g-f}{2}(\beta^2+\beta^{*2})+a^\dagger a\Big]\vdots\\&=2\sqrt{\frac{fg}{D}}\vdots\exp\left[\frac{g-f}{2D}(a^{\dagger2}+a^2)+\frac{4fg-f-g}{D}a^\dagger a\right]\vdots,\end{aligned}\tag{8.62}$$

where $D=(2f-1)(2g-1)$. Using the operator identity $e^{\alpha a^\dagger a}=e^{-\alpha}\vdots\exp\left[(1-e^{-\alpha})aa^\dagger\right]\vdots$, we have

$$\begin{aligned}\vdots\exp\left(\frac{4fg-f-g}{D}a^\dagger a\right)\vdots&=\vdots\exp\left[\left(1-\frac{1-f-g}{D}\right)a^\dagger a\right]\vdots\\&=\frac{D}{1-f-g}\exp\left(a^\dagger a\ln\frac{D}{1-f-g}\right),\end{aligned}\tag{8.63}$$

thus Eq. (8.62) becomes

$$\rho(0)=\frac{2\sqrt{fgD}}{1-f-g}\exp\left(\frac{g-f}{2D}a^2\right)\exp\left(a^\dagger a\ln\frac{D}{1-f-g}\right)\exp\left(\frac{g-f}{2D}a^{\dagger2}\right).\tag{8.64}$$

In the amplitude decay channel, the relation between the time-dependent density operator and the initial density operator is given by Eq. (6.20). So, inserting Eq. (8.62) into Eq. (6.20) leads to

$$\begin{aligned}\rho(t)&=2\sqrt{\frac{fg}{D}}\sum_{n=0}^{\infty}\frac{T^n}{n!}e^{-\kappa ta^\dagger a}\vdots a^n\exp\left[\frac{g-f}{2D}(a^{\dagger2}+a^2)\right.\\&\quad\left.+\frac{4fg-f-g}{D}a^\dagger a\right]a^{\dagger n}\vdots e^{-\kappa ta^\dagger a}.\end{aligned}\tag{8.65}$$

Now, we derive the normal ordering product of $\rho(t)$. For this, noting that the coherent state $|\beta\rangle = \exp\left(-|\beta|^2/2 + \beta a^\dagger\right)|0\rangle$ yields

$$e^{-\kappa t a^\dagger a}|\beta\rangle = e^{-\frac{|\beta|^2}{2}} e^{\beta e^{-kt} a^\dagger}|0\rangle, \tag{8.66}$$

where we have used the operator identity

$$e^{-\beta a^\dagger a} f\left(a, a^\dagger\right) e^{\beta a^\dagger a} = f\left(a e^{\beta}, a^\dagger e^{-\beta}\right). \tag{8.67}$$

Further, using the P-representation in the coherent state representation and the completeness relation of coherent states, that is $\pi^{-1}\int d^2\beta\,|\beta\rangle\langle\beta| = 1$, as well as the normal ordering product $|0\rangle\langle 0| =: \exp\left(-a^\dagger a\right):$, we give

$$\begin{aligned}\rho(t) &= 2\sqrt{\frac{fg}{D}}\int\frac{d^2\beta}{\pi}\sum_{n=0}^{\infty}\frac{T^n}{n!}|\beta|^{2n} : \exp\left(\frac{4fg-f-g}{D}-1\right)|\beta|^2 \\ &\quad + e^{-kt}a^\dagger\beta + e^{-kt}a\beta^* + \frac{g-f}{2D}\left(\beta^2+\beta^{*2}\right) - a^\dagger a : \\ &= 2\sqrt{\frac{fg}{D}}\int\frac{d^2\beta}{\pi} : \exp\left[-\mu|\beta|^2 + e^{-kt}\left(a^\dagger\beta + a\beta^*\right)\right. \\ &\quad \left. + \frac{g-f}{2D}\left(\beta^2+\beta^{*2}\right) - a^\dagger a\right] : ,\end{aligned} \tag{8.68}$$

where $\mu = (1 - f - g - DT)/D$. Further, using the integral formula in Eq. (1.57), we can obtain the normal ordering product of $\rho(t)$ as

$$\rho(t) = 2\sqrt{\frac{fgD}{F_\mu}} : \exp\left\{\frac{D^2 e^{-2kt}}{F_\mu}\left[\mu a^\dagger a + E\left(a^2 + a^{\dagger 2}\right)\right] - a^\dagger a\right\} : , \tag{8.69}$$

where

$$F_\mu = \mu^2 D^2 - (g-f)^2, \quad E = (g-f)/2D. \tag{8.70}$$

For $t = 0$, $e^{-2\kappa t} = 1$, noting that

$$\begin{aligned}(4fg - f - g - D)^2 - (g-f)^2 &= D \\ D - (4fg - f - g) &= 1 - f - g,\end{aligned} \tag{8.71}$$

so $\rho(t) \to \rho(0) = 2\sqrt{fg} : \exp\left[\frac{1}{2}(g-f)\left(a^2 + a^{\dagger 2}\right) - (f+g)a^\dagger a\right] :$, as expected. To further confirm that Eq. (8.69) is indeed a density operator, we need to prove $\mathtt{tr}\rho(t) = 1$. Actually, using the completeness relation, i.e., $\pi^{-1}\int d^2\beta\,|\beta\rangle\langle\beta| = 1$ and Eq. (1.57), we have

$$
\begin{aligned}
\mathrm{tr}\rho(t) &= 2\sqrt{\frac{fgD}{F_\mu}}\int\frac{d^2\beta}{\pi}\langle\beta|:\exp\left\{\frac{D^2\mathrm{e}^{-2kt}}{F_\mu}\left[\mu a^\dagger a+E\left(a^2+a^{\dagger 2}\right)\right]-a^\dagger a\right\}:|\beta\rangle \\
&= 2\sqrt{\frac{fgD}{F_\mu}}\int\frac{d^2\beta}{\pi}\exp\left[\left(\frac{\mu D^2\mathrm{e}^{-2kt}-F_\mu}{F_\mu}\right)|\beta|^2+\frac{ED^2\mathrm{e}^{-2kt}}{F_\mu}\left(\beta^2+\beta^{*2}\right)\right] \\
&= 1.
\end{aligned} \tag{8.72}
$$

In fact, Eq. (8.69) can also be written in the following equivalent form, that is

$$
\begin{aligned}
\rho(t) = 2\sqrt{\frac{fgD}{F_\mu}}&\exp\left(\frac{ED^2\mathrm{e}^{-2kt}}{F_\mu}a^{\dagger 2}\right) \\
&\times\exp\left(a^\dagger a\ln\frac{\mu D^2\mathrm{e}^{-2kt}}{F_\mu}\right)\exp\left(\frac{ED^2\mathrm{e}^{-2kt}}{F_\mu}a^2\right),
\end{aligned} \tag{8.73}
$$

which is Gaussian quadratic form. Noting that $\rho_c(\lambda)=\left(1-\mathrm{e}^\lambda\right)\mathrm{e}^{\lambda a^\dagger a}$, so $\rho(t)$ can be viewed as a photon-added thermal state with the average photon number $n''=\left[F_\mu/\left(\mu D^2\mathrm{e}^{-2kt}\right)-1\right]^{-1}=1/(f_1+g_1)-1$.

On the other hand, by comparing Eq. (8.73) and the squeezed chaotic field in Eq. (3.97), or the following form

$$
\rho(0)=2\sqrt{fg}\exp\left[\frac{1}{2}(g-f)a^{\dagger 2}\right]\exp\left\{a^\dagger a\ln[1-(f+g)]\right\}\exp\left[\frac{1}{2}(g-f)a^2\right], \tag{8.74}
$$

we find that, when the squeezed chaotic light passes through the amplitude decay channel, the form of the density operator $\rho(t)$ is exactly the same as that of $\rho(0)$. This indicates that at any given time t, $\rho(t)$ can still be viewed as a squeezed chaotic field. Introducing the following parameters:

$$
g'-f'=2\frac{ED^2\mathrm{e}^{-2kt}}{F_\mu},\quad g'+f'=1-\frac{\mu D^2\mathrm{e}^{-2kt}}{F_\mu}, \tag{8.75}
$$

which leads to

$$
\begin{aligned}
g' &= \frac{1}{2}\left(1-\frac{\mu-2E}{F_\mu}D^2\mathrm{e}^{-2kt}\right)=\frac{g}{2g-(2g-1)\,\mathrm{e}^{-2\kappa t}}, \\
f' &= \frac{1}{2}\left(1-\frac{\mu+2E}{F_\mu}D^2\mathrm{e}^{-2kt}\right)=\frac{f}{2f-(2f-1)\,\mathrm{e}^{-2\kappa t}}.
\end{aligned} \tag{8.76}
$$

So Eq. (8.73) can be rewritten as

$$\begin{aligned}\rho(t) &= 2\sqrt{g'f'}\exp\left[\frac{1}{2}(g'-f')(a^{\dagger 2}-a^2)-(g'+f')a^\dagger a\right]: \\ &= 2\sqrt{g'f'}\exp\left[\frac{1}{2}(g'-f')a^{\dagger 2}\right] \\ &\quad\times\exp\left\{a^\dagger a\ln\left[1-(f'+g')\right]\right\}\exp\left[\frac{1}{2}(g'-f')a^{\dagger 2}\right].\end{aligned} \tag{8.77}$$

Further, using Eq. (8.74), Eq. (8.77) can be rewritten as a squeezed chaotic state, that is

$$\rho(t) = S(r')\rho_c' S^\dagger(r'), \tag{8.78}$$

where the average photon number of ρ_c' is

$$\begin{aligned} n(t) &= \frac{1}{2}(uv-1), \quad \lambda'(t) = \ln\frac{uv-1}{uv+1}, \quad r'(t) = \frac{1}{2}\ln\frac{u}{v}, \\ u &= \sqrt{\frac{1-f_1}{f_1}} = \sqrt{(2n'+1)\mathrm{e}^{-2\kappa t}\mathrm{e}^{2r}+T}, \\ v &= \sqrt{\frac{1-g_1}{g_1}} = \sqrt{(2n'+1)\mathrm{e}^{-2\kappa t}\mathrm{e}^{-2r}+T}, \end{aligned} \tag{8.79}$$

which shows that the average photon number and the squeezed parameters depend on the time-evolution factor $\mathrm{e}^{-2\kappa t}$. From Eqs. (8.78) and (8.79), it can be clearly seen that, when the squeezed chaotic state undergoes the amplitude decay channel, the output state can still be regarded as a squeezed chaotic state, and its average photon number $n(t)$ and the squeezing parameter $r'(t)$ are given by Eq. (8.79). Using Eqs. (8.78) and (8.79), it is convenient to calculate some distribution functions, such as photon number distribution, Q function, and Wigner function, etc.

Next, we continue to discuss the decay of the squeezed chaotic light field by virtue of the parameters g' and f'. Comparing the initial state $\rho(0)$ in Eq. (8.74) with the evolved state $\rho(t)$ for amplitude decay in Eq. (8.77), it can be concluded that the evolution law of the state $\rho(0)$ during amplitude decay process (photon number decay) can be fully determined by the time-evolution of two parameters, that is

$$g \to g', \qquad f \to f', \tag{8.80}$$

where g' and f' determined by Eq. (8.76) are two effective parameters that can be used to describe the evolution of the squeezed chaotic field in the amplitude decay channel. Obviously, when $t = 0$, Eq. (8.77) reduces to Eq. (8.74). If $t \to \infty$, thus $g' \to 1/2$, $f' \to 1/2$, Eq. (8.77) becomes $\rho(t) \to: \exp(-a^\dagger a): = |0\rangle\langle 0|$, which shows that the state $\rho(0)$ ultimately evolves into vacuum.

(i) Time invariant in decay. Noting that

$$g' - f' = \frac{g}{2g - (2g-1)\,\mathrm{e}^{-2\kappa t}} - \frac{f}{2f - (2f-1)\,\mathrm{e}^{-2\kappa t}}$$
$$= \frac{(g-f)\,\mathrm{e}^{-2\kappa t}}{\left[2g - (2g-1)\,\mathrm{e}^{-2\kappa t}\right]\left[2f - (2f-1)\,\mathrm{e}^{-2\kappa t}\right]}, \tag{8.81}$$

and

$$g' + f' - 4g'f' = \frac{(g + f - 4fg)\,\mathrm{e}^{-2\kappa t}}{\left[2g - (2g-1)\,\mathrm{e}^{-2\kappa t}\right]\left[2f - (2f-1)\,\mathrm{e}^{-2\kappa t}\right]}, \tag{8.82}$$

thus we have

$$\frac{g' - f'}{g' + f' - 4g'f'} = \frac{g - f}{g + f - 4fg}, \tag{8.83}$$

which means that the factor $(g - f)/(g + f - 4fg)$ cannot change with time t during the decay process, and this quantity will appear in the calculation of second-order coherence.

(ii) Decay of the average photon number of the state $\rho(0)$. Using the completeness relation of coherent states, that is

$$\int \frac{\mathrm{d}^2 z}{\pi} |z\rangle\langle z| = \int \frac{\mathrm{d}^2 z}{\pi} \colon \exp\left(-|z|^2 + za^\dagger + z^* a - a^\dagger a\right) \colon = 1, \tag{8.84}$$

the average photon number of the initial state $\rho(0)$ reads

$$\langle N\rangle_0 \equiv \mathrm{tr}\left(\rho(0)\, a^\dagger a\right) = \mathrm{tr}\left(\rho(0)\, aa^\dagger\right) - 1$$
$$= 2\sqrt{gf}\,\mathrm{tr}\left\{ \colon a^\dagger a \exp\left[\frac{1}{2}(g-f)\left(a^{\dagger 2} + a^2\right) - (g+f)\, a^\dagger a\right] \colon \right\} - 1$$
$$= 2\sqrt{gf} \int \frac{\mathrm{d}^2 z}{\pi} |z|^2 \exp\left[\frac{1}{2}(g-f)\left(z^{*2} + z^2\right) - (g+f)\,|z|^2\right] - 1$$
$$= \frac{g+f}{4gf} - 1. \tag{8.85}$$

Clearly, using Eqs. (8.77) and (8.85), we immediately obtain the evolved average photon number at arbitrary time t in the amplitude decay channel, that is

$$\langle N\rangle_t = \mathrm{tr}\left(\rho(t)\, a^\dagger a\right) = \frac{g' + f'}{4g'f'} - 1$$
$$= \frac{g(1-2f) + f(1-2g)}{4gf}\mathrm{e}^{-2\kappa t}$$
$$= \left(\frac{g+f}{4gf} - 1\right)\mathrm{e}^{-2\kappa t}$$
$$= \langle N\rangle_0\, \mathrm{e}^{-2\kappa t}, \tag{8.86}$$

that is, the average photon number $\langle N\rangle_t$ decays exponentially with time t.

(iii) Time evolution of the quantum fluctuation. Using Eqs. (8.77) and (8.85), we immediately write the anti-normal ordering product of the density operator $\rho(t)$, i.e.,

$$\rho(t) = 2\sqrt{\frac{g'f'}{D'}}\vdots \exp\left[\frac{g'-f'}{2D'}\left(a^{\dagger 2}+a^2\right)+\frac{4g'f'-g'-f'}{D'}a^{\dagger}a\right]\vdots \tag{8.87}$$

where $D' = \left(2f'-1\right)\left(2g'-1\right)$.

Letting

$$u = \frac{g'-f'}{2D'},\quad v = \frac{4g'f'-g'-f'}{D'}, \tag{8.88}$$

thus

$$2\sqrt{\frac{g'f'}{D'}} = \sqrt{v^2-4u^2}, \tag{8.89}$$

so we have

$$\rho(t) = \sqrt{v^2-4u^2}\vdots \exp\left[u\left(a^{\dagger 2}+a^2\right)+va^{\dagger}a\right]\vdots. \tag{8.90}$$

To investigate the quantum fluctuation of the state $\rho(t)$, we first calculate $\langle N^2\rangle_t$ as

$$\langle N^2\rangle_t = \mathtt{tr}\left(\rho(t)a^{\dagger}aa^{\dagger}a\right) = \mathtt{tr}\left(\rho(t)a^{\dagger 2}a^2\right)+\langle N\rangle_t. \tag{8.91}$$

Using Eq. (8.90) and the completeness relation of coherent states, we have

$$\begin{aligned}
\mathtt{tr}\left(\rho(t)a^{\dagger 2}a^2\right) &= \sqrt{v^2-4u^2}\,\mathtt{tr}\left\{a^2\vdots \exp\left[u\left(a^{\dagger 2}+a^2\right)+va^{\dagger}a\right]\vdots a^{\dagger 2}\right\}\\
&= \sqrt{v^2-4u^2}\,\mathtt{tr}\left\{\int\frac{d^2z}{\pi}a^2\vdots \exp\left[u\left(a^{\dagger 2}+a^2\right)+va^{\dagger}a\right]\vdots |z\rangle\langle z|\,a^{\dagger 2}\right\}\\
&= \sqrt{v^2-4u^2}\,\mathtt{tr}\left\{\int\frac{d^2z}{\pi}|z|^4\exp\left[u\left(z^{*2}+z^2\right)+v|z|^2\right]|z\rangle\langle z|\right\}\\
&= \sqrt{v^2-4u^2}\,\mathtt{tr}\left[\int\frac{d^2z}{\pi}:\ |z|^4\exp\left(v-1\right)|z|^2+za^{\dagger}+z^{*}a\right.\\
&\quad\left.+u\left(z^{*2}+z^2\right)-a^{\dagger}a\right]:\\
&= \frac{2v^2+4u^2}{\left(v^2-4u^2\right)^2} = \frac{2\left(4g'f'-g'-f'\right)^2+\left(g'-f'\right)^2}{16\left(g'f'\right)^2},
\end{aligned} \tag{8.92}$$

it follows from Eq. (8.92) that

$$\begin{aligned}
\langle N^2\rangle_t &= \mathtt{tr}\left(\rho(t)a^{\dagger 2}a^2\right)+\langle N\rangle_t\\
&= \frac{2\left(4g'f'-g'-f'\right)^2+\left(g'-f'\right)^2}{16\left(g'f'\right)^2}+\frac{g'+f'}{4g'f'}-1,
\end{aligned} \tag{8.93}$$

and

$$\langle N^2 \rangle_t - \langle N \rangle_t^2 = \frac{g'^2 + f'^2 - 2g'f'\left(g' + f'\right)}{8\left(g'f'\right)^2}. \tag{8.94}$$

Therefore, the relative fluctuation of photon number is

$$\frac{\Delta N_t}{\langle N \rangle_t} = \frac{\sqrt{\langle N^2 \rangle_t - \langle N \rangle_t^2}}{\langle N \rangle_t} = \frac{\sqrt{2\left[g'^2 + f'^2 - 2g'f'\left(g' + f'\right)\right]}}{g' + f' - 4g'f'}. \tag{8.95}$$

Substituting Eq. (8.76) into Eq. (8.95) leads to

$$\frac{\Delta N_t}{\langle N \rangle_t} = \frac{\sqrt{2}\sqrt{g^2\left(2f-1\right)^2 + f^2\left(2g-1\right)^2 + 2gf\left(g + f - 4gf\right)e^{2\kappa t}}}{g + f - 4gf}, \tag{8.96}$$

which means that the relative fluctuation of photon number almost increases with $e^{\kappa t}$, corresponding to the rapid decay.

The second-order coherence reads

$$G^{(2)} \equiv \frac{\langle N^2 \rangle_t - \langle N \rangle_t}{\langle N \rangle_t^2} = 2 + \frac{\left(g' - f'\right)^2}{\left(g' + f' - 4g'f'\right)^2} = 2 + \frac{\left(g - f\right)^2}{\left(g + f - 4gf\right)^2}, \tag{8.97}$$

which shows that the photon statistical distribution is a super-Poisson distribution and the light field with the density operator $\rho\left(t\right)$ exhibits photon bunching effect.

To sum up, we found two parameters g and f in the decay process of the squeezed chaotic light, and they can well reflect the essence of decay. The result is that the squeezed chaotic light evolves into a new squeezed chaotic light only by changing the two parameters g and f into g' and f', respectively. Acording to this evolution law, we calculated the quantum fluctuation and found that $\left(g - f\right)/\left(g + f - 4gf\right)$ closely related to the second-order coherence cannot change with time t, i.e., the time invariance.

8.3 Evolution of Degenerate Parametric Amplifier for Amplitude Decay

8.3.1 Quantum State $\rho\left(0\right)$ Describing Degenerate Parametric Amplifier

The Hamiltonian corresponding to the degenerate parametric amplification process is [6]

$$H = \omega a^\dagger a + n^* a^{\dagger 2} + n a^2, \tag{8.98}$$

where n is the dielectric constant, its corresponding normalized density operator reads

$$\rho(0) = \frac{e^{-\beta H}}{\mathtt{tr}e^{-\beta H}} = \frac{1}{\mathtt{tr}e^{-\beta H}} \exp[-\beta(\omega a^\dagger a + n^* a^{\dagger 2} + n a^2)], \tag{8.99}$$

where $\beta = 1/kT$, k is Boltzmann constant. Noting that the operators $\frac{1}{2}\left(a^\dagger a + \frac{1}{2}\right)$, $\frac{1}{2}a^{\dagger 2}$ and $\frac{1}{2}a^2$ yield SU(1,1) Lie Algebra, we obtain

$$\begin{aligned}
&\exp\left(f a^\dagger a + g a^{\dagger 2} + k a^2\right) \\
&= e^{-f/2} \exp\left(\frac{g a^{\dagger 2}}{\mathcal{D}\coth\mathcal{D} - f}\right) \\
&\times \exp\left\{\left(a^\dagger a + \frac{1}{2}\right) \ln \frac{\mathcal{D}\mathrm{sech}\mathcal{D}}{\mathcal{D} - f\tanh\mathcal{D}}\right\} \exp\left(\frac{k a^2}{\mathcal{D}\coth\mathcal{D} - f}\right),
\end{aligned} \tag{8.100}$$

which is also obtained directly from the simple operator algebra without using Lie algebra, $\mathcal{D}^2 = f^2 - 4kg$. Using Eq. (8.100), the exponent on the right side of Eq. (8.99) can be decomposed into

$$e^{-\beta H} = \rho(0)\mathtt{tr}e^{-\beta H} = \sqrt{\lambda e^{\beta\omega}} \exp(E^* a^{\dagger 2}) \exp(a^\dagger a \ln\lambda) \exp(E a^2), \tag{8.101}$$

where

$$\begin{aligned}
\lambda &= \frac{D}{\omega\sinh(\beta D) + D\cosh(\beta D)}, \\
E &= \frac{-\lambda n}{D}\sinh(\beta D), \\
D^2 &= \omega^2 - 4|n|^2.
\end{aligned} \tag{8.102}$$

On the other hand, using the operator identity $\exp(a^\dagger a \ln\lambda) =: \exp[(\lambda - 1)a^\dagger a]:$, we can calculate its partition function as

$$Z(\beta) = \mathtt{tr}e^{-\beta H} = \sqrt{\lambda e^{\beta\omega}}\mathtt{tr}\left\{\exp(E^* a^{\dagger 2})\colon \exp[(\lambda - 1)a^\dagger a]\colon \exp(E a^2)\right\}. \tag{8.103}$$

Using the completeness relation of coherent states in Eq. (8.84) and the integral formula in Eq. (1.57), we obtain

$$\begin{aligned}
Z(\beta) &= \sqrt{\lambda e^{\beta\omega}}\mathtt{tr}\left\{\exp(E^* a^{\dagger 2})\colon \exp[(\lambda - 1)a^\dagger a]\colon \exp(E a^2) \int \frac{d^2 z}{\pi}|z\rangle\langle z|\right\} \\
&= \sqrt{\lambda e^{\beta\omega}} \int \frac{d^2 z}{\pi}\langle z|\exp(E^* a^{\dagger 2})\colon \exp[(\lambda - 1)a^\dagger a]\colon \exp(E a^2)|z\rangle \\
&= \sqrt{\lambda e^{\beta\omega}} \int \frac{d^2 z}{\pi} \exp[E^* z^{*2} + E z^2 + (\lambda - 1)|z|^2]
\end{aligned}$$

$$= \frac{\sqrt{\lambda e^{\beta\omega}}}{\sqrt{(1-\lambda)^2 - 4|E|^2}} = \frac{e^{\beta\omega/2}}{2\sinh(\beta D/2)}, \tag{8.104}$$

where

$$\sqrt{(1-\lambda)^2 - 4|E|^2} = 2\sqrt{\lambda}\sinh(\beta D/2) \equiv A. \tag{8.105}$$

The proof of Eq. (8.105) is as follows:

$$\begin{aligned}
\sqrt{(1-\lambda)^2 - 4|E|^2} &= \sqrt{(1-\lambda)^2 - \frac{\lambda^2}{D^2}(\omega^2 - D^2)\sinh^2(\beta D)} \\
&= \sqrt{\lambda^2[\frac{D^2\cosh^2(\beta D) - \omega^2\sinh^2(\beta D)}{D^2}] - 2\lambda + 1} \\
&= \sqrt{\frac{D\cosh(\beta D) - \omega\sinh(\beta D)}{\omega\sinh(\beta D) + D\cosh(\beta D)} - \frac{2D}{\omega\sinh(\beta D) + D\cosh(\beta D)} + 1} \\
&= \sqrt{\frac{2D[\cosh(\beta D) - 1]}{\omega\sinh(\beta D) + D\cosh(\beta D)}} \\
&= 2\sqrt{\lambda}\sinh(\beta D/2).
\end{aligned} \tag{8.106}$$

Finally, we have

$$\begin{aligned}
\rho(0) &= \frac{\sqrt{\lambda e^{\beta\omega}}\exp(E^* a^{\dagger 2})\exp(a^\dagger a \ln\lambda)\exp(E a^2)}{\frac{\beta\omega/2}{2\sinh(\beta D/2)}} \\
&= A\exp(E^* a^{\dagger 2})\exp(a^\dagger a \ln\lambda)\exp(E a^2),
\end{aligned} \tag{8.107}$$

which yields $\mathtt{tr}\rho(0) = 1$.

8.3.2 Thermal Vacuum State Corresponding to the State ρ (0)

Using Eqs. (3.5) and (8.101), we have [6]

$$\begin{aligned}
\rho(0)\,\mathtt{tr}e^{-\beta H} &= \sqrt{\lambda e^{\beta\omega}}\exp\left(E^* a^{\dagger 2}\right) : \exp\left[(\lambda - 1)\,a^\dagger a\right] : \exp\left(E a^2\right) \\
&= \sqrt{\lambda e^{\beta\omega}}\int \frac{d^2 z}{\pi} e^{E^* a^{\dagger 2} + \sqrt{\lambda} z^* a^\dagger} |0\rangle\langle 0| e^{E a^2 + \sqrt{\lambda} z a} \langle \tilde{z} \left|\tilde{0}\right\rangle \langle \tilde{0}\,|\tilde{z}\rangle \\
&= \sqrt{\lambda e^{\beta\omega}}\int \frac{d^2 z}{\pi} \langle \tilde{z}| e^{E^* a^{\dagger 2} + \sqrt{\lambda} z^* a^\dagger} \left|0\tilde{0}\right\rangle \langle 0\tilde{0} e^{E a^2 + \sqrt{\lambda} z a} |\tilde{z}\rangle \\
&= \sqrt{\lambda e^{\beta\omega}}\,\widetilde{\mathtt{tr}}\left(e^{E^* a^{\dagger 2} + \sqrt{\lambda} a^\dagger \tilde{a}^\dagger} \left|0\tilde{0}\right\rangle \langle 0\tilde{0} e^{E a^2 + \sqrt{\lambda} a\tilde{a}}\right) \\
&\equiv \left(\mathtt{tr}e^{-\beta H}\right)\widetilde{\mathtt{tr}}\left(|\phi(\beta)\rangle\langle\phi(\beta)|\right),
\end{aligned} \tag{8.108}$$

which shows that, the quantum state resulting from the Hamiltonian in Eq. (8.98) can be written in pure state form in the expanded Fock space, i.e.,

$$|\phi(\beta)\rangle = \sqrt{\frac{\lambda^{1/2}e^{\beta\omega/2}}{Z(\beta)}} e^{E^* a^{\dagger 2} + \sqrt{\lambda} a^\dagger \tilde{a}^\dagger} \left|0\tilde{0}\right\rangle, \tag{8.109}$$

where $Z(\beta)$ is the partition function, so the normalized quantum state $|\phi(\beta)\rangle$ is

$$|\phi(\beta)\rangle = \sqrt{2\lambda^{1/2}\sinh(\beta D/2)} e^{E^* a^{\dagger 2} + \sqrt{\lambda} a^\dagger \tilde{a}^\dagger} \left|0\tilde{0}\right\rangle. \tag{8.110}$$

Also, the internal energy of the system is

$$\langle H\rangle_e = -\frac{\partial}{\partial\beta}\ln Z(\beta) = \frac{D\coth(\beta D/2) - \omega}{2}, \tag{8.111}$$

and the quanutum entropy of the system is

$$\begin{aligned} S &= -k\mathtt{tr}(\rho\ln\rho) = \frac{1}{T}\langle H\rangle_e + k\ln Z(\beta) \\ &= \frac{D}{2T}\coth(\beta D/2) - k\ln[2\sinh(\beta D/2)]. \end{aligned} \tag{8.112}$$

Especially, for $k = 0$, $D = \omega$, Eq. (8.110) reduces to $|0(\beta)\rangle$ ($\omega \to \hbar\omega$), thus Eqs. (8.111) and (8.112) respectively become $\frac{\omega}{2}\left[\coth\left(\frac{\beta\omega}{2}\right) - 1\right]$ and $\frac{\omega}{2T}\coth\left(\frac{\beta\omega}{2}\right)$, which are the expected results. It can be seen that using IWOP method, we can use the partial tracing method to transform the density operator of the mixed state into a pure state that is convenient for calculation.

As an application of Eq. (8.110), let's calculate the contributions of each term in the Hamiltonian to the internal energy of the system. As mentioned earlier, the average value of the system operator A can be replaced by the average value of A in the extended pure state, i.e., $\langle A\rangle_e = \langle\phi(\beta)|A|\phi(\beta)\rangle$. Using the completeness relation of coherent states, and the inner product $\langle\phi(\beta)|\phi(\beta)\rangle = 1$, $(1-\lambda)^2 - 4|E|^2 = 4\lambda\sinh^2\left(\frac{\beta D}{2}\right)$, thus we have

$$\begin{aligned} \left\langle\omega a^\dagger a\right\rangle_e &= \omega\langle\phi(\beta)|\left(aa^\dagger - 1\right)|\phi(\beta)\rangle \\ &= 2\omega\lambda^{1/2}\sinh\left(\frac{\beta D}{2}\right)\frac{\partial}{\partial\lambda}\int\frac{d^2z}{\pi}e^{-(1-\lambda)|z|^2 + Ez^2 + E^*z^{*2}} - \omega \\ &= 2\omega\lambda^{1/2}\sinh\left(\frac{\beta D}{2}\right)\frac{\partial}{\partial\lambda}\frac{1}{\sqrt{(1-\lambda)^2 - 4|E|^2}} - \omega \\ &= \frac{\omega}{2}\left[\frac{\omega}{D}\coth\left(\frac{\beta D}{2}\right) - 1\right], \end{aligned} \tag{8.113}$$

and

$$\begin{aligned}\left\langle \kappa^* a^{\dagger 2} \right\rangle_e &= 2\kappa^* \lambda^{1/2} \sinh\left(\frac{\beta D}{2}\right) \frac{\partial}{\partial E^*} \frac{1}{\sqrt{(1-\lambda)^2 - 4EE^*}} \\ &= -\frac{|\kappa|^2}{D} \coth\left(\frac{\beta D}{2}\right),\end{aligned} \tag{8.114}$$

as well as

$$\left\langle \kappa a^2 \right\rangle_e = -\frac{|\kappa|^2}{D} \coth\left(\frac{\beta D}{2}\right). \tag{8.115}$$

According to Eqs. (8.114) and (8.115), we find that $\kappa^* a^{\dagger 2}$ and κa^2 contribute equally to the system energy. In terms of Eqs. (8.113) and (8.115), we can verify Eq. (8.111).

8.3.3 Decay of the State $\rho(0)$

Substituting Eq. (8.107) into Eq. (6.20) leads to

$$\rho(t) = A \sum_{n=0}^{+\infty} \frac{T^n}{n!} e^{-\kappa t a^\dagger a} a^n \exp(E^* a^{\dagger 2}) \exp(a^\dagger a \ln \lambda) \exp(E a^2) a^{\dagger n} e^{-\kappa t a^\dagger a}. \tag{8.116}$$

According to the formulas

$$\begin{aligned}\exp\left(a^\dagger a \ln B\right) f\left(a^\dagger\right) \exp\left(-a^\dagger a \ln B\right) &= f\left(B a^\dagger\right), \\ \exp\left(a^\dagger a \ln B\right) f\left(a\right) \exp\left(-a^\dagger a \ln B\right) &= f\left(a/B\right),\end{aligned} \tag{8.117}$$

thus $\rho(t)$ becomes

$$\rho(t) = A \sum_{n=0}^{+\infty} \frac{T^n e^{2\kappa t n}}{n!} a^n \exp\left(E^* e^{-2\kappa t} a^{\dagger 2}\right) \exp[(\ln\lambda - 2\kappa t) a^\dagger a] \exp\left(E e^{-2\kappa t} a^2\right) a^{\dagger n}, \tag{8.118}$$

where

$$\begin{aligned}\exp[(\ln\lambda - 2\kappa t) a^\dagger a] &= :\exp\left[\left(e^{\ln\lambda - 2\kappa t} - 1\right) a^\dagger a\right]: \\ &= :\exp\left[\left(\lambda e^{-2\kappa t} - 1\right) a^\dagger a\right]:.\end{aligned} \tag{8.119}$$

Due to the presence of the part $\exp(E^* e^{-2\kappa t} a^{\dagger 2}) \exp[(\ln\lambda - 2\kappa t) a^\dagger a] \exp(E e^{-2\kappa t} a^2)$ between a^n and $a^{\dagger n}$, we encounter difficulties when summing n, and the way to solve this problem is to convert it into the anti-normal ordering form [6]. For this, we use the following formula

$$\rho = \int \frac{d^2\beta}{\pi} \vdots \langle -\beta|\rho|\beta\rangle \exp\left(|\beta|^2 + \beta^* a - \beta a^\dagger + a^\dagger a\right) \vdots \tag{8.120}$$

to calculate

$$\begin{aligned}
&\exp(E^* e^{-2\kappa t} a^{\dagger 2}) \exp[(\ln\lambda - 2\kappa t) a^\dagger a] \exp(E e^{-2\kappa t} a^2) \\
&= \int \frac{d^2\beta}{\pi} \vdots \langle -\beta| \exp(E^* e^{-2\kappa t} a^{\dagger 2}) \colon \exp\left[\left(\lambda e^{-2\kappa t} - 1\right) a^\dagger a\right] \colon \\
&\times \exp(E e^{-2\kappa t} a^2)|\beta\rangle \exp\left(|\beta|^2 + \beta^* a - \beta a^\dagger + a^\dagger a\right) \vdots \\
&= \int \frac{d^2\beta}{\pi} \vdots \exp\left(-\lambda e^{-2\kappa t}|\beta|^2 + \beta^* a - \beta a^\dagger + E e^{-2\kappa t}\beta^2 + E^* e^{-2\kappa t}\beta^{*2} + a^\dagger a\right) \vdots \\
&= \frac{e^{2\kappa t}}{\sqrt{\lambda^2 - 4|E|^2}} \vdots \exp\left(\frac{-\lambda a a^\dagger + E a^2 + E^* a^{\dagger 2}}{\lambda^2 - 4|E|^2} e^{2\kappa t} + a^\dagger a\right) \vdots .
\end{aligned} \tag{8.121}$$

Obviously, after inserting Eq. (8.121) into Eq. (8.118), we can find that Eq. (8.118) is arranged in an anti-normal ordering manner. Since the operators a and $a^\dagger$ are commutative in the anti-normal ordering symbol $\vdots\ \vdots$, we can sum n within $\vdots\ \vdots$, and obtain the anti-normal ordering product of the density operator $\rho(t)$, that is

$$\begin{aligned}
\rho(t) &= \frac{A e^{2\kappa t}}{\sqrt{\lambda^2 - 4|E|^2}} \vdots \sum_{n=0}^{+\infty} \frac{(T e^{2\kappa t} a a^\dagger)^n}{n!} \exp\left(\frac{-\lambda a a^\dagger + E a^2 + E^* a^{\dagger 2}}{\lambda^2 - 4|E|^2} e^{2\kappa t} + a^\dagger a\right) \vdots \\
&= \frac{A e^{2\kappa t}}{\sqrt{\lambda^2 - 4|E|^2}} \vdots \exp(e^{2\kappa t} a a^\dagger) \exp\left(\frac{-\lambda a a^\dagger + E a^2 + E^* a^{\dagger 2}}{\lambda^2 - 4|E|^2} e^{2\kappa t}\right) \vdots \\
&= A' \vdots \exp\left(e^{2\kappa t} \frac{(\lambda^2 - 4|E|^2 - \lambda) a a^\dagger + E a^2 + E^* a^{\dagger 2}}{\lambda^2 - 4|E|^2}\right) \vdots ,
\end{aligned} \tag{8.122}$$

where $A' = A e^{2\kappa t}/\sqrt{\lambda^2 - 4|E|^2}$ and $T = 1 - e^{-2\kappa t}$. Using Eq. (8.122) and the P-representation formula in the coherent state representation

$$\rho(t) = \int \frac{d^2 z}{\pi} |z\rangle\langle z| \mathrm{P}_t(z), \tag{8.123}$$

we can calculate the trace of $\rho(t)$ as

$$\begin{aligned}
\mathtt{tr}\rho(t) &= \int \frac{d^2 z}{\pi} \langle z| \mathrm{P}_t(z) |z\rangle \\
&= \int \frac{d^2 z}{\pi} \mathrm{P}_t(z) \\
&= A' \int \frac{d^2 z}{\pi} \exp\left[e^{2\kappa t} \frac{(\lambda^2 - 4|E|^2 - \lambda)|z|^2 + E z^2 + E^* z^{*2}}{\lambda^2 - 4|E|^2}\right]
\end{aligned}$$

$$
\begin{aligned}
&= \frac{A\sqrt{\lambda^2 - 4|E|^2}}{\sqrt{(\lambda^2 - 4|E|^2 - \lambda)^2 - 4|E|^2}} \\
&= \frac{A}{\sqrt{(\lambda - 1)^2 - 4|E|^2}} = 1,
\end{aligned}
\tag{8.124}
$$

which is consistent with expected result.

Now, using the anti-normal ordering product of $\rho(t)$ in Eq. (8.122) and its P-representation in Eq. (8.123) to calculate the energy of the system at time t, we finally obtain

$$
\begin{aligned}
\mathtt{tr}\left(\rho(t)a^\dagger a\right) &= \mathtt{tr}\left(\int \frac{d^2 z}{\pi} \mathrm{P}_t\,(z)\,|z\rangle\langle z|a^\dagger a\right) = \int \frac{d^2 z}{\pi} \mathrm{P}_t\,(z)\,|z|^2 \\
&= A' \int \frac{d^2 z}{\pi} |z|^2 \exp\left(e^{2\kappa t}|z|^2 + \frac{Ez^2 + E^* z^{*2} - \lambda|z|^2}{\lambda^2 - 4|E|^2} e^{2\kappa t}\right) \\
&= A' e^{-4\kappa t} \int \frac{d^2 z}{\pi} |z|^2 \exp\left(|z|^2 + \frac{Ez^2 + E^* z^{*2} - \lambda|z|^2}{\lambda^2 - 4|E|^2}\right) \\
&= A' e^{-4\kappa t} \frac{\partial}{\partial f} \int \frac{d^2 z}{\pi} \exp\left(f|z|^2 + \frac{Ez^2 + E^* z^{*2} - \lambda|z|^2}{\lambda^2 - 4|E|^2}\right)\Bigg|_{f=1} \\
&= A' e^{-4\kappa t} \frac{\partial}{\partial f} \int \frac{d^2 z}{\pi} \exp\left\{\frac{Ez^2 + E^* z^{*2} + \left[f\left(\lambda^2 - 4|E|^2\right) - \lambda\right]|z|^2}{\lambda^2 - 4|E|^2}\right\}\Bigg|_{f=1} \\
&= A' e^{-4\kappa t} \frac{\partial}{\partial f} \frac{\lambda^2 - 4|E|^2}{\sqrt{\left[f\left(\lambda^2 - 4|E|^2\right) - \lambda\right]^2 - 4|E|^2}}\Bigg|_{f=1} \\
&= A' e^{-4\kappa t} \frac{\sqrt{\lambda^2 - 4|E|^2}[4|E|^2 - \lambda(\lambda - 1)]}{[(\lambda - 1)^2 - 4|E|^2]^{3/2}} \\
&= e^{-2\kappa t} \frac{A[4|E|^2 - \lambda(\lambda - 1)]}{[(\lambda - 1)^2 - 4|E|^2]^{3/2}} \\
&= e^{-2\kappa t} \frac{4|E|^2 - \lambda(\lambda - 1)}{(\lambda - 1)^2 - 4|E|^2}.
\end{aligned}
\tag{8.125}
$$

In the last step, we have used the identity $A \equiv \sqrt{(1 - \lambda)^2 - 4|E|^2} = 2\sqrt{\lambda} \sinh\left(\frac{\beta D}{2}\right)$.

Noting that

$$
\begin{aligned}
e^{-2\kappa t} \frac{4|E|^2 - \lambda(\lambda - 1)}{(\lambda - 1)^2 - 4|E|^2} &= e^{-2\kappa t} \left[\frac{1 - \lambda}{(\lambda - 1)^2 - 4|E|^2} - 1\right] \\
&= e^{-2\kappa t} \left[\frac{1 - \lambda}{4\lambda \sinh^2(\beta D/2)} - 1\right]
\end{aligned}
\tag{8.126}
$$

and inserting $\lambda = D/\left[\omega \sinh(\beta D) + D\cosh(\beta D)\right]$ into Eq. (8.126), we get

$$
\mathtt{tr}\left(\rho(t)a^\dagger a\right) = \frac{e^{-2\kappa t}}{2}\left[\frac{\omega}{D}\coth(\beta D/2) - 1\right]. \tag{8.127}
$$

Since

$$\mathtt{tr}\left(\rho(0)a^{\dagger}a\right)=\frac{\omega}{2D}\coth\frac{\beta D}{2}-\frac{1}{2}, \tag{8.128}$$

we obtain

$$\mathtt{tr}\left(\rho(t)a^{\dagger}a\right)=\mathrm{e}^{-2\kappa t}\mathtt{tr}\left(\rho(0)a^{\dagger}a\right)=\mathrm{e}^{-Rt/(2L)}\mathtt{tr}\left(\rho(0)a^{\dagger}a\right), \tag{8.129}$$

which shows that the energy decays with time t according to the exponent $\mathrm{e}^{-Rt/(2L)}$.

8.4 Evolutions of Two-Mode Quantum States for Single-Mode Amplitude Decay

8.4.1 Decay of Two-Mode Squeezed Vacuum State

Now, we investigate the decay of a two-mode squeezed vacuum state in the single-mode amplitude decay channel. We point out that when the signal mode or idle mode of a two-mode squeezed vacuum state (generated by a parametric amplifier) undergoes a single-mode amplitude decay channel (denoted as $a_2^{\dagger}$ mode) described by the following master equation [8, 9]:

$$\frac{\mathrm{d}\rho(t)}{\mathrm{d}t}=\kappa\left(2a_2\rho(t)a_2^{\dagger}-a_2^{\dagger}a_2\rho(t)-\rho(t)a_2^{\dagger}a_2\right), \tag{8.130}$$

κ is the decay constant, thus the two-mode squeezed vacuum state will evolve into a definite form, and let's derive this form as follows. The evolved density operator $\rho(t)$ in the amplitude decay channel yields the following identity:

$$\rho(t)=\sum_{n=0}^{\infty}\frac{T^n}{n!}\mathrm{e}^{-\kappa t a_2^{\dagger}a_2}a_2^n\rho(0)a_2^{\dagger n}\mathrm{e}^{-\kappa t a_2^{\dagger}a_2},\quad T=1-\mathrm{e}^{-2\kappa t}. \tag{8.131}$$

Substituting a two-mode squeezed vacuum state

$$\rho(0)=\operatorname{sech}^2\lambda\mathrm{e}^{a_1^{\dagger}a_2^{\dagger}\tanh\lambda}|00\rangle\langle 00|\,\mathrm{e}^{a_1a_2\tanh\lambda} \tag{8.132}$$

into Eq. (8.131) leads to

$$\begin{aligned}\rho(t)&=\operatorname{sech}^2\lambda\sum_{n=0}^{\infty}\frac{T^n}{n!}\mathrm{e}^{-\kappa t a_2^{\dagger}a_2}a_2^n\mathrm{e}^{a_1^{\dagger}a_2^{\dagger}\tanh\lambda}\\&\quad|00\rangle\langle 00|\,\mathrm{e}^{a_1a_2\tanh\lambda}a_2^{\dagger n}\mathrm{e}^{-\kappa t a_1^{\dagger}a_1}\\&=\operatorname{sech}^2\lambda\sum_{n=0}^{\infty}\frac{T^n\tanh^{2n}\lambda}{n!}a_1^{\dagger n}\mathrm{e}^{\mathrm{e}^{-\kappa t}a_1^{\dagger}a_2^{\dagger}\tanh\lambda}\end{aligned}$$

$$|00\rangle\langle 00|\, e^{e^{-\kappa t}a_1 a_2 \tanh\lambda} a_1^n. \tag{8.133}$$

Using the normal ordering product of vacuum projector, that is

$$|00\rangle\langle 00| =: e^{-a_1^\dagger a_1 - a_2^\dagger a_2}:,\ |0\rangle_{11}\langle 0| =: e^{-a_1^\dagger a_1}:,\ |0\rangle_{22}\langle 0| =: e^{-a_2^\dagger a_2}: \tag{8.134}$$

and the operator identity $: e^{a^\dagger a(e^\lambda - 1)}: = e^{\lambda a^\dagger a}$, the density operator $\rho(t)$ can be expressed as

$$\begin{aligned}\rho(t) &= \mathrm{sech}^2\lambda \sum_{n=0}^{\infty} \frac{T^n \tanh^{2n}\lambda}{n!} : a_1^{\dagger n} a_1^n e^{e^{-\kappa t} a_1^\dagger a_2^\dagger \tanh\lambda} e^{e^{-\kappa t} a_1 a_2 \tanh\lambda - a_1^\dagger a_1 - a_2^\dagger a_2}: \\ &= \mathrm{sech}^2\lambda : e^{a_1^\dagger a_1 (1 - e^{-2\kappa t})\tanh^2\lambda + e^{-\kappa t}\tanh\lambda\left(a_1^\dagger a_2^\dagger + a_1 a_2\right) - a_1^\dagger a_1 - a_2^\dagger a_2}: \\ &= \mathrm{sech}^2\lambda e^{e^{-\kappa t} a_1^\dagger a_2^\dagger \tanh\lambda} e^{a_1^\dagger a_1 \ln\left[(1 - e^{-2\kappa t})\tanh^2\lambda\right]} |0\rangle_{22}\langle 0|\, e^{e^{-\kappa t} a_1 a_2 \tanh\lambda} \\ &= \mathrm{sech}^2\lambda e^{e^{-\kappa t} a_1^\dagger a_2^\dagger \tanh\lambda} \left[(1 - e^{-2\kappa t})\tanh^2\lambda\right]^{a_1^\dagger a_1} |0\rangle_{22}\langle 0|\, e^{e^{-\kappa t} a_1 a_2 \tanh\lambda}.\end{aligned} \tag{8.135}$$

Equation (8.135) indicates that during the process of single-mode amplitude decay, the squeezed parameter $\tanh\lambda$ of the initial squeezed state $\rho(0)$ decreases to $e^{-\kappa t}\tanh\lambda$, and the first mode without decay evolves into a chaotic state $\left[(1 - e^{-2\kappa t})\tanh^2\lambda\right]^{a_1^\dagger a_1}$. That is to say, the output state $\rho(t)$ is no longer a pure state, but an entangled mixed state.

Referring to Eq. (8.135), we introduce a new two-mode light field state described by the following identity:

$$\rho = \mathrm{sech}^2\sigma\,\mathrm{sech}^2\tau\, e^{a_2^\dagger a_1^\dagger \tanh\sigma} \left(\mathrm{sech}^2\sigma \tanh^2\tau\right)^{a_1^\dagger a_1} |0\rangle_{22}\langle 0|\, e^{a_2 a_1 \tanh\sigma}, \quad \mathrm{tr}\rho = 1. \tag{8.136}$$

Obviously, the above state ρ contains both squeezing information and chaotic behavior, where the term $\left(\mathrm{sech}^2\sigma \tanh^2\tau\right)^{a_1^\dagger a_1}$ corresponding to mode 1 represents the chaotic state, and σ, τ are two independent parameters, $|0\rangle_{22}\langle 0|$ represents the vacuum field of mode 2, and $e^{a_2^\dagger a_1^\dagger \tanh\sigma}$ represents the two-mode squeezed operator that is only related to σ. Therefore, ρ is a mixed state that includes squeezing and noise. Specifically, when $\tau = 0$, $\left(\mathrm{sech}^2\sigma \tanh^2\tau\right)^{a_1^\dagger a_1} \to |0\rangle_{11}\langle 0|$, then $\rho \to \mathrm{sech}^2\sigma e^{a_2^\dagger a_1^\dagger \tanh\sigma} |0\rangle_1 |0\rangle_{22}\langle 0|_1 \langle 0| e^{a_2 a_1 \tanh\sigma}$, which is precisely the two-mode squeezed vacuum state. However, it should be noted that ρ is not equivalent to the result of the two-mode squeezing transformation of the product state of chaotic and vacuum fields, i.e., $S_2\left(\mathrm{sech}^2\sigma \tanh^2\tau\right)^{a_1^\dagger a_1} |0\rangle_{22}\langle 0|\, S_2^{-1}$, where $S_2 \equiv e^{\sigma\left(a_2^\dagger a_1^\dagger - a_2 a_1\right)}$ is the two-mode squeezing operator. Therefore, the density operator in Eq. (8.136) is neither trivial nor obvious, and even the coefficient $\mathrm{sech}^2\sigma\,\mathrm{sech}^2\tau$ is determined by the normalization condition $\mathrm{tr}\rho(t) = 1$ ($\mathrm{tr} = \mathrm{tr}_1\mathrm{tr}_2$). To confirm the validity of $\mathrm{tr}\rho(t) = 1$, we use the completeness relation of two-mode coherent states, that is

$$\int \frac{\mathrm{d}^2 z_1 \mathrm{d}^2 z_2}{\pi^2} |z_1, z_2\rangle \langle z_1, z_2| = 1, \tag{8.137}$$

where $|z_1, z_2\rangle$ is a two-mode coherent state, i.e.,

$$|z_1, z_2\rangle = \exp\left[-\frac{1}{2}\left(|z_1|^2 + |z_2|^2\right) + z_1 a^\dagger + z_2 b^\dagger\right] |00\rangle , \tag{8.138}$$

to verify the traceability of $\rho(t)$, i.e.,

$$\begin{aligned} \mathrm{tr}\rho(t) &= \mathrm{sech}^2\lambda \int \frac{\mathrm{d}^2 z_1 \mathrm{d}^2 z_2}{\pi^2} \langle z_1, z_2| \, 0\rangle_{a_2 a_2} \langle 0| \\ &: \mathrm{e}^{a_1^\dagger a_1\left[\left(1-\mathrm{e}^{-2\kappa t}\right)\tanh^2\lambda - 1\right]} : |z_1, z_2\rangle \, \mathrm{e}^{\mathrm{e}^{-\kappa t}\left(z_1 z_2 + z_1^* z_2^*\right)\tanh\lambda} \\ &= \mathrm{sech}^2\theta \int \frac{\mathrm{d}^2 z_2}{\pi} \exp\left[\left(\tanh^2\theta - 1\right)|z_2|^2\right] = 1. \end{aligned} \tag{8.139}$$

In sum, we propose a new light field state that includes both squeezing and chaotic behavior, which can be generated by one mode of the two-mode squeezed state passing through a single-mode amplitude decay channel. We found that measuring a mode of a two-mode light field state can obtain the chaotic fields at different temperatures. This means that the measurement of different modes of this new light field state will lead to the temperature effects that are reflected in the different average photon numbers of different modes.

8.4.2 Decay of *l*-Photon-Added Two-Mode Squeezed State

Quantum control has attracted widespread attention. The generation of non-classical quantum states will be beneficial for quantum state engineering and quantum information processing. From a theoretical perspective, the simplest way to generate non-classical quantum states is to subtract or add photons from traditional light fields, which results in quantum states exhibiting rich non-classical properties. By applying the photon creation or annihilation operators to classical states, such as thermal and coherent states, the resulting states are called non-Gaussian states (having non-Gaussian distributions of the Wigner functions). Currently, photon-added coherent states and thermal states have been experimentally achieved. In addition, photon addition and subtraction can be used to improve the entanglement of Gaussian states.

In this section, we investigate how the l-photon-added two-mode squeezed vacuum state $C_l a^{\dagger l} S_2(\lambda)|00\rangle \equiv |\psi(\lambda)\rangle_l$ (a type of non-Gaussian state) evolves in the amplitude decay channel? Here, C_l is the undetermined normalization coefficient, and $S_2 = \exp\left[\lambda\left(a^\dagger b^\dagger - ab\right)\right]$ is the two-mode squeezed operator. This non-classical state is theoretically obtained by interacting the operator $a^{\dagger l}$ with the two mode squeezed vacuum

$$S_2 |00\rangle = \text{sech}\lambda e^{a^\dagger b^\dagger \tanh\lambda} |00\rangle . \tag{8.140}$$

Let the state $|\psi(\lambda)\rangle_l$ pass through the amplitude decay channel again and see what happens? Next, using the IWOP method, it will be demonstrated that the initial l-photon-added two-mode squeezed state

$$\rho(0) = |C_l|^2 a^{\dagger l} S_2 |00\rangle \langle 00| S_2^\dagger a^l \tag{8.141}$$

evolves into a Laguerre-polynomial-weighted squeezed state $\rho(t)$ with strong decoherence. In particular, for the case of $l = 0$, the squeezing of one mode of the state $\rho(0)$ decreases during the decay process, and the other mode becomes a chaotic light field.

Firstly, we calculate the normalization coefficient C_l. Using the two-mode coherent state [10, 11]

$$\left|z, \tilde{z}'\right\rangle = \exp\left(-\frac{|z|^2}{2} - \frac{|z'|^2}{2} + za^\dagger + z'b^\dagger\right)|00\rangle \tag{8.142}$$

and its completeness relation

$$\int \frac{d^2z d^2z'}{\pi^2} \left|z, \tilde{z}'\right\rangle\left\langle z, \tilde{z}'\right| = 1, \tag{8.143}$$

we obtain

$$\begin{aligned}
\langle 00| S_2^\dagger a^l a^{\dagger l} S_2 |00\rangle &= \text{sech}^2\lambda \langle 00| e^{ab\tanh\lambda} a^l a^{\dagger l} e^{a^\dagger b^\dagger \tanh\lambda} |00\rangle \\
&= \text{sech}^2\lambda \langle 00| e^{ab\tanh\lambda} a^l \int \frac{d^2z d^2z'}{\pi^2} \left|z, \tilde{z}'\right\rangle\left\langle z, \tilde{z}'\right| a^{\dagger l} e^{a^\dagger b^\dagger \tanh\lambda} |00\rangle \\
&= \text{sech}^2\lambda \int \frac{d^2z d^2z'}{\pi^2} |z|^{2l} e^{-|z|^2 - |z'|^2 + (z^* z'^* + zz')\tanh\lambda} \\
&= \text{sech}^2\lambda \int \frac{d^2z}{\pi^2} |z|^{2l} e^{-|z|^2 \text{sech}^2\lambda} = l! \cosh^{2l}\lambda.
\end{aligned} \tag{8.144}$$

So, the normalization coefficient reads $C_l = \text{sech}^l\lambda/\sqrt{l!}$, and the normalized quantum state $C_l a^{\dagger l} S_2(\lambda)|00\rangle$ is given by

$$|\psi(\lambda)\rangle_l \equiv C_l a^{\dagger l} S_2(\lambda)|00\rangle = \frac{\text{sech}^{l+1}\lambda}{\sqrt{l!}} a^{\dagger l} e^{a^\dagger b^\dagger \tanh\lambda} |00\rangle . \tag{8.145}$$

Letting $\text{sech}^2\lambda = \gamma$, thus Eq. (8.145) can be represented as

$$\begin{aligned} |\psi(\lambda)\rangle_l &= \sqrt{\frac{\gamma^{l+1}}{l!}} a^{\dagger l} \mathrm{e}^{a^\dagger b^\dagger \sqrt{1-\gamma}} |0,0\rangle \\ &= \sqrt{\frac{\gamma^{l+1}}{l!}} \sum_{n=0}^{\infty} \sqrt{(1-\gamma)^n} \frac{a^{\dagger n+l} b^{\dagger n}}{n!} |0,0\rangle \\ &= \sum_{n=0}^{\infty} \sqrt{\binom{n+l}{l} \gamma^{l+1} (1-\gamma)^n} |l+n, n\rangle, \end{aligned} \tag{8.146}$$

where $\binom{n+l}{l}\gamma^{l+1}(1-\gamma)^n$ is the negative binomial coefficient, so the l-photon-added two-mode squeezed vacuum state can also be viewed as a negative binomial state.

When the state $\rho(0)$ in Eq. (8.141) undergoes the single-mode amplitude decay channel, the evolution of the density operator $\rho(0)$ is determined by (8.131) [12]. Substituting Eq. (8.141) into Eq. (8.131), we obtain the following operator-sum representation of $\rho(t)$, that is

$$\begin{aligned} \rho(t) &= \frac{\operatorname{sech}^{2l}\lambda}{l!} \sum_{n=0}^{\infty} \frac{T^n}{n!} \mathrm{e}^{-\kappa t a^\dagger a} a^n a^{\dagger l} S_2 |00\rangle \langle 00| S_2^\dagger a^l a^{\dagger n} \mathrm{e}^{-\kappa t a^\dagger a} \\ &= \frac{\operatorname{sech}^{2(l+1)}\lambda}{l!} \sum_{n=0}^{\infty} \frac{T^n}{n!} \mathrm{e}^{-\kappa t a^\dagger a} a^n a^{\dagger l} \exp\left(a^\dagger b^\dagger \tanh\lambda\right) \\ &\quad |00\rangle \langle 00| \exp(ab\tanh\lambda) a^l a^{\dagger n} \mathrm{e}^{-\kappa t a^\dagger a}, \end{aligned} \tag{8.147}$$

where the normal ordering product of $a^n a^{\dagger l}$ is

$$\begin{aligned} a^n a^{\dagger l} &= (-\mathrm{i})^{n+l} : \mathrm{H}_{l,n}\left(\mathrm{i}a^\dagger, \mathrm{i}a\right) : \\ &= (-\mathrm{i})^{n+l} \sum_{k=0}^{\min(l,n)} \frac{l!n!(-1)^l (\mathrm{i}a^\dagger)^{l-k} (\mathrm{i}a)^{n-k}}{k!(l-k)!(n-k)!}, \end{aligned} \tag{8.148}$$

where $\mathrm{H}_{l,n}(\cdot,\cdot)$ are two-variable Hermite polynomials. In fact, Eq. (8.148) is proved as follows:

$$\begin{aligned} \sum_{m,n=0}^{\infty} \frac{g^m a^m f^n a^{\dagger n}}{m!n!} &= \mathrm{e}^{fa} \mathrm{e}^{ga^\dagger} = \mathrm{e}^{ga^\dagger} \mathrm{e}^{fa} \mathrm{e}^{fg} =: \mathrm{e}^{ga^\dagger + fa + fg} : \\ &=: \mathrm{e}^{(-\mathrm{i}g)\mathrm{i}a^\dagger + (-\mathrm{i}f)\mathrm{i}a - (-\mathrm{i}f)(-\mathrm{i}g)} : \\ &= \sum_{m,n=0}^{\infty} \frac{(-\mathrm{i}g)^m (-\mathrm{i}f)^n}{m!n!} : \mathrm{H}_{m,n}\left(\mathrm{i}a^\dagger, \mathrm{i}a\right) :, \end{aligned} \tag{8.149}$$

which leads to the operator identity

$$a^m a^{\dagger n} = (-\mathrm{i})^{m+n} : \mathrm{H}_{m,n}\left(\mathrm{i}a^\dagger, \mathrm{i}a\right) : . \tag{8.150}$$

Noting that the action of the annihilation operator on the two-mode squeezed vacuum state is equivalent to the action of the creation operator, that is

$$a^{n-k}\exp\left(a^{\dagger}b^{\dagger}\tanh\lambda\right)|00\rangle=\left(b^{\dagger}\tanh\lambda\right)^{n-k}\exp\left(a^{\dagger}b^{\dagger}\tanh\lambda\right)|00\rangle\,, \tag{8.151}$$

so using Eqs. (8.148) and (8.151), we obtain

$$\begin{aligned}&a^{n}a^{\dagger l}\exp\left(a^{\dagger}b^{\dagger}\tanh\lambda\right)|00\rangle\\&=(-\mathrm{i})^{n+l}\sum_{k=0}^{\min(l,n)}\frac{l!n!\,(-1)^{k}\,(\mathrm{i}a^{\dagger})^{l-k}(\mathrm{i}b^{\dagger}\tanh\lambda)^{n-k}}{k!\,(l-k)!\,(n-k)!}\exp\left(a^{\dagger}b^{\dagger}\tanh\lambda\right)|00\rangle\\&=(-\mathrm{i})^{n+l}\,\mathrm{H}_{l,n}\left(\mathrm{i}a^{\dagger},\mathrm{i}b^{\dagger}\tanh\lambda\right)\exp\left(a^{\dagger}b^{\dagger}\tanh\lambda\right)|00\rangle\,.\end{aligned} \tag{8.152}$$

Substituting Eq. (8.152) into Eq. (8.147), and using the operator identity $\mathrm{e}^{-\kappa t a^{\dagger}a}a^{\dagger}\mathrm{e}^{\kappa t a^{\dagger}a}=a^{\dagger}\mathrm{e}^{-\kappa t}$ and the normal ordering product $|00\rangle\langle 00|=:\mathrm{e}^{-a^{\dagger}a-b^{\dagger}b}:$, we can write the density operator $\rho(t)$ as

$$\begin{aligned}\rho(t)=&\frac{\operatorname{sech}^{2(l+1)}\lambda}{l!}\sum_{n=0}^{\infty}\frac{T^{n}}{n!}(-\mathrm{i})^{n+l}\,\mathrm{e}^{-\kappa t a^{\dagger}a}\\&\times\exp\left(a^{\dagger}b^{\dagger}\tanh\lambda\right)\mathrm{H}_{l,n}\left(\mathrm{i}a^{\dagger},\mathrm{i}b^{\dagger}\tanh\lambda\right)|00\rangle\langle 00|\\&\times\mathrm{H}_{l,n}\left(-\mathrm{i}a,-\mathrm{i}b\tanh\lambda\right)\mathrm{i}^{n+l}\exp\left(ab\tanh\lambda\right)\mathrm{e}^{-\kappa t a^{\dagger}a}\\=&\frac{\operatorname{sech}^{2(l+1)}\lambda}{l!}\mathrm{e}^{a^{\dagger}b^{\dagger}\mathrm{e}^{-\kappa t}\tanh\lambda}\sum_{n=0}^{\infty}\frac{T^{n}}{n!}\mathrm{H}_{l,n}\left(\mathrm{i}\mathrm{e}^{-\kappa t}a^{\dagger},\mathrm{i}b^{\dagger}\tanh\lambda\right)\\&\times:\mathrm{e}^{-a^{\dagger}a-b^{\dagger}b}:\mathrm{H}_{l,n}\left(-\mathrm{i}\mathrm{e}^{-\kappa t}a,-\mathrm{i}b\tanh\lambda\right)\mathrm{e}^{ab\mathrm{e}^{-\kappa t}\tanh\lambda}\\=&\frac{\operatorname{sech}^{2(l+1)}\lambda}{l!}\mathrm{e}^{a^{\dagger}b^{\dagger}\mathrm{e}^{-\kappa t}\tanh\lambda}\sum_{n=0}^{\infty}\frac{T^{n}}{n!}:\mathrm{H}_{l,n}\left(\mathrm{i}\mathrm{e}^{-\kappa t}a^{\dagger},\mathrm{i}b^{\dagger}\tanh\lambda\right)\\&\times\mathrm{H}_{l,n}\left(-\mathrm{i}\mathrm{e}^{-\kappa t}a,-\mathrm{i}b\tanh\lambda\right)\mathrm{e}^{-a^{\dagger}a-b^{\dagger}b}:\mathrm{e}^{ab\mathrm{e}^{-\kappa t}\tanh\lambda}.\end{aligned} \tag{8.153}$$

Using the IWOP method and the summation formula

$$\begin{aligned}&\sum_{n=0}^{\infty}\frac{z^{n}}{n!}\mathrm{H}_{m,n}(x,y)\,\mathrm{H}_{m,n}\left(x',y'\right)\\&=(-z)^{m}\,\mathrm{e}^{zyy'}\mathrm{H}_{m,m}\left(\mathrm{i}(\sqrt{z}y'-\frac{x}{\sqrt{z}}),\mathrm{i}(\sqrt{z}y-\frac{x'}{\sqrt{z}})\right),\end{aligned} \tag{8.154}$$

we have

$$
\begin{aligned}
&\sum_{n=0}^{\infty}\frac{T^n}{n!}:\mathrm{H}_{l,n}\left(\mathrm{i}\mathrm{e}^{-\kappa t}a^{\dagger},\mathrm{i}b^{\dagger}\tanh\lambda\right)\mathrm{H}_{l,n}\left(-\mathrm{i}\mathrm{e}^{-\kappa t}a,-\mathrm{i}b\tanh\lambda\right)\mathrm{e}^{-a^{\dagger}a-b^{\dagger}b}:\\
&=:\ (-T)^l\,\mathrm{e}^{Tb^{\dagger}b\tanh^2\lambda}\mathrm{H}_{l,l}\left(\sqrt{T}b\tanh\lambda+\frac{a^{\dagger}\mathrm{e}^{-\kappa t}}{\sqrt{T}},\right.\\
&\left.-\sqrt{T}b^{\dagger}\tanh\lambda-\frac{a\mathrm{e}^{-\kappa t}}{\sqrt{T}}\right)\mathrm{e}^{-a^{\dagger}a-b^{\dagger}b}:\\
&=(-T)^l:\mathrm{e}^{Tb^{\dagger}b\tanh^2\lambda}\mathrm{H}_{l,l}\left(\sqrt{T}b\tanh\lambda+\frac{a^{\dagger}\mathrm{e}^{-\kappa t}}{\sqrt{T}},\right.\\
&\left.-\sqrt{T}b^{\dagger}\tanh\lambda-\frac{a\mathrm{e}^{-\kappa t}}{\sqrt{T}}\right)|00\rangle\langle 00|:.
\end{aligned}
\tag{8.155}
$$

Finally, inserting Eq. (8.155) into Eq. (8.153), and using the relations between two-variable Hermite polynomials and Laguerre polynomials in Eq. (2.81), we obtain

$$
\begin{aligned}
\rho(t)=&\frac{\mathrm{sech}^{2(l+1)}\lambda}{l!}(-T)^l\exp\left(a^{\dagger}b^{\dagger}\mathrm{e}^{-\kappa t}\tanh\lambda\right)\\
&\times:\mathrm{e}^{Tb^{\dagger}b\tanh^2\lambda}\mathrm{H}_{l,l}\left(\sqrt{T}b\tanh\lambda+\frac{a^{\dagger}\mathrm{e}^{-\kappa t}}{\sqrt{T}},\right.\\
&\left.-\sqrt{T}b^{\dagger}\tanh\lambda-\frac{a\mathrm{e}^{-\kappa t}}{\sqrt{T}}\right)|00\rangle\langle 00|:\\
&\times\exp\left(ab\mathrm{e}^{-\kappa t}\tanh\lambda\right)\\
=&\,T^l\mathrm{sech}^{2(l+1)}\lambda\mathrm{e}^{a^{\dagger}b^{\dagger}\mathrm{e}^{-\kappa t}\tanh\lambda}:\mathrm{e}^{b^{\dagger}b\left(1-\mathrm{e}^{-2\kappa t}\right)\tanh^2\lambda}\\
&\times\mathrm{L}_l\left[-\left(\sqrt{T}b\tanh\lambda+\frac{a^{\dagger}\mathrm{e}^{-\kappa t}}{\sqrt{T}}\right)\right.\\
&\left.\times\left(\sqrt{T}b^{\dagger}\tanh\lambda+\frac{a\mathrm{e}^{-\kappa t}}{\sqrt{T}}\right)\right]|00\rangle\langle 00|:\mathrm{e}^{ab\mathrm{e}^{-\kappa t}\tanh\lambda},
\end{aligned}
\tag{8.156}
$$

which is a Laguerre-polynomial-weighted two-mode squeezed state. It can be seen that after the l-photon-added two-mode squeezed state $C_l a^{\dagger l} S_2|00\rangle$ passing through a single-mode amplitude decay channel, a Laguerre-polynomial-weighted two-mode squeezed state can be generated [12].

To verify the correctness of the above results, it is necessary to prove that $\mathtt{tr}\rho(t)=1$. In fact, the normal ordering product of the exponential operator in Eq. (8.156) can be expressed as

$$
\begin{aligned}
&:\mathrm{e}^{\left(a^{\dagger}b^{\dagger}+ab\right)\mathrm{e}^{-\kappa t}\tanh\lambda+b^{\dagger}b\left(1-\mathrm{e}^{-2\kappa t}\right)\tanh^2\lambda}:\\
&=:\exp\left[\left(\sqrt{T}b\tanh\lambda+\frac{a^{\dagger}\mathrm{e}^{-\kappa t}}{\sqrt{T}}\right)\right.\\
&\left.\times\left(\sqrt{T}b^{\dagger}\tanh\lambda+\frac{a\mathrm{e}^{-\kappa t}}{\sqrt{T}}\right)-\frac{a^{\dagger}a\mathrm{e}^{-2\kappa t}}{T}\right]:,
\end{aligned}
\tag{8.157}
$$

thus Eq. (8.156) becomes

$$\rho(t) = T^l \operatorname{sech}^{2(l+1)} \lambda \colon \mathrm{e}^{-F} \mathrm{L}_l(F)\, \mathrm{e}^{-\frac{a^\dagger a e^{-2\kappa t}}{T}} |00\rangle\langle 00| \colon, \tag{8.158}$$

where

$$F = -\left(\sqrt{T} b \tanh\lambda + \frac{a^\dagger \mathrm{e}^{-\kappa t}}{\sqrt{T}}\right)\left(\sqrt{T} b^\dagger \tanh\lambda + \frac{a \mathrm{e}^{-\kappa t}}{\sqrt{T}}\right). \tag{8.159}$$

It follows from Eq. (8.158) that

$$\begin{aligned}
\mathtt{tr}\rho(t) &= \mathtt{tr}\left(\rho(t) \iint \frac{\mathrm{d}^2 z \mathrm{d}^2 z'}{\pi^2} \left|z, z'\right\rangle\left\langle z, z'\right|\right) \\
&= T^l \operatorname{sech}^{2(l+1)} \lambda \iint \frac{\mathrm{d}^2 z \mathrm{d}^2 z'}{\pi^2} \left\langle z, z'\right| \colon \mathrm{e}^{-F} \\
&\times \mathrm{L}_l(F)\, \mathrm{e}^{-\frac{a^\dagger a e^{-2\kappa t}}{T}} |00\rangle\langle 00| \colon \left|z, z'\right\rangle \\
&= T^l \operatorname{sech}^{2(l+1)} \lambda \iint \frac{\mathrm{d}^2 z \mathrm{d}^2 z'}{\pi^2} \mathrm{e}^{\frac{\mathrm{e}^{-2\kappa t}}{T}\left(\mathrm{e}^{\kappa t} T z' \tanh\lambda + z^*\right)\left(\mathrm{e}^{\kappa t} T z'^* \tanh\lambda + z\right)} \\
&\times \mathrm{L}_l\left[-\frac{\mathrm{e}^{-2\kappa t}}{T}\left(\mathrm{e}^{\kappa t} T z' \tanh\lambda + z^*\right)\left(\mathrm{e}^{\kappa t} T z'^* \tanh\lambda + z\right)\right] \\
&\times \mathrm{e}^{-\left(\frac{\mathrm{e}^{-2\kappa t}}{T}+1\right)|z|^2 - |z'|^2} \\
&= T^l \operatorname{sech}^{2(l+1)} \lambda \iint \frac{\mathrm{d}^2 z'' \mathrm{d}^2 z'}{\pi^2} \mathrm{e}^{\frac{\mathrm{e}^{-2\kappa t}}{T}|z''|^2} \\
&\times \mathrm{L}_l\left(-\frac{\mathrm{e}^{-2\kappa t}|z''|^2}{T}\right) \mathrm{e}^{-\left(\frac{\mathrm{e}^{-2\kappa t}}{T}+1\right)|z'' - \mathrm{e}^{\kappa t} T z'^* \tanh\lambda|^2 - |z'|^2}.
\end{aligned} \tag{8.160}$$

Noting $T = 1 - \mathrm{e}^{-2\kappa t}$, $\mathrm{e}^{-2\kappa t}/T + 1 = 1/T$, thus the integral of the variable z' in Eq. (8.160) reads

$$\begin{aligned}
&\int \frac{\mathrm{d}^2 z'}{\pi} \mathrm{e}^{-\left(1+\mathrm{e}^{2\kappa t} T \tanh^2\lambda\right)|z'|^2 - \tanh\lambda \mathrm{e}^{\kappa t}\left(z' z'' + z'^* z''^*\right)} \\
&= \frac{1}{1 + \mathrm{e}^{2\kappa t} T \tanh^2\lambda} \exp\left(\frac{\tanh^2\lambda \mathrm{e}^{2\kappa t}}{1 + \mathrm{e}^{2\kappa t} T \tanh^2\lambda}|z''|^2\right),
\end{aligned} \tag{8.161}$$

and the remaining integral in Eq. (8.160) is

$$\begin{aligned}
&\int \frac{d^2 z''}{\pi} L_l\left(-\frac{e^{-2\kappa t}|z''|^2}{T}\right) \exp\left[\left(\frac{\tanh^2\lambda e^{2\kappa t}}{1+e^{2\kappa t}T\tanh^2\lambda}-1\right)|z''|^2\right] \\
&= \int \frac{d^2 z''}{\pi} L_l\left(-\frac{e^{-2\kappa t}|z''|^2}{T}\right) \exp\left(-\frac{\mathrm{sech}^2\lambda}{1+e^{2\kappa t}T\tanh^2\lambda}|z''|^2\right) \\
&= Te^{2\kappa t}\int \frac{d^2 z''}{\pi} L_l\left(-|z''|^2\right) \exp\left(-\frac{Te^{2\kappa t}\mathrm{sech}^2\lambda}{1+e^{2\kappa t}T\tanh^2\lambda}|z''|^2\right) \\
&= \frac{1+e^{2\kappa t}T\tanh^2\lambda}{T^l \mathrm{sech}^{2(l+1)}\lambda},
\end{aligned} \tag{8.162}$$

where we have used the integral formula in Eq. (5.73). Substituting Eq. (8.162) into Eq. (8.160) leads to

$$\mathtt{tr}\rho(t) = \frac{1+e^{2\kappa t}T\tanh^2\lambda}{T^l\mathrm{sech}^{2(l+1)}\lambda}\frac{T^l\mathrm{sech}^{2(l+1)}\lambda}{1+e^{2\kappa t}T\tanh^2\lambda} = 1. \tag{8.163}$$

Finally, we want to calculate the average photon number of a-mode in the state $\rho_{l=0}(t)$. For $l = 0$, using the operator identity $\exp[(f-1)]b^\dagger b]: = \exp\left(b^\dagger b\ln f\right)$, we have

$$\rho_{l=0}(t) = \mathrm{sech}^2\lambda e^{a^\dagger b^\dagger e^{-\kappa t}\tanh\lambda}|0\rangle\langle 0| e^{b^\dagger b\ln\left[\left(1-e^{-2\kappa t}\right)\tanh^2\lambda\right]}e^{abe^{-\kappa t}\tanh\lambda}, \tag{8.164}$$

where $\exp\left\{b^\dagger b\ln\left[\left(1-e^{-2\kappa t}\right)\tanh^2\lambda\right]\right\}$ is a time-dependent chaotic field of b-mode. Comparing Eq. (8.164) with two-mode squeezed vacuum state $S_2|00\rangle\langle 00|S_2^\dagger = \mathrm{sech}^2\lambda e^{a^\dagger b^\dagger\tanh\lambda}|00\rangle\langle 00|e^{ab\tanh\lambda}$, we can clearly see how a pure state evolves into a mixed state, that is, the squeezing parameter $\tanh\lambda$ evolves into $e^{-\kappa t}\tanh\lambda$, and the vacuum state $|00\rangle\langle 00|$ evolves into

$$|0\rangle_{aa}\langle 0|\exp\left\{b^\dagger b\ln\left[\left(1-e^{-2\kappa t}\right)\tanh^2\lambda\right]\right\}. \tag{8.165}$$

Using the completeness relation of two-mode coherent states, we obtain

$$\begin{aligned}
\mathtt{tr}\left(\rho_{l=0}(t)a^\dagger a\right) &= \mathrm{sech}^2\lambda\mathtt{tr}\int\frac{d^2zd^2z'}{\pi^2}\left|z,z'\right\rangle\left\langle z,z'\right|a^\dagger a e^{a^\dagger b^\dagger e^{-\kappa t}\tanh\lambda} \\
&|0\rangle\langle 0|e^{b^\dagger b\ln\left[\left(1-e^{-2\kappa t}\right)\tanh^2\lambda\right]}e^{abe^{-\kappa t}\tanh\lambda} \\
&= \mathrm{sech}^2\lambda\int\frac{d^2zd^2z'}{\pi^2}\left\langle z,z'\right|a^\dagger a e^{a^\dagger b^\dagger e^{-\kappa t}\tanh\lambda} \\
&|0\rangle\langle 0|e^{b^\dagger b\ln\left[\left(1-e^{-2\kappa t}\right)\tanh^2\lambda\right]}e^{abe^{-\kappa t}\tanh\lambda}\left|z,z'\right\rangle \\
&= \mathrm{sech}^2\lambda\int\frac{d^2zd^2z'}{\pi^2}z^*\left\langle z,z'\right|e^{a^\dagger b^\dagger e^{-\kappa t}\tanh\lambda}e^{-a^\dagger b^\dagger e^{-\kappa t}\tanh\lambda} \\
&\times ae^{a^\dagger b^\dagger e^{-\kappa t}\tanh\lambda}|0\rangle\langle 0|e^{b^\dagger b\ln\left[\left(1-e^{-2\kappa t}\right)\tanh^2\lambda\right]}e^{zz'e^{-\kappa t}\tanh\lambda}\left|z,z'\right\rangle
\end{aligned}$$

$$
\begin{aligned}
&= \operatorname{sech}^2\lambda \int \frac{d^2z d^2z'}{\pi^2} z^* e^{z^* z'^* e^{-\kappa t}\tanh\lambda} \langle z, z'| \left(a + b^\dagger e^{-\kappa t}\tanh\lambda\right) \\
&|0\rangle\langle 0| e^{b^\dagger b \ln[(1-e^{-2\kappa t})\tanh^2\lambda]} |z, z'\rangle e^{zz' e^{-\kappa t}\tanh\lambda} \\
&= \operatorname{sech}^2\lambda e^{-\kappa t}\tanh\lambda \int \frac{d^2z d^2z'}{\pi^2} z^* z'^* \\
&\times e^{(z^* z'^* + zz')e^{-\kappa t}\tanh\lambda - |z|^2} \langle z'| e^{b^\dagger b \ln[(1-e^{-2\kappa t})\tanh^2\lambda]} |z'\rangle, \quad (8.166)
\end{aligned}
$$

where we have used the identity

$$
\begin{aligned}
\langle z'| e^{b^\dagger b \ln[(1-e^{-2\kappa t})\tanh^2\lambda]} |z'\rangle &= \langle z'| : e^{b^\dagger b[(1-e^{-2\kappa t})\tanh^2\lambda - 1]} : |z'\rangle \\
&= e^{[(1-e^{-2\kappa t})\tanh^2\lambda - 1]|z'|^2}. \quad (8.167)
\end{aligned}
$$

By carrying out the integral of the variable z in Eq. (8.166), we obtain

$$
\int \frac{d^2z}{\pi} z^* e^{(z^* z'^* + zz')e^{-\kappa t}\tanh\lambda - |z|^2} = z' e^{-\kappa t}\tanh\lambda e^{|z'|^2 \tanh^2\lambda e^{-2\kappa t}}. \quad (8.168)
$$

Substituting Eq. (8.168) into Eq. (8.166) and using the integral formula

$$
\int \frac{d^2z}{\pi} |z|^2 e^{-\lambda|z|^2} = \frac{1}{\lambda^2}, \quad (8.169)
$$

we thus have

$$
\begin{aligned}
\mathrm{tr}\left(\rho_{l=0}(t) a^\dagger a\right) &= \operatorname{sech}^2\lambda \tanh^2\lambda e^{-2\kappa t} \int \frac{d^2z'^*}{\pi} z'^* z' e^{(\tanh^2\lambda - 1)|z'|^2} \\
&= e^{-2\kappa t}\sinh^2\lambda. \quad (8.170)
\end{aligned}
$$

Comparing the average photon number $\mathrm{tr}\left(\rho_{l=0}(t) a^\dagger a\right)$ at the initial time, a decay factor $e^{-2\kappa t}$ is added in $\mathrm{tr}\left(\rho_{l=0}(t) a^\dagger a\right)$. For $l \neq 0$, the results are more complicated and will not be discussed in detail here.

References

1. Fan HY (2004) Entangled states, squeezed states gained via the route of developing Dirac's symbolic method and their applications. Int J Mod Phys B 18(10–11):1387–1455
2. Meng XG, Wang JS, Liang BL (2023) Entangled state representations in quantum optics. Springer, Singapore
3. Wan ZL, Fan HY (2016) Dynamic evolution of squeezing maintenance. Int J Theor Phys 55(8):3682–3688
4. Da C, Fan HY (2019) Time evolution of the variance of squeezed chaotic state in amplitude dissipation channel. Rep Math Phys 83(3):329–338

5. Xu XL, Li HQ, Fan HY (2015) Time evolution of a squeezed chaotic field in an amplitude damping channel when used as a generating field for a squeezed number state. Chin Phys B 24(7):070306
6. Chen F, Fang BL, He R, Fan HY (2014) Evolution law of the optical field of degenerate parametric amplifier in dissipative channel. Int J Theor Phys 53(8):2846–2854
7. Fan HY, He R (2015) Photon-added chaotic field and its damping in a thermo enviroment. Can J Phys 93(4):456–459
8. Zhou NR, Hu LY, Fan HY (2011) Dissipation of a two-mode squeezed vacuum state in the single-mode amplitude damping channel. Chin Phys B 20(12):120301
9. Meng XG, Wang JS, Liang BL, Han CX (2018) Evolution of a two-mode squeezed vacuum for amplitude decay via continuous-variable entangled state approach. Front Phys 13(5):130322
10. Meng XG, Goan XS, Wang JS, Zhang R (2018) Nonclassical thermal-state superpositions: Analytical evolution law and decoherence behavior. Opt Commun 411(7):15–20
11. Wu WF (2015) Analytical evolution of displaced thermal states for amplitude damping. Chin J Phys 53(4):17–24
12. He R, Fan HY, Song J, Zhou J (2016) Laguerre-polynomial-weighted two-mode squeezed state. Int J Theor Phys 55(7):3329–3337

Chapter 9
Photon Counting Formula and Wigner Function Formula

This chapter introduces the photon counting formula and Wigner function formula in the thermal field dynamics, and investigates how their evolutions in the amplitude decay channel.

9.1 Integral Form of Photon Counting Formula

The formula for photon counting distribution in quantum mechanics was first derived by Kelley and Kleiner [1]. For a single-mode field with the density operator ρ, the probability $p(m, \Delta\tau)$ of detecting m photoelectrons within a time interval $\Delta\tau$ is

$$p(m, \Delta\tau) = \mathtt{tr}\left(\rho\colon \frac{(\xi a^{\dagger}a)^{m}}{m!}\mathrm{e}^{-\xi a^{\dagger}a}\colon\right), \tag{9.1}$$

where $\xi \propto \Delta\tau$ is the measurement efficiency of the detector. In the Fock state representation, the density operator $\rho = \sum_{n=0}^{\infty}\mathrm{P}_n \left|n\right\rangle \left\langle n\right|$, $\left|n\right\rangle = a^{\dagger n}\left|0\right\rangle/\sqrt{n!}$, thus Eq. (9.1) represents the Bernoulli distribution of the field, that is

$$p(m, \Delta\tau) = \sum_{n=m}^{\infty}\mathrm{P}_n\binom{n}{m}\xi^{m}(1-\xi)^{n-m}. \tag{9.2}$$

For example, when ρ is the coherent state light field, i.e., $\rho \equiv \left|\alpha\right\rangle\left\langle\alpha\right|$, $\left|\alpha\right\rangle = \exp\left(-|\alpha|^2/2 + \alpha a^{\dagger}\right)\left|0\right\rangle$, thus Eq. (9.1) becomes

$$p(m, \Delta\tau) = \mathtt{tr}\left(\left|\alpha\right\rangle\left\langle\alpha\right| : \frac{(\xi a^{\dagger}a)^{n}}{n!}\mathrm{e}^{-\xi a^{\dagger}a}\colon\right) = \frac{(\xi|\alpha|^2)^{n}}{n!}\mathrm{e}^{-\xi|\alpha|^2}. \tag{9.3}$$

H. Fan and L. Hu, *Dissipation and Diffusion of Quantum Light Field*,
https://doi.org/10.1007/978-981-95-1528-8_9

Obviously, for the ideal detection case, i.e., $\xi \to 1$, Eq. (9.3) refers to the Poisson distribution.

Using the thermal vacuum state $|\psi\rangle_\beta$ whose density operator is $\rho = \mathtt{tr}_{\tilde{a}}\left(|\psi\rangle_{\beta\beta}\langle\psi|\right)$, thus Eq. (9.1) becomes

$$p(m, \Delta\tau) =_\beta \langle\psi| : \frac{\left(\xi a^\dagger a\right)^m}{m!} e^{-\xi a^\dagger a} : |\psi\rangle_\beta . \tag{9.4}$$

For example, the thermal vacuum state belonging to the negative binomial state is [2]

$$|\psi(\beta)\rangle_s = \sum_{n=0}^{\infty} \sqrt{\binom{n+s}{n} \gamma^{s+1} (1-\gamma)^n} \, |n, \tilde{s} + \tilde{n}\rangle , \tag{9.5}$$

thus the corresponding photon counting distribution reads

$$\begin{aligned} p(m, \Delta\tau) &= \sum_{n=0}^{\infty} \binom{n+s}{n} \gamma^{s+1} (1-\gamma)^n \frac{\xi^m}{m!} \langle n| a^{\dagger m} : e^{-\xi a^\dagger a} : a^m |n\rangle \\ &= \sum_{n=0}^{\infty} \binom{n+s}{n} \gamma^{s+1} (1-\gamma)^n \frac{n! \xi^m}{m! (n-m)!} \langle n-m| (1-\xi)^{a^\dagger a} |n-m\rangle \\ &= \sum_{n=0}^{\infty} \gamma^{s+1} (1-\gamma)^n \frac{\xi^m (1-\xi)^{n-m}}{m!} \frac{(n+s)!}{s! (n-m)!} . \end{aligned} \tag{9.6}$$

Next, we will derive the integral form of the photon counting formula. For this, we first get the anti-normal ordering product of the normal ordering operator in Eq. (9.1). Using the operator formula in Eq. (2.76), we give

$$\begin{aligned} &: \left(a^\dagger a\right)^m e^{-\xi a^\dagger a} : \\ &= \int \frac{d^2 z}{\pi} \langle -z| : \left(a^\dagger a\right)^m e^{-\xi a^\dagger a} : |z\rangle \,\vdots \exp\left(|z|^2 + z^* a - z a^\dagger + a a^\dagger\right) \vdots \\ &= \int \frac{d^2 z}{\pi} (-1)^m |z|^{2m} \vdots \exp\left[-(1-\xi)|z|^2 + z^* a - z a^\dagger + a a^\dagger\right] \vdots , \end{aligned} \tag{9.7}$$

where $|z\rangle$ is a coherent state. Using the mathematical integral formula [3]

$$\begin{aligned} &\int \frac{d^2 z}{\pi} z^n z^{*m} e^{k|z|^2 + Bz + Cz^*} \\ &= e^{-BC/k} \sum_{l=0}^{\min(m,n)} \frac{n! m! B^{m-l} C^{n-l}}{l! (n-l)! (m-l)! (-k)^{n+m-l+1}}, \quad \mathrm{Re} k < 0 \end{aligned} \tag{9.8}$$

to complete the integral in Eq. (9.7), we have

$$: \left(a^{\dagger}a\right)^{m} \mathrm{e}^{-\xi a^{\dagger}a} : \, = \vdots \mathrm{e}^{\xi aa^{\dagger}/(\xi-1)} \sum_{l=0}^{m} \frac{(-)^{l} (m!)^{2} a^{m-l} \left(a^{\dagger}\right)^{m-l}}{l! \left[(m-l)!\right]^{2} (1-\xi)^{2m-l+1}} \vdots \tag{9.9}$$

Using the definition of Laguerre polynomials

$$\begin{aligned} \mathrm{L}_m(x) &= \sum_{l=0}^{m} (-1)^{l} \binom{m}{l} \frac{x^{l}}{l!} \\ &= \sum_{l'=0}^{m} (-1)^{m-l'} \binom{m}{m-l'} \frac{x^{m-l'}}{(m-l')!}, \end{aligned} \tag{9.10}$$

Equation (9.9) becomes

$$: \left(a^{\dagger}a\right)^{m} \mathrm{e}^{-\xi a^{\dagger}a} : \, = \frac{(-)^{m} m!}{(1-\xi)^{m+1}} \vdots \mathrm{e}^{\xi aa^{\dagger}/(\xi-1)} \mathrm{L}_m \left(\frac{aa^{\dagger}}{1-\xi}\right) \vdots, \tag{9.11}$$

which is a new operator identity.

When the anti-normal ordering of the operator is known, using the P-representation, it can be inferred that

$$\vdots A\left(a, a^{\dagger}\right) \vdots = \int \frac{\mathrm{d}^2 z}{\pi} A\left(z, z^{*}\right) |z\rangle \langle z| \,. \tag{9.12}$$

Inserting Eq. (9.11) into Eq. (9.1) and using Eq. (9.12), we obtain

$$\begin{aligned} p(m, \Delta\tau) &= \frac{(-\xi)^{m}}{(1-\xi)^{m+1}} \mathrm{tr} \left[\rho \vdots \mathrm{e}^{\xi aa^{\dagger}/(\xi-1)} \mathrm{L}_m \left(\frac{aa^{\dagger}}{1-\xi}\right) \vdots\right] \\ &= \left(\frac{\xi}{\xi-1}\right)^{m} \mathrm{tr} \left[\rho \int \frac{\mathrm{d}^2 z}{\pi} \frac{\mathrm{e}^{-\xi|z|^2/(1-\xi)}}{(1-\xi)} \mathrm{L}_m \left(\frac{|z|^2}{1-\xi}\right) |z\rangle \langle z|\right] \\ &= \left(\frac{\xi}{\xi-1}\right)^{m} \int \frac{\mathrm{d}^2 z}{\pi} \mathrm{e}^{-\xi|z|^2} \mathrm{L}_m \left(|z|^2\right) \left\langle \sqrt{1-\xi} z \right| \rho \left| \sqrt{1-\xi} z \right\rangle, \end{aligned} \tag{9.13}$$

which is a new formula for calculating photon counting distribution [4]. Letting

$$: \frac{\left(\xi a^{\dagger}a\right)^{n}}{n!} \mathrm{e}^{-\xi a^{\dagger}a} : \, \equiv M_a(\xi), \tag{9.14}$$

we call it the counting operator of a detector with quantum efficiency ξ.

9.2 Photon Counting Formula in Thermal Field Dynamics

Here, we transform the calculation of the thermal statistical mean value in Eq. (9.1) into a matrix element in a pure state, that is, we hope to explain the photon counting distribution $p(m, \Delta\tau)$ from a new perspective. Using the thermal field dynamics theory proposed by Takahashi and Umezawa, we provide a new explanation for $p(m, \Delta\tau)$. The basic point of this theory is the double expansion the true real state space, that is, the single-mode number space $|m\rangle$ is expanded to the two-mode number state space $|m, \tilde{m}\rangle$, where $|\tilde{m}\rangle = \frac{\tilde{a}^{\dagger m}}{\sqrt{m!}}|0\rangle$ is the number state of the fictitious mode corresponding to the real mode space, $\tilde{a}^\dagger$ is the creation operator of fictitious mode corresponding to the real mode $a^\dagger$, satisfying $[\tilde{a}, \tilde{a}^\dagger] = 1$, $\tilde{a}\,|00\rangle = 0$. Based on the theory of thermal field dynamics, we replace the field operator

$$:\mathrm{e}^{-\xi a^\dagger a}: \; = (1-\xi)^{a^\dagger a} \tag{9.15}$$

by the thermal squeezed operator $\mathrm{e}^{(1-\xi)a^\dagger \tilde{a}^\dagger}$, and replace the density operator ρ by the state vector $\langle\rho|$, where

$$\langle\rho| = \langle I\rho|\,, \quad \langle I| = \sum_l^\infty \left\langle l, \tilde{l}\right| = \left\langle 0.\tilde{0}\right| \mathrm{e}^{a\tilde{a}}, \tag{9.16}$$

thus the photon counting formula $p(m, \Delta\tau)$ can be explained as the matrix elements of the thermal squeezed operator $\mathrm{e}^{(1-\xi)a^\dagger \tilde{a}^\dagger}$ in the pure states $\langle\rho|$ and $|m, \tilde{m}\rangle = |m\rangle\,|\tilde{m}\rangle$, i.e.,

$$p(m, \Delta\tau) = \xi^m \langle\rho|\, \mathrm{e}^{(1-\xi)a^\dagger \tilde{a}^\dagger}\, |m, \tilde{m}\rangle\,. \tag{9.17}$$

Using the thermal entangled state representation in Eq. (5.49), thus Eq. (9.17) also becomes

$$p(m, \Delta\tau) = \xi^m \int \frac{\mathrm{d}^2\eta}{\pi} \chi(\eta)\, \langle\eta|\, \mathrm{e}^{(1-\xi)a^\dagger \tilde{a}^\dagger}\, |m, \tilde{m}\rangle\,. \tag{9.18}$$

Next, we will prove Eqs. (9.17) and (9.18), and calculate the matrix elements $\langle\eta|\mathrm{e}^{(1-\xi)a^\dagger \tilde{a}^\dagger}\,|m, \tilde{m}\rangle$. Actually, using the completeness relation of Fock states, that is, $\sum_{k=0}^{\infty} |k\rangle\,\langle k| = 1$ and the inner product $\langle\tilde{k}|\tilde{l}\rangle = \delta_{l,k}$, thus Eq. (9.1) can be rewritten as

$$\begin{aligned}
p(m, \Delta\tau) &= \mathtt{tr}\left(\sum_{k=0}^{\infty} |k\rangle\,\langle k|\,\rho\colon \frac{\left(\xi a^\dagger a\right)^m}{m!}\mathrm{e}^{-\xi a^\dagger a}\right) \\
&= \sum_{k=0}^{\infty} \langle k|\,\rho\colon \frac{\left(\xi a^\dagger a\right)^m}{m!}\mathrm{e}^{-\xi a^\dagger a}\colon\, |k\rangle \\
&= \frac{\xi^m}{m!}\sum_{k=0}^{\infty}\left\langle k, \tilde{k}\right| \rho a^{\dagger m} : \mathrm{e}^{-\xi a^\dagger a} : a^m \sum_l^\infty \left|l, \tilde{l}\right\rangle
\end{aligned}$$

$$= \frac{\xi^m}{m!} \langle\rho| a^{\dagger m} (1-\xi)^{a^\dagger a} a^m |I\rangle , \tag{9.19}$$

it follows from Eq. (9.16) that

$$a^m |I\rangle = a^m \sum_{l=0}^{\infty} \left| l, \tilde{l} \right\rangle = a^m \mathrm{e}^{a^\dagger \tilde{a}^\dagger} \left| 0, \tilde{0} \right\rangle = \tilde{a}^\dagger |I\rangle . \tag{9.20}$$

Therefore, Eq. (9.19) finally becomes

$$\begin{aligned} p(m, \Delta\tau) &= \frac{\xi^m}{m!} \langle\rho| a^{\dagger m} (1-\xi)^{a^\dagger a} \tilde{a}^{\dagger m} |I\rangle \\ &= \frac{\xi^m}{m!} \langle\rho| a^{\dagger m} \tilde{a}^{\dagger m} \sum_{l}^{\infty} (1-\xi)^l \frac{a^{\dagger l} \tilde{a}^{\dagger l}}{l!} \left| 0, \tilde{0} \right\rangle \\ &= \xi^m \langle\rho| \mathrm{e}^{(1-\xi) a^\dagger \tilde{a}^\dagger} |m, \tilde{m}\rangle , \end{aligned} \tag{9.21}$$

which is just Eq. (9.17). Further, using Eqs. (4.108) and (4.109), and noting $\left[\left(a - \tilde{a}^\dagger \right), \left(a^\dagger - \tilde{a} \right) \right] = 0$, which leads to the commum eigen-states $|\eta\rangle$ of the operators $a - \tilde{a}^\dagger$ and $a^\dagger - \tilde{a}$ [5, 6], we can prove that the state $|\eta\rangle$ is orthogonal, that is

$$\left\langle \eta' | \eta \right\rangle = \pi \delta(\eta' - \eta) \delta(\eta'^* - \eta^*). \tag{9.22}$$

Using the normal ordering product $\left| 0\tilde{0} \right\rangle \left\langle 0\tilde{0} \right| = :\ \exp(-a^\dagger a - \tilde{a}^\dagger \tilde{a})$ and the IWOP method, we can directly prove that the state $|\eta\rangle$ is complete, that is

$$\begin{aligned} \int \frac{\mathrm{d}^2\eta}{\pi} |\eta\rangle \langle\eta| = \int \frac{\mathrm{d}^2\eta}{\pi} &:\ \exp(-|\eta|^2 + \eta a^\dagger - \eta^* \tilde{a}^\dagger + \eta^* a^\dagger \\ &- \eta \tilde{a} + a^\dagger \tilde{a}^\dagger + a\tilde{a} - a^\dagger a - \tilde{a}^\dagger \tilde{a}): \\ &= 1. \end{aligned} \tag{9.23}$$

Using the identity

$$|\eta\rangle = D(\eta) |I\rangle , \quad D(\eta) = \mathrm{e}^{\eta a^\dagger - \eta^* a}, \tag{9.24}$$

we obtain

$$\begin{aligned} \langle\rho |\eta\rangle = \langle\rho| D(\eta) |I\rangle &= \sum_{n,m}^{\infty} \langle n, \tilde{n}| \rho D(\eta) |m, \tilde{m}\rangle \\ &= \sum_{n}^{\infty} \langle n| D(\eta) \rho |n\rangle = \mathtt{tr} (\rho D(\eta)) = \chi(\eta) . \end{aligned} \tag{9.25}$$

Clearly, the inner product $\langle\rho\,|\eta\rangle$ happens to be the characteristic function of the density operator ρ. Further, using Eqs. (9.23) and (9.25), Eq. (9.21) can be written as the form in Eq. (9.18), that is

$$
\begin{aligned}
p(m, \Delta\tau) &= \xi^m \int \frac{d^2\eta}{\pi} \langle\rho\,|\eta\rangle \langle\eta|\, e^{(1-\xi)a^\dagger\tilde{a}^\dagger}\, |m, \tilde{m}\rangle \\
&= \xi^m \int \frac{d^2\eta}{\pi} \chi(\eta) \langle\eta|\, e^{(1-\xi)a^\dagger\tilde{a}^\dagger}\, |m, \tilde{m}\rangle\,. \qquad (9.26)
\end{aligned}
$$

Finally, let's calculate the matrix element $\langle\eta|e^{(1-\xi)a^\dagger\tilde{a}^\dagger}\,|m, \tilde{m}\rangle$. To calculate the above matrix element, the conjugate entangled state $|\xi\rangle$ of the thermal entangled state $|\eta\rangle$ is introduced [5, 6], that is

$$
|\xi\rangle = \exp\left(-\frac{1}{2}|\xi|^2 + \xi a^\dagger + \xi^* \tilde{a}^\dagger - a^\dagger\tilde{a}^\dagger\right)\left|0\tilde{0}\right\rangle, \qquad (9.27)
$$

which is complete, that is

$$
\int \frac{d^2\xi}{\pi} |\xi\rangle\langle\xi| = 1, \qquad (9.28)
$$

and the inner product of the states $\langle\xi|$ and $|\eta\rangle$ is

$$
\langle\xi\,|\eta\rangle = \frac{1}{2}\exp\left[\frac{1}{2}(\eta\xi^* - \eta^*\xi)\right]. \qquad (9.29)
$$

On the other hand, the thermal squeezing operator $S = e^{\lambda(a^\dagger\tilde{a}^\dagger - a\tilde{a})}$ in the above thermal entangled state representations can be naturally expressed as [7]

$$
S = \int \frac{d^2\eta}{\pi\mu} |\eta/\mu\rangle\langle\eta| = \mu \int \frac{d^2\xi}{\pi} |\mu\xi\rangle\langle\xi|\,, \quad \mu = e^\lambda. \qquad (9.30)
$$

Actually, using the IWOP method to carry out the integral in Eq. (9.30), we obtain

$$
\begin{aligned}
S &= \int \frac{d^2\eta}{\pi\mu} |\eta/\mu\rangle\langle\eta| \\
&= \int \frac{d^2\eta}{\mu} :\, \exp\left[-\frac{|\eta|^2}{2}\left(1 + \frac{1}{\mu^2}\right) + \frac{\eta}{\mu}a^\dagger - \frac{\eta^*}{\mu}\tilde{a}^\dagger\right. \\
&\quad \left. + \eta^* a - \eta\tilde{a} + a^\dagger\tilde{a}^\dagger + a\tilde{a} - a^\dagger a - a^\dagger\tilde{a}^\dagger\right] : \\
&= \exp\left(-a^\dagger\tilde{a}^\dagger \tanh\lambda\right) \exp\left[(a^\dagger a + \tilde{a}^\dagger a + 1)\ln \operatorname{sech}\lambda\right] \exp\left(a\tilde{a}\tanh\lambda\right), \qquad (9.31)
\end{aligned}
$$

it follows that the thermal vacuum state is obtained as

$$S\left|0\tilde{0}\right\rangle = \mathrm{sech}\lambda \mathrm{e}^{-a^\dagger\tilde{a}^\dagger \tanh\lambda}\left|0\tilde{0}\right\rangle = \mu\int\frac{\mathrm{d}^2\xi}{\pi}\left|\mu\xi\right\rangle\left\langle\xi\right|0\tilde{0}\rangle$$
$$= \mu\int\frac{\mathrm{d}^2\xi}{\pi}\left|\mu\xi\right\rangle \mathrm{e}^{-|\xi|^2/2}. \tag{9.32}$$

Letting $1-\xi=f$ and introducing the real parameter g, thus the matrix element $\langle\eta|\mathrm{e}^{(1-\xi)a^\dagger\tilde{a}^\dagger}|m,\tilde{m}\rangle$ becomes

$$\langle\eta|\,\mathrm{e}^{(1-\xi)a^\dagger\tilde{a}^\dagger}\,|m,\tilde{m}\rangle = \frac{1}{m!}\frac{\mathrm{d}^m}{\mathrm{d}g^m}\langle\eta|\,\mathrm{e}^{(f+g)a^\dagger\tilde{a}^\dagger}\left|0\tilde{0}\right\rangle\Big|_{g=0}. \tag{9.33}$$

Comparing $\mathrm{e}^{(f+g)a^\dagger\tilde{a}^\dagger}\left|0\tilde{0}\right\rangle$ with $\mathrm{e}^{-a^\dagger\tilde{a}^\dagger\tanh\lambda}\left|0\tilde{0}\right\rangle$ in Eq. (9.33), and letting

$$f+g=g'=-\tanh\lambda=\frac{\mathrm{e}^{-\lambda}-\mathrm{e}^{\lambda}}{\mathrm{e}^{-\lambda}+\mathrm{e}^{\lambda}}=\frac{1-\mu^2}{1+\mu^2}, \tag{9.34}$$

and $\mathrm{sech}^2\lambda = 1-g'^2$, thus we have

$$\mu^2=\frac{1-g'}{1+g'}, \tag{9.35}$$

so using Eq. (9.32) to obtain the identity

$$\mathrm{e}^{(f+g)a^\dagger\tilde{a}^\dagger}\left|0\tilde{0}\right\rangle = \mathrm{e}^{g'a^\dagger\tilde{a}^\dagger}\left|0\tilde{0}\right\rangle$$
$$=\sqrt{\frac{1}{1-g'^2}}\sqrt{\frac{1-g'}{1+g'}}\int\frac{\mathrm{d}^2\zeta}{\pi}\left|\sqrt{\frac{1-g'}{1+g'}}\zeta\right\rangle \mathrm{e}^{-|\zeta|^2/2}$$
$$=\frac{1}{1+g'}\int\frac{\mathrm{d}^2\zeta}{\pi}\left|\sqrt{\frac{1-g'}{1+g'}}\zeta\right\rangle \mathrm{e}^{-|\zeta|^2/2}. \tag{9.36}$$

Further, using the completeness relation in Eq. (9.28) and the inner product in Eq. (9.29), the matrix element $\langle\eta|\mathrm{e}^{g'a^\dagger\tilde{a}^\dagger}\left|0\tilde{0}\right\rangle$ is obtained as

$$\langle\eta|\,\mathrm{e}^{g'a^\dagger\tilde{a}^\dagger}\left|0\tilde{0}\right\rangle = \frac{1}{1+g'}\int\frac{\mathrm{d}^2\zeta}{\pi}\left\langle\eta\left|\sqrt{\frac{1-g'}{1+g'}}\zeta\right.\right\rangle \mathrm{e}^{-|\zeta|^2/2}$$
$$=\frac{1}{2(1+g')}\int\frac{\mathrm{d}^2\zeta}{\pi}\exp\left(\sqrt{\frac{1-g'}{1+g'}}\frac{\eta^*\zeta-\zeta^*\eta}{2}-\frac{|\zeta|^2}{2}\right)$$
$$=\frac{1}{1+g'}\mathrm{e}^{-|\eta|^2/2}\exp\left(\frac{g'}{1+g'}|\eta|^2\right). \tag{9.37}$$

Noting that

$$\frac{g'}{1+g'} = \frac{f+g}{1+f+g} = \frac{g}{1+f+g}\frac{1}{1+f} + \frac{f}{1+f} \tag{9.38}$$

and the generating function for the Laguerre polynomial $\mathrm{L}_n(x)$, that is

$$\mathrm{L}_n(x) = \frac{1}{n!}\frac{\partial^n}{\partial t^n}\left[(1-t)^{-1}\exp\left(\frac{-xt}{1-t}\right)\right]\bigg|_{t=0} \tag{9.39}$$

thus Eq. (9.33) becomes

$$\begin{aligned}
&\langle\eta|\, \mathrm{e}^{(1-\xi)a^\dagger a}\, |m,\tilde{m}\rangle \\
&= \frac{1}{m!}\frac{\mathrm{d}^m}{\mathrm{d}g^m}\, \langle\eta|\, \mathrm{e}^{(f+g)a^\dagger\tilde{a}^\dagger}\left|0\tilde{0}\right\rangle\Big|_{g=0} \\
&= \frac{1}{m!}\frac{\mathrm{d}^m}{\mathrm{d}g^m}\left[\frac{\mathrm{e}^{\frac{f|\eta|^2}{1+f}-|\eta|^2/2}}{1+f+g}\exp\left(\frac{g|\eta|^2}{(1+f+g)(1+f)}\right)\right]\bigg|_{g=0} \\
&= \frac{\mathrm{e}^{\frac{f-1}{2(1+f)}|\eta|^2}}{m!}\frac{\mathrm{d}^m}{\mathrm{d}g^m}\left[\frac{1}{1+f+g}\exp\left(\frac{g|\eta|^2}{(1+f+g)(1+f)}\right)\right]\bigg|_{g=0} \\
&= \frac{\mathrm{e}^{\frac{f-1}{2(1+f)}|\eta|^2}}{m!\,(1+f)^{m+1}}\frac{\mathrm{d}^m}{\mathrm{d}\left(\frac{g}{1+f}\right)^m}\left[\frac{1}{1+\frac{g}{1+f}}\exp\left(\frac{\frac{g}{1+f}}{1+\frac{g}{1+f}}\frac{|\eta|^2}{1+f}\right)\right]\Bigg|_{g=0} \\
&= \frac{(-1)^m\mathrm{e}^{\frac{f-1}{2(1+f)}|\eta|^2}}{(1+f)^{m+1}}\mathrm{L}_m\left(\frac{-|\eta|^2}{1+f}\right).
\end{aligned} \tag{9.40}$$

Substituting $1-\xi = f$ and Eq. (9.40) into Eq. (9.18) leads to

$$\begin{aligned}
p(m,\Delta\tau) &= \frac{(-\xi)^m}{(1+f)^{m+1}}\int\frac{\mathrm{d}^2\eta}{\pi}\langle\rho\,|\eta\rangle\,\mathrm{e}^{\frac{f-1}{2(1+f)}|\eta|^2}\mathrm{L}_m(\frac{-|\eta|^2}{1+f}) \\
&= \frac{(-\xi)^m}{(2-\xi)^{m+1}}\int\frac{\mathrm{d}^2\eta}{\pi}\chi(\eta)\,\mathrm{e}^{\frac{-\xi}{2(2-\xi)}|\eta|^2}\mathrm{L}_m\left(\frac{-|\eta|^2}{2-\xi}\right),
\end{aligned} \tag{9.41}$$

which is a new integral formula for calculating the photon counting distribution when the characteristic function denoted as $\chi(\eta)$ of the density operator ρ is known [8]. For example, for the pure coherent state $|\alpha\rangle\langle\alpha|$, its characteristic function is

$$\chi(\eta) = \mathtt{tr}(|\alpha\rangle\langle\alpha|\, D(\eta)) = \langle\alpha|\, D(\eta)\,|\alpha\rangle = \mathrm{e}^{-|\eta|^2/2+\alpha^*\eta-\alpha\eta^*}, \tag{9.42}$$

inserting Eq. (9.42) into Eq. (9.41), we obtain

$$\begin{aligned} p_{|\alpha\rangle\langle\alpha|}(m, \Delta\tau) &= \frac{(-\xi)^m}{(2-\xi)^{m+1}} \int \frac{d^2\eta}{\pi} e^{-|\eta|^2/2+\alpha^*\eta-\alpha\eta^*} e^{\frac{-\xi}{2(2-\xi)}|\eta|^2} L_m\left(\frac{-|\eta|^2}{2-\xi}\right) \\ &= \frac{\left(\xi|\alpha|^2\right)^m e^{-\xi|\alpha|^2}}{m!}, \end{aligned} \tag{9.43}$$

which is completely consistent with the previous result.

In short, based on the theory of thermal field dynamics, we transform the photon counting distribution formula into the matrix element of the thermal entangled operator (squeezed operator) $e^{(1-\xi)a^\dagger \tilde{a}^\dagger}$ in the pure states $\langle\rho|$ and $|m, \tilde{m}\rangle$. A new integral formula for calculating photon counting distribution of a quantum state was obtained by introducing the thermal entangled state representation when the characteristic function of the density operator of the state is known. This is convenient for calculating the photon counting distributions of quantum states, especially quantum states with complex density operators.

9.3 Single-Mode Photon Counting Distribution in Two-Mode Squeezed State

An interesting question is: when counting photons on one mode of a two-mode squeezed vacuum state, what state will the other mode be in? Or if one of the two modes has n photon counts, how will the other mode collapse? The two-mode squeezed vacuum state is [9]

$$\rho_{a,b} \equiv \operatorname{sech}^2\lambda e^{a^\dagger b^\dagger \tanh\lambda} |00\rangle \langle 00| e^{ab\tanh\lambda}, \quad (|00\rangle \equiv |0\rangle_a |0\rangle_b) \tag{9.44}$$

If only counting n photons on a-mode in the two-mode states $\rho_{a,b}$, then Eq. (9.1) naturally becomes a partial trace, that is

$$p_a(n) \to \mathtt{tr}_a\left(\rho_{a,b} \colon \frac{\left(\xi a^\dagger a\right)^n}{n!} e^{-\xi a^\dagger a} \colon\right) = \rho_b, \tag{9.45}$$

which means that the counting operation on a-mode is equivalent to an operator of b-mode. Substituting Eq. (9.13) into Eq. (9.45) and using the normal ordering product, that is $|0\rangle_{22}\langle 0| = \colon e^{-b^\dagger b}\colon$, we obtain

$$\begin{aligned} \rho_b &= \frac{\xi^n \operatorname{sech}^2\lambda}{(\xi-1)^{n+1}} \int \frac{d^2\beta}{\pi} \langle -\beta| e^{a^\dagger b^\dagger \tanh\lambda} |00\rangle \langle 00| \\ &\quad e^{ab\tanh\lambda} |\beta\rangle e^{\frac{\xi-2}{\xi-1}|\beta|^2} L_n\left(\frac{|\beta|^2}{\xi-1}\right) \\ &= \frac{\xi^n \operatorname{sech}^2\lambda}{(\xi-1)^{n+1}} \int \frac{d^2\beta}{\pi} e^{-\beta^* b^\dagger \tanh\lambda} |0\rangle_{22}\langle 0| \end{aligned}$$

$$\begin{aligned}
& e^{\beta b \tanh\lambda} e^{\left(\frac{\xi-2}{\xi-1}-1\right)|\beta|^2} \mathrm{L}_n\left(\frac{|\beta|^2}{\xi-1}\right) \\
&= \frac{\xi^n \operatorname{sech}^2\lambda}{(\xi-1)^{n+1}} \int \frac{d^2\beta}{\pi} : e^{\frac{-1}{\xi-1}|\beta|^2 + (\beta b - \beta^* b^\dagger)\tanh\lambda - b^\dagger b} \mathrm{L}_n\left(\frac{|\beta|^2}{\xi-1}\right) : \\
&= \frac{\xi^n \operatorname{sech}^2\lambda}{n!\,(\xi-1)^{n+1}} \frac{\partial^n}{\partial t^n} : \int \frac{d^2\beta}{\pi(1-t)} e^{\frac{|\beta|^2}{(\xi-1)(t-1)} + (\beta b - \beta^* b^\dagger)\tanh\lambda - b^\dagger b}\bigg|_{t=0} :,
\end{aligned} \tag{9.46}$$

where we have used the generating function for the Laguerre polynomial, that is

$$\sum_{n=0}^{\infty} \mathrm{L}_n(x)\, t^n = \frac{\exp\left(\frac{-xt}{1-t}\right)}{1-t}. \tag{9.47}$$

Or, using Eq. (9.39) and the IWOP method to complete the integral in Eq. (9.46), we obtain

$$\begin{aligned}
\rho_b &= \frac{\xi^n \operatorname{sech}^2\lambda}{n!\,(\xi-1)^n} \frac{\partial^n}{\partial t^n} : e^{\left[(\xi-1)(t-1)\tanh^2\lambda - 1\right] b^\dagger b} : |_{t=0} \\
&= \frac{\xi^n \operatorname{sech}^2\lambda}{n!\,(\xi-1)^n} \frac{\partial^n}{\partial t^n} : e^{\left[(\xi-1)t\tanh^2\lambda - \xi\tanh^2\lambda - \operatorname{sech}^2\lambda\right] b^\dagger b} : |_{t=0} \\
&= \xi^n \operatorname{sech}^2\lambda \tanh^{2n}\lambda : \frac{1}{n!} b^{\dagger n} b^n e^{\left(-\xi\tanh^2\lambda - \operatorname{sech}^2\lambda\right) b^\dagger b} : .
\end{aligned} \tag{9.48}$$

Letting

$$\xi' = \xi\tanh^2\lambda + \operatorname{sech}^2\lambda, \tag{9.49}$$

thus Eq. (9.48) can be written as

$$\begin{aligned}
\rho_b &= \frac{\left(\xi\tanh^2\lambda\right)^n \operatorname{sech}^2\lambda}{\xi'^n} : \frac{\left(\xi' b^\dagger b\right)^n}{n!} e^{-\xi' b^\dagger b} : \\
&= \frac{\left(\xi\tanh^2\lambda\right)^n \operatorname{sech}^2\lambda}{\xi'^n} M_b\left(\xi'\right).
\end{aligned} \tag{9.50}$$

Therefore, when counting photons on a-mode in the state $\rho_{a,b}$, the b-mode of the state $\rho_{a,b}$ collapses into a new counting operator with a counting efficiency ξ' determined by Eq. (9.49) [9].

Besides, using the operator identity

$$b^{\dagger k+n} b^{k+n} = \vdots \mathrm{H}_{k+n,k+n}\left(b, b^\dagger\right) \vdots \tag{9.51}$$

and the summation formula

$$\sum_{k=0}^{\infty} \frac{f^k}{k!} \mathrm{H}_{k+m,k+n}(x, y) = \frac{e^{fxy/(f+1)}}{(f+1)^{(m+n+2)/2}} \mathrm{H}_{m,n}\left(\frac{x}{\sqrt{f+1}}, \frac{y}{\sqrt{f+1}}\right), \tag{9.52}$$

thus Eq. (9.48) can be expressed as

$$\begin{aligned}
\rho_b &= \xi^n \mathrm{sech}^2\lambda \tanh^{2n}\lambda \colon \frac{1}{n!} b^{\dagger n} b^n \mathrm{e}^{\left(-\xi \tanh^2\lambda - \mathrm{sech}^2\lambda\right) b^\dagger b} \colon \\
&= \frac{\xi^n}{n!} \mathrm{sech}^2\lambda \tanh^{2n}\lambda \sum_{k=0}^{\infty} b^{\dagger k+n} b^{k+n} \frac{\left(-\xi \tanh^2\lambda - \mathrm{sech}^2\lambda\right)^k}{k!} \\
&= \frac{\xi^n}{n!} \mathrm{sech}^2\lambda \tanh^{2n}\lambda \sum_{k=0}^{\infty} \vdots \mathrm{H}_{k+n,k+n}\left(b, b^\dagger\right) \vdots \frac{\left(-\xi \tanh^2\lambda - \mathrm{sech}^2\lambda\right)^k}{k!} \\
&= \frac{\xi^n}{n! \sinh^2\lambda \left(1-\xi\right)^n} \vdots \mathrm{e}^{\left(-\xi \tanh^2\lambda - \mathrm{sech}^2\lambda\right) b b^\dagger / \left[(1-\xi)\tanh^2\lambda\right]} \\
&\quad \times \mathrm{H}_{n,n}\left(\frac{b}{\tanh\lambda\sqrt{1-\xi}}, \frac{b^\dagger}{\tanh\lambda\sqrt{1-\xi}}\right) \vdots, \qquad (9.53)
\end{aligned}$$

which is the anti-normal ordering product of the density operator ρ_b of b-mode. So, the P-representation of ρ_b is

$$\mathrm{e}^{-\frac{\xi \tanh^2\lambda + \mathrm{sech}^2\lambda}{(1-\xi)\tanh^2\lambda}|\alpha|^2} \mathrm{L}_n\left(\frac{|\alpha|^2}{\tanh^2\lambda\left(1-\xi\right)}\right), \qquad (9.54)$$

which is useful for further calculating the related quantities of the density operator ρ_b of b-mode.

9.4 Photon Counting Formula for Amplitude Decay

When the density operator $\rho(0)$ passes through the amplitude decay channel, the evolved density operator $\rho(t)$ of the system is given by the operator-sum representation in Eq. (6.21) [10]. So, substituting Eq. (6.21) into Eq. (9.1) leads to

$$\begin{aligned}
p(n,t) &= \mathtt{tr}\left(\rho(t) \colon \frac{\left(\xi a^\dagger a\right)^n}{n!} \mathrm{e}^{-\xi a^\dagger a} \colon\right) \\
&= \mathtt{tr}\left(\rho(0)\, O(t)\right), \qquad (9.55)
\end{aligned}$$

where

$$O(t) = \sum_{m=0}^{\infty} M_m^\dagger \colon \frac{\left(\xi a^\dagger a\right)^n}{n!} \mathrm{e}^{-\xi a^\dagger a} \colon M_m. \qquad (9.56)$$

Using the operator identities

$$\colon \mathrm{e}^{-\xi a^\dagger a} \colon = \exp\left[a^\dagger a \ln\left(1-\xi\right)\right] = \left(1-\xi\right)^{a^\dagger a},$$

$$\begin{aligned} a^{\dagger} e^{\lambda} &= e^{\lambda a^{\dagger} a} a^{\dagger} e^{-\lambda a^{\dagger} a}, \\ a e^{-\lambda} &= e^{\lambda a^{\dagger} a} a e^{-\lambda a^{\dagger} a}, \end{aligned} \tag{9.57}$$

thus Eq. (9.56) can be rewritten as its normal ordering product, that is

$$\begin{aligned} O(t) &= \sum_{m=0}^{\infty} \frac{T^m}{m!} a^{\dagger m} e^{-\kappa t a^{\dagger} a} : \frac{\left(\xi a^{\dagger} a\right)^n}{n!} e^{-\xi a^{\dagger} a} : e^{-\kappa t a^{\dagger} a} a^m \\ &= \sum_{m=0}^{\infty} \frac{T^m}{m!} a^{\dagger m} e^{-\kappa t a^{\dagger} a} \frac{\xi^n a^{\dagger n}}{n!} (1-\xi)^{a^{\dagger} a} a^n e^{-\kappa t a^{\dagger} a} a^m \\ &= \frac{\xi^n}{n!} e^{-2n\kappa t} \sum_{m=0}^{\infty} \frac{T^m}{m!} a^{\dagger n+m} e^{-\kappa t a^{\dagger} a} e^{a^{\dagger} a \ln(1-\xi)} e^{-\kappa t a^{\dagger} a} a^{n+m} \\ &= \frac{\xi^n}{n!} e^{-2\kappa t n} \sum_{m=0}^{\infty} \frac{T^m}{m!} a^{\dagger n+m} : e^{\left[e^{[\ln(1-\xi)-2\kappa t]}-1\right] a^{\dagger} a} : a^{n+m}. \end{aligned} \tag{9.58}$$

Noting $T = 1 - e^{-2\kappa t}$ and the summation method within the normal ordering (a and $a^{\dagger}$ are commutative within the normal ordering symbol $: \, :$), so we have

$$\begin{aligned} O(t) &= \frac{\xi^n}{n!} e^{-2\kappa t n} a^{\dagger n} : \sum_{m=0}^{\infty} \frac{\left(T a^{\dagger} a\right)^m}{m!} e^{\left[e^{[\ln(1-\xi)-2\kappa t]}-1\right] a^{\dagger} a} : a^n \\ &= \frac{\left(\xi e^{-2\kappa t}\right)^n}{n!} a : \exp\left(-\xi e^{-2\kappa t} a^{\dagger} a\right) : a^n \\ &= : \frac{\left(\xi e^{-2\kappa t} a^{\dagger} a\right)^n}{n!} e^{-\xi e^{-2\kappa t} a^{\dagger} a} : . \end{aligned} \tag{9.59}$$

Inserting Eq. (9.59) into Eq. (9.55), thus we get

$$p(n,t) = \mathrm{tr}\left(\rho(0) : \frac{\left(\xi e^{-2\kappa t} a^{\dagger} a\right)^n}{n!} e^{-\xi e^{-2\kappa t} a^{\dagger} a} : \right). \tag{9.60}$$

Combining Eqs. (9.1) and (9.60), we finally find that the correspondence between the initial counting distribution and the evolved counting distribution at time t, i,e.,

$$\begin{aligned} & p(n,0) = \mathrm{tr}\left(\rho(0) : \frac{\left(\xi a^{\dagger} a\right)^n}{n!} e^{-\xi a^{\dagger} a} : \right) \\ \rightarrow & \; p(n,t) = \mathrm{tr}\left(\rho(0) : \frac{\left(\xi e^{-2\kappa t} a^{\dagger} a\right)^n}{n!} e^{-\xi e^{-2\kappa t} a^{\dagger} a} : \right). \end{aligned} \tag{9.61}$$

Therefore, the law presented by the formula (9.61) can be seen as the detection efficiency of the detector changing from ξ to $\xi e^{-2\kappa t}$, that is, the detection efficiency gradually decreases over time t. In sum, in the amplitude decay channel, the photon counting distribution is related to the decay of quantum detection efficiency, i.e., $\xi' = \xi e^{-2\kappa t}$.

Next, we will use an examples to verify the correctness of the formula (9.60). When the initial state is a number state, $\rho(0) = |m\rangle\langle m|$, using the operator identity, i.e., $a^{\dagger m}a^m = N(N-1)\cdots(N-m+1)$ $(N = a^\dagger a)$, we obtain the photon counting distribution as

$$\begin{aligned}
p(n,0) &= \langle m| : \frac{(\xi a^\dagger a)^n}{n!} e^{-\xi a^\dagger a} : |m\rangle \\
&= \frac{\xi^n}{n!}\sum_{l=0}^{\infty}\frac{(-\xi)^l}{l!}\langle m| a^{\dagger n+l}a^{n+l}|m\rangle \\
&= \frac{\xi^n}{n!}\sum_{l=0}^{\infty}\frac{(-\xi)^l}{l!}\langle m| N(N-1)\cdots(N-n-l+1)|m\rangle \\
&= \sum_{l=0}^{m-n}\frac{(-\xi)^l}{l!}\frac{\xi^n}{n!}\frac{m!}{(m-n-l)!} \\
&= \frac{m!}{n!(m-n)!}\xi^n(1-\xi)^{m-n},
\end{aligned} \tag{9.62}$$

where ξ represents the probability of a photon being detected within the time interval of $\Delta\tau$, so the probability of n out of m photons being detected is proportional to the probability ξ^n of detecting n photons multiplied by the probability $(1-\xi)^{m-n}$ of $m-n$ photons not being detected. In terms of the new formula in Eq. (9.61), we get the evolved photon counting formula of the number state in the amplitude decay channel, i.e.,

$$\begin{aligned}
p(n,t) &= \mathtt{tr}\left(\rho(0) : \frac{(\xi e^{-2\kappa t}a^\dagger a)^n}{n!} e^{-\xi e^{-2\kappa t}a^\dagger a} :\right) \\
&= \frac{m!(\xi e^{-2\kappa t})^n}{n!(m-n)!}\left(1-\xi e^{-2\kappa t}\right)^{m-n}.
\end{aligned} \tag{9.63}$$

In fact, we can also verify the correctness of Eq. (9.63) using the following method. Using Eq. (6.58), we find that the initial number state $|m\rangle\langle m|$ evolves into a binomial state, that is

$$|m\rangle\langle m| \to e^{-2m\kappa t}\sum_{l=0}^{m}\frac{m!\left(e^{2\kappa t}-1\right)^l}{l!(m-l)!}|m-l\rangle\langle m-l| = \rho(t), \tag{9.64}$$

substituting Eq. (9.64) into Eq. (9.1) leads to

$$\begin{aligned}
p(n,t) &= \sum_{l=0}^{m} \frac{m!\left(\mathrm{e}^{2\kappa t}-1\right)^{l}}{l!\,(m-l)!\mathrm{e}^{2m\kappa t}} \langle m-l| : \frac{\left(\xi a^{\dagger}a\right)^{n}}{n!}\mathrm{e}^{-\xi a^{\dagger}a} : |m-l\rangle \\
&= \mathrm{e}^{-2m\kappa t}\sum_{l=0}^{m-n} \frac{m!\left(\mathrm{e}^{2\kappa t}-1\right)^{l}}{l!n!\,(m-n-l)!}\xi^{n}\,(1-\xi)^{m-l-n} \\
&= \frac{m!\xi^{n}\mathrm{e}^{-2m\kappa t}}{n!\,(m-n)!}\sum_{l=0}^{m}\binom{m-n}{m-n-l}\left(\mathrm{e}^{2\kappa t}-1\right)^{l}(1-\xi)^{m-n-l} \\
&= \frac{m!\left(\xi \mathrm{e}^{-2\kappa t}\right)^{n}}{n!\,(m-n)!}\left(1-\xi \mathrm{e}^{-2\kappa t}\right)^{m-n},
\end{aligned} \tag{9.65}$$

which is the same as the result in Eq. (9.63).

In short, for the amplitude decay channel, the photon counting distribution at time t is exactly the same as the initial photon counting formula, except that the detection efficiency changes from ξ to $\xi \mathrm{e}^{-2\kappa t}$, that is, making the substitution $\xi \to \xi \mathrm{e}^{-2\kappa t}$ in the initial counting distribution leads to the counting distribution at time t. This law greatly simplifies the theoretical study of photon counting distributions of quantum light fields.

9.5 Evolution of Wigner Function for Amplitude Decay

The section examines the thermal entangled state representation of the Wigner function, and further investigates the evolution of the Wigner function in the amplitude decay channel [5, 11, 12].

Noting that the definition of the Wigner function for the density operator ρ is

$$W\left(\alpha,\alpha^{*}\right) = \mathrm{tr}\left(\Delta\left(\alpha,\alpha^{*}\right)\rho\right), \tag{9.66}$$

where $\Delta(\alpha,\alpha^{*})$ is the Wigner operator in the coherent state representation, that is

$$\Delta\left(\alpha,\alpha^{*}\right) = \frac{1}{\pi}D(2\alpha)(-1)^{-a^{\dagger}a}, \tag{9.67}$$

and using the thermal field dynamics theory, the Wigner function in Eq. (9.66) can be expressed as

$$\begin{aligned}
W\left(\alpha,\alpha^{*}\right) &= \sum_{m,n=0}^{\infty}\langle n,\tilde{n}|\,\Delta\left(\alpha,\alpha^{*}\right)\rho\,|m,\tilde{m}\rangle \\
&= \frac{1}{\pi}\langle \eta=0|\,D(2\alpha)(-1)^{a^{\dagger}a}|\rho\rangle \\
&= \frac{1}{\pi}\langle \eta=-2\alpha|(-1)^{a^{\dagger}a}|\rho\rangle
\end{aligned}$$

$$= \frac{1}{\pi} \langle \zeta = -2\alpha | \rho \rangle, \tag{9.68}$$

where $|\rho\rangle \equiv \rho |I\rangle$, and $\langle \zeta = -2\alpha|$ is the conjugate state of the thermal entangled state $\langle \eta|$, seen its specific form in Eq. (9.27). Using the inner product of entangled states in Eq. (9.29) and its completeness relation in Eq. (9.23), we have

$$\begin{aligned} W\left(\alpha, \alpha^*\right) &= \int \frac{d^2\eta}{\pi^2} \langle \upsilon = 2\alpha | \eta \rangle \langle \eta | \rho \rangle \\ &= \int \frac{d^2\eta}{2\pi^2} \exp(\alpha^* \eta - \alpha \eta^*) \langle \eta | \rho \rangle . \end{aligned} \tag{9.69}$$

So, once the inner product $\langle \eta | \rho \rangle$ is known, the Wigner function $W(\alpha, \alpha^*)$ of the system can be obtained by virtue of the Fourier transform of $\langle \eta | \rho \rangle$.

The evolution of the density operator ρ in the amplitude decay channel can be described by the master equation:

$$\frac{d\rho(t)}{dt} = \kappa \left(2a\rho(t)a^\dagger - a^\dagger a\rho(t) - \rho(t)a^\dagger a\right), \tag{9.70}$$

where κ is the decay rate. To obtain the relation between the evolved Wigner function $W(\alpha, \alpha^*, t)$ with time t and the initial Wigner function $W(\alpha, \alpha^*, 0)$, acting Eq. (9.70) on the state $|I\rangle$ leads to

$$\frac{d}{dt} |\rho(t)\rangle = \kappa(2a\rho(t)a^\dagger - a^\dagger a\rho(t) - \rho(t)a^\dagger a) |I\rangle . \tag{9.71}$$

Noting that the operators $a^\dagger a$ and $\tilde{a}^\dagger \tilde{a}$ are commutative, and using Eq. (6.9), thus Eq. (9.71) can be rewritten as

$$\frac{d}{dt} |\rho(t)\rangle = \kappa(2a\tilde{a} - a^\dagger a - \tilde{a}^\dagger \tilde{a}) |\rho(t)\rangle, \tag{9.72}$$

its formal solution reads

$$\begin{aligned} |\rho(t)\rangle &= \exp\left[\kappa t(2a\tilde{a} - a^\dagger a - \tilde{a}^\dagger \tilde{a})\right] |\rho(0)\rangle \\ &= \exp\left\{\kappa t\left[-(a^\dagger - \tilde{a})(a - \tilde{a}^\dagger) + a\tilde{a} - \tilde{a}^\dagger a^\dagger + 1\right]\right\} |\rho(0)\rangle, \end{aligned} \tag{9.73}$$

where $|\rho(0)\rangle = \rho(0)|I\rangle$, $\rho(0)$ is the initial density operator. Using the commutation relation

$$\left[a\tilde{a} - \tilde{a}^\dagger a^\dagger, -(a^\dagger - \tilde{a})(a - \tilde{a}^\dagger)\right] = -2\left[-(a^\dagger - \tilde{a})(a - \tilde{a}^\dagger)\right] \tag{9.74}$$

and the operator identity

$$\mathrm{e}^{\lambda(A+\sigma B)} = \mathrm{e}^{\lambda A} \exp\left[\sigma(1-\mathrm{e}^{-\lambda\tau})B/\tau\right], \tag{9.75}$$

which is valid for $[A, B] = \tau B$, thus Eq. (9.73) is disentangled into

$$|\rho(t)\rangle = \mathrm{e}^{\kappa t(a\tilde{a}-\tilde{a}^\dagger a^\dagger+1)} \exp\left[\frac{1-\mathrm{e}^{2\kappa t}}{2}(a^\dagger-\tilde{a})(a-\tilde{a}^\dagger)\right] |\rho(0)\rangle. \tag{9.76}$$

Projecting Eq. (9.76) onto the entangled state $\langle\eta|$, and noting that $\exp\left[\kappa t\left(a\tilde{a}-\tilde{a}^\dagger a^\dagger\right)\right]$ is actually a two-mode squeezed operator, thus we obtain

$$\langle\eta| \exp\left[\kappa t\left(a\tilde{a}-\tilde{a}^\dagger a^\dagger\right)\right] = \mathrm{e}^{-\kappa t}\left\langle\eta\mathrm{e}^{-\kappa t}\right|. \tag{9.77}$$

Combining Eqs. (9.76) and (9.77), we have

$$\langle\eta|\rho(t)\rangle = \mathrm{e}^{-\frac{1}{2}T|\eta|^2}\left\langle\eta\mathrm{e}^{-\kappa t}\right|\rho(0)\rangle, \tag{9.78}$$

where $T = 1-\mathrm{e}^{-2\kappa t}$. Substituting Eq. (9.78) into Eq. (9.69), thus the Wigner function at time t can be expressed as

$$W\left(\alpha,\alpha^*,t\right) = \int \frac{\mathrm{d}^2\eta}{2\pi^2} \exp(\alpha^*\eta-\alpha\eta^*)\mathrm{e}^{-\frac{1}{2}T|\eta|^2}\left\langle\eta\mathrm{e}^{-\kappa t}\right|\rho(0)\rangle. \tag{9.79}$$

Noting that the initial Wigner function can be represented as $W(\beta,\beta^*,0) = \langle\eta = 2\beta|\rho(0)\rangle$, so using the completeness relation in Eq. (9.23), we obtain

$$\begin{aligned}
W\left(\alpha,\alpha^*,t\right) &= \int \frac{\mathrm{d}^2 v'}{\pi} \int \frac{\mathrm{d}^2\eta}{2\pi^2} \exp\left(\alpha^*\eta-\alpha\eta^*-\frac{T}{2}|\eta|^2\right) \\
&\quad \times \left\langle\eta\mathrm{e}^{-\kappa t}\right| v' = 2\beta\rangle \langle v' = 2\beta|\rho(0)\rangle \\
&= \int \frac{\mathrm{d}^2\beta}{\pi} \int \frac{\mathrm{d}^2\eta}{\pi} \exp\left[-\frac{T}{2}|\eta|^2+\eta\left(\alpha^*-\beta^*\mathrm{e}^{-\kappa t}\right)\right. \\
&\quad \left.+\eta^*\left(\beta\mathrm{e}^{-\kappa t}-\alpha\right)\right] W\left(\beta,\beta^*,0\right) \\
&= \frac{2}{T}\int \frac{\mathrm{d}^2\beta}{\pi} \exp\left(-\frac{2}{T}\left|\alpha-\beta\mathrm{e}^{-\kappa t}\right|^2\right) W\left(\beta,\beta^*,0\right).
\end{aligned} \tag{9.80}$$

For convenience, making the variable substitution, i.e., $\left(\alpha-\beta\mathrm{e}^{-\kappa t}\right)/\sqrt{T} \to \beta$ in Eq. (9.80), thus we have

$$\begin{aligned}
W\left(\alpha,\alpha^*,t\right) &= \frac{2}{T}\int \frac{\mathrm{d}^2\beta}{\pi} \exp\left(-\frac{2}{T}\left|\alpha-\beta\mathrm{e}^{-\kappa t}\right|^2\right) W\left(\beta,\beta^*,0\right) \\
&= 2\mathrm{e}^{2\kappa t}\int \mathrm{d}^2\beta W_{|0\rangle\langle 0|}\left(\beta,\beta^*\right) W\left(\alpha\mathrm{e}^{\kappa t}-\sqrt{T}\beta\mathrm{e}^{\kappa t}0\right),
\end{aligned} \tag{9.81}$$

where $W_{|0\rangle\langle 0|}(\beta, \beta^*)$ is the Wigner function for vacuum, i.e., $W_{|0\rangle\langle 0|}(\beta, \beta^*) = \pi^{-1}\mathrm{e}^{-2|\beta|^2}$. For the systems in other decoherence channels, the time-evolutions of the Wigner functions are also obtained using similar methods as mentioned above.

References

1. Kelley KL, Kleiner WH (1964) Theory of electromagnetic field measurement and photoelectron counting. Phys Rev 136(2A):A316
2. Wan ZL, Fan HY (2016) Expectation value theorem for thermo vacuum states of optical chaotic field and negative-binomial field. Int J Theor Phys 55(7):3164–3172
3. Meng XG, Wang JS, Liang BL (2023) Entangled state representations in quantum optics. Springer, Singapore
4. Fan HY, Hu LY (2008) Two quantum-mechanical photocount formulas. Opt Lett 33(5):443–445
5. Fan HY, Hu LY (2008) Operator-sum representation of density operators as solution to master equations obtained via the entangled states approach. Mod Phys Lett B 22(25):2435–2468
6. Meng XG, Wang JS, Liang BL, Han CX (2018) Evolution of a two-mode squeezed vacuum for amplitude decay via continuous-variable entangled state approach. Front Phys 13(5):130322
7. Fan HY (2004) Entangled states, squeezed states gained via the route of developing Dirac's symbolic method and their applications. Int J Mod Phys B 18(10–11):1387–1455
8. Ren G, Du JM, Yu HJ, Fan HY (2017) New photon-counting formula in thermo field dynamics. J Mod Opt 64(7):681–684
9. Zhou NR, Hu LY, Fan HY (2011) Dissipation of a two-mode squeezed vacuum state in the single-mode amplitude damping channel. Chin Phys B 20(12):120301
10. Fan HY, Lou SY, Hu LY (2013) Damping law of photocount distribution in a dissipative channel. Chin Phys Lett 30(9):090304
11. Yu ZS, Ren GH, Yu ZY, Wei CHN, Fan HY (2018) Time evolution of the wigner operator as a quasi-density operator in amplitude dessipative channel. Int J Theor Phys 57(6):1888–1893
12. Chen F, Fan HY (2012) A new approach to the time evolution of characteristic function of the density operator obtained by virtue of thermal entangled state representation. Sci China Phys Mech Astron 55(11):2076–2080

Chapter 10
Operator Hermite Polynomials and Their Applications

In the course of "Mathematical and Physical Equations", we have learned that the set of Hermite polynomials can form a Hilbert space and also have complete orthogonality. Here, we would like to point out that the operator Hermite polynomials have the unique convenience for describing certain quantum light fields, finding new generating functions for Hermite polynomials and proposing some new special functions.

10.1 Introducing Operator Hermite Polynomials

According to the generating function for the Hermite polynomial [1–4], i.e.,

$$\mathrm{e}^{2\lambda x-\lambda^2}=\sum_{n=0}^{\infty}\frac{\lambda^n}{n!}\mathrm{H}_n(x), \tag{10.1}$$

and using the identity of summation, i.e.,

$$\sum_{n=0}^{\infty}\sum_{m=0}^{\infty}A(m,n)=\sum_{n=0}^{\infty}\sum_{m=0}^{[n/2]}A(m,n-2m), \tag{10.2}$$

we obtain

$$\begin{aligned}\sum_{n=0}^{\infty}\frac{(2\lambda)^n}{n!}x^n=\mathrm{e}^{2\lambda x}&=\sum_{m=0}^{\infty}\frac{\lambda^{2m}}{m!}\sum_{n=0}^{\infty}\frac{\lambda^n\mathrm{H}_n(x)}{n!}\\&=\sum_{n=0}^{\infty}\sum_{m=0}^{[n/2]}\frac{\lambda^n\mathrm{H}_{n-2m}(x)}{m!(n-2m)!}.\end{aligned} \tag{10.3}$$

H. Fan and L. Hu, *Dissipation and Diffusion of Quantum Light Field*,
https://doi.org/10.1007/978-981-95-1528-8_10

Comparing the coefficients of λ^n on both sides of Eq. (10.3), we can see that, using the polynomials $\mathrm{H}_n(x)$, x^n can be expanded as

$$x^n = \sum_{m=0}^{[n/2]} \frac{n!\mathrm{H}_{n-2m}(x)}{2^n m!(n-2m)!}. \tag{10.4}$$

So what is its inverse relation? To conveniently solve this problem, we turn to consider the operator Hermite polynomial $\mathrm{H}_n(X)$ with the coordinate operator X as the variable. Noting that the relation between X and $(a, a^\dagger)$ is

$$X = \sqrt{\frac{\hbar}{2m\omega}}\left(a + a^\dagger\right), \quad \left[a, a^\dagger\right] = 1 \tag{10.5}$$

and, for convenience, taking $\hbar = 1$, $m = 1$ and $\omega = 1$, so we have

$$\begin{aligned}\sum_{n=0}^{\infty} \frac{\lambda^n}{n!}\mathrm{H}_n(X) &= \mathrm{e}^{2\lambda X - \lambda^2} = \mathrm{e}^{\sqrt{2}(a+a^\dagger)\lambda - \lambda^2} \\ &= \mathrm{e}^{\sqrt{2}a^\dagger\lambda}\mathrm{e}^{\sqrt{2}a\lambda} =: \mathrm{e}^{\sqrt{2}a^\dagger\lambda}\mathrm{e}^{\sqrt{2}a\lambda}: \\ &= :\mathrm{e}^{2\lambda X}: =: \sum_{n=0}^{\infty}\frac{(2\lambda)^n}{n!}X^n :. \end{aligned}\tag{10.6}$$

Comparing the coefficients of λ^n on both sides of Eq. (10.6), we obtain

$$\mathrm{H}_n(X) = 2^n : X^n :. \tag{10.7}$$

Combining Eqs. (10.4) and (10.7) leads to

$$X^n = \sum_{m=0}^{[n/2]} \frac{n!\mathrm{H}_{n-2m}(X)}{2^n m!(n-2m)!} = \sum_{m=0}^{[n/2]} \frac{n!}{2^{2m} m!(n-2m)!} : X^{n-2m} :. \tag{10.8}$$

Noting that

$$\sum_{n=0}^{\infty} \frac{(-\lambda)^n X^n}{n!} = \mathrm{e}^{-\lambda X} =: \mathrm{e}^{\lambda^2/4 - \lambda X}: = \sum_{n=0}^{\infty} \frac{(\mathrm{i}\lambda/2)^n}{n!} : \mathrm{H}_n(\mathrm{i}X): \tag{10.9}$$

and comparing the coefficients of λ^n on both sides of Eq. (10.9), we have

$$X^n = (2\mathrm{i})^{-n} : \mathrm{H}_n(\mathrm{i}X) :. \tag{10.10}$$

Using Eqs. (10.8) and (10.10), we obtain

$$(2\mathrm{i})^n \sum_{m=0}^{[n/2]} \frac{n!\!: X^{n-2m}:}{2^{2m}m!\,(n-2m)!} =: \mathrm{H}_n\,(\mathrm{i}X)\,:, \tag{10.11}$$

which means

$$\mathrm{H}_n\,(x) = 2^n \sum_{k=0}^{[n/2]} \frac{(-1)^k\, n!}{2^{2k}k!\,(n-2k)!} x^{n-2k}, \tag{10.12}$$

which is the inverse relation of Eq. (10.4) (some literatures define it as a Hermite polynomial). As an application, using the operator Hermite polynomial as a function of the orthogonal component of the field (e.g., electric field operator) to excite vacuum, i.e.,

$$C_n X^n \left|0\right\rangle = (2\mathrm{i})^{-n} : \mathrm{H}_n\left(\mathrm{i}\frac{a+a^\dagger}{\sqrt{2}}\right) : \left|0\right\rangle = (2\mathrm{i})^{-n}\,\mathrm{H}_n\left(\frac{\mathrm{i}a^\dagger}{\sqrt{2}}\right)\left|0\right\rangle, \tag{10.13}$$

we obtain a Hermite state, where the normalization coefficient C_n is determined by

$$\begin{aligned} 1 &= |C_n|^2 \left\langle 0\right| X^{2n} \left|0\right\rangle = |C_n|^2\,(2\mathrm{i})^{-2n} \left\langle 0\right| : \mathrm{H}_{2n}\,(\mathrm{i}X) : \left|0\right\rangle \\ &= |C_n|^2\,(2\mathrm{i})^{-2n}\,\mathrm{H}_{2n}\,(0)\,. \end{aligned} \tag{10.14}$$

On the other hand, it is worth noting that annihilating m photons in a squeezed vacuum state can also lead to a Hermite state of light field. To illustrate this point, we first use the IWOP method and Eq. (1.57) to derive an operator identity:

$$\begin{aligned} a^m \mathrm{e}^{\nu a^{\dagger 2}} &= \int \frac{\mathrm{d}^2 z}{\pi} z^m \mathrm{e}^{\nu z^{*2}} \left|z\right\rangle\left\langle z\right| \\ &= \int \frac{\mathrm{d}^2 z}{\pi} z^m : \exp\left(-|z|^2 + z a^\dagger + z^* a + \nu z^{*2} - a^\dagger a\right) : \\ &= \frac{\partial^m}{\partial\lambda^m} \int \frac{\mathrm{d}^2 z}{\pi} : \exp\left[-|z|^2 + z\left(a^\dagger + \lambda\right) + z^* a + \nu z^{*2} - a^\dagger a\right] : \Big|_{\lambda=0} \\ &= \frac{\partial^m}{\partial\lambda^m} : \exp\left[\left(a^\dagger + \lambda\right) a + \nu\left(a^\dagger + \lambda\right)^2 - a^\dagger a\right] : \Big|_{\lambda=0} \\ &= \mathrm{e}^{\nu a^{\dagger 2}} \frac{\partial^m}{\partial\lambda^m} : \exp\left[\lambda\left(2a^\dagger \nu + a\right) + \nu\lambda^2\right] : \Big|_{\lambda=0} \\ &= \left(\mathrm{i}\sqrt{\nu}\right)^m \mathrm{e}^{\nu a^{\dagger 2}} \frac{\partial^m}{\partial\left(\mathrm{i}\sqrt{\nu}\lambda\right)^m} : \exp\left[2\left(\mathrm{i}\sqrt{\nu}\lambda\right)\frac{2a^\dagger \nu + a}{\mathrm{i}2\sqrt{\nu}} - \left(\mathrm{i}\sqrt{\nu}\lambda\right)^2\right] : \Big|_{\lambda=0} \\ &= \left(\mathrm{i}\sqrt{\nu}\right)^m \mathrm{e}^{\nu a^{\dagger 2}} : \mathrm{H}_m\left(\frac{2a^\dagger \nu + a}{\mathrm{i}2\sqrt{\nu}}\right) :. \end{aligned} \tag{10.15}$$

So, the final state of annihilating m photons in a squeezed vacuum state is

$$a^m \mathrm{e}^{\nu a^{\dagger 2}} \left|0\right\rangle = \left(\mathrm{i}\sqrt{\nu}\right)^m \mathrm{e}^{\nu a^{\dagger 2}} \mathrm{H}_m\left(-\mathrm{i}\sqrt{\nu} a^\dagger\right)\left|0\right\rangle, \tag{10.16}$$

which is a Hermite state of light field.

Using Eqs. (1.78) and (10.7), we have

$$\sum_{m=0}^{\infty}\frac{t^m}{m!}\mathrm{H}_{2m}(X)=\sum_{m=0}^{\infty}\frac{t^m}{m!}4^m:X^{2m}:=:\mathrm{e}^{4tX^2}:=\sqrt{\frac{1}{1+4t}}\mathrm{e}^{\frac{4t}{1+4t}X^2}. \tag{10.17}$$

Since

$$\mathrm{H}_{2m}(X)|0\rangle=2^{2m}:X^{2m}:|0\rangle=2^m a^{\dagger 2m}|0\rangle, \tag{10.18}$$

thus we have

$$\sum_{m=0}^{\infty}\frac{t^m}{m!}\mathrm{H}_{2m}(X)|0\rangle=\sum_{m=0}^{\infty}\frac{t^m}{m!}2^m a^{\dagger 2m}|0\rangle=\mathrm{e}^{2ta^{\dagger 2}}|0\rangle. \tag{10.19}$$

By comparing Eq. (1.77), we can see that $\mathrm{e}^{2ta^{\dagger 2}}|0\rangle$ in Eq. (10.19) refers to a squeezed state that is the superposition of even oder Hermite states $\mathrm{H}_{2m}(X)|0\rangle$. Using the IWOP method, we find that the normalized form of the state $\mathrm{e}^{2ta^{\dagger 2}}|0\rangle$ reads

$$\sqrt{1-16t^2}\mathrm{e}^{2ta^{\dagger 2}}|0\rangle. \tag{10.20}$$

Letting $t=(\tanh\lambda)/4$, thus

$$\operatorname{sech}^{1/2}\lambda\mathrm{e}^{\frac{1}{2}a^{\dagger 2}\tanh\lambda}|0\rangle=\operatorname{sech}^{1/2}\lambda\sum_{m=0}^{\infty}\frac{1}{m!}\left(\frac{1}{4}\tanh\lambda\right)^m\mathrm{H}_{2m}(X)|0\rangle, \tag{10.21}$$

it follows that its wave function reads

$$\begin{aligned}&\operatorname{sech}^{1/2}\lambda\langle x|\mathrm{e}^{\frac{1}{2}a^{\dagger 2}\tanh\lambda}|0\rangle\\&=\operatorname{sech}^{1/2}\lambda\mathrm{e}^{-x^2/2}\sum_{m=0}^{\infty}\frac{1}{m!}\left(\frac{1}{4}\tanh\lambda\right)^m\mathrm{H}_{2m}(x)\\&=\mathrm{e}^{-\lambda/2}\mathrm{e}^{-x^2/2}\mathrm{e}^{\frac{\tanh\lambda}{1+\tanh\lambda}x^2}.\end{aligned} \tag{10.22}$$

Besides, we want to discuss the two-variable Hermite polynomials in the coherent state representation [4]. We know that the normal ordering product of the coherent state projection operator $|z\rangle\langle z|$ is

$$|z\rangle\langle z|=:\exp\left(-|z|^2+za^\dagger+z^*a-a^\dagger a\right):. \tag{10.23}$$

The generating function for the two-variable Hermite polynomial $\mathrm{H}_{m,n}(\zeta,\zeta^*)$ is

$$\exp(-tt' + t\zeta + t'\zeta^*) = \sum_{m,n=0}^{\infty} \frac{t^m t'^n}{m!n!} \mathrm{H}_{m,n}\left(\zeta, \zeta^*\right), \tag{10.24}$$

which is a non-degenerate extension of the generating function for the single-variable Hermite polynomial in Eq. (10.1), so $\mathrm{H}_{m,n}(\zeta, \zeta^*)$ is called the two-variable Hermite polynomial. By direct differentiation, the series representation of $\mathrm{H}_{m,n}(\zeta, \zeta^*)$ is obtained as:

$$\begin{aligned} \mathrm{H}_{m,n}\left(\zeta, \zeta^*\right) &= \frac{\partial^{n+m}}{\partial t^m \partial t'^n} \exp(-tt' + t\zeta + t'\zeta^*)|_{t=t'=0} \\ &= \sum_{l=0}^{\min(m,n)} \frac{n!m!\,(-1)^l}{l!\,(m-l)!\,(n-l)!} \zeta^{m-l} \zeta^{*n-l}. \end{aligned} \tag{10.25}$$

Using Eqs. (10.24), (10.23) can be expanded as

$$|z\rangle \langle z| = \mathrm{e}^{-|z|^2} \sum_{m,n=0}^{\infty} : \frac{a^{\dagger m} a^n}{m!n!} \mathrm{H}_{m,n}\left(z, z^*\right) :, \tag{10.26}$$

it follows that the completeness relation of coherent states $|z\rangle$ can be written as

$$\begin{aligned} 1 &= \int \frac{\mathrm{d}^2 z}{\pi} |z\rangle \langle z| = \int \frac{\mathrm{d}^2 z}{\pi} : \mathrm{e}^{-(z^* - a^\dagger)(z-a)} : \\ &= \int \frac{\mathrm{d}^2 z}{\pi} \mathrm{e}^{-|z|^2} \sum_{m,n=0}^{\infty} : \frac{a^{\dagger m} a^n}{m!n!} \mathrm{H}_{m,n}\left(z, z^*\right) :, \end{aligned} \tag{10.27}$$

which leads to a new integration formula

$$\int \frac{\mathrm{d}^2 z}{\pi} \mathrm{e}^{-|z|^2} \mathrm{H}_{m,n}\left(z, z^*\right) = \delta_{m,0}\delta_{n,0}. \tag{10.28}$$

Using Eq. (10.26) and

$$a^n |l\rangle = \sqrt{\frac{l!}{(l-n)!}} |l-n\rangle, \quad |l\rangle = \frac{a^{\dagger l}}{\sqrt{l!}} |0\rangle, \tag{10.29}$$

we can calculate

$$\begin{aligned} \langle l | z\rangle \langle z | l\rangle &= \mathrm{e}^{-|z|^2} \langle l| \sum_{m,n=0}^{\infty} \frac{a^{\dagger m} a^n}{m!n!} \mathrm{H}_{m,n}\left(z, z^*\right) |l\rangle \\ &= \mathrm{e}^{-|z|^2} \langle l-m| \sum_{m,n=0}^{l} \frac{1}{m!n!} \sqrt{\frac{l!l!}{(l-m)!\,(l-n)!}} \mathrm{H}_{m,n}\left(z, z^*\right) |l-n\rangle \end{aligned}$$

$$= e^{-|z|^2} \sum_{n=0}^{l} \frac{l!}{(l-n)!n!n!} H_{n,n}\left(z, z^*\right), \tag{10.30}$$

where

$$\begin{aligned} H_{n,n}\left(z, z^*\right) &= n! \sum_{k=0}^{n} \frac{n!\,(-1)^k}{k!\,(n-k)!\,(n-k)!} |z|^{2(n-k)} \\ &= n! \sum_{k=0}^{n} \frac{n!\,(-1)^{n-k}}{k!k!\,(n-k)!} |z|^{2k} \\ &= n!\,(-1)^n L_n\left(|z|^2\right), \end{aligned} \tag{10.31}$$

and $L_n(x)$ is the Laugerre polynomial with the specific form

$$L_n(x) = \sum_{k=0}^{n} (-1)^k \binom{n}{n-k} \frac{x^k}{k!}. \tag{10.32}$$

On the other hand, we easily know

$$\langle l\,|z\rangle\,\langle z|\,l\rangle = e^{-|z|^2} \frac{|z|^{2l}}{l!}. \tag{10.33}$$

Comparing Eqs. (10.30), (10.31) and (10.33), we have

$$\sum_{n=0}^{l} (-1)^n \binom{l}{n} L_n\left(|z|^2\right) = \frac{|z|^{2l}}{l!}. \tag{10.34}$$

Letting $x = |z|^2$, Eq. (10.34) becomes

$$\sum_{n=0}^{l} (-1)^n \binom{l}{n} L_n(x) = \frac{x^l}{l!}, \tag{10.35}$$

which is the inverse relation of Eq. (10.32) and is not recorded in previous mathematical manuals. So, from the coherent state representation, we can find some new properties of Laguerre polynomials. Using them, we can also obtain the new relation of Hermite polynomials:

$$\sum_{n=0}^{l} (-1)^n \binom{l}{n} \sum_{k=0}^{n} \binom{n}{k} \frac{(-1)^k}{2^k k!} H_k(x) = \frac{1}{2^l l!} H_l(x). \tag{10.36}$$

Proof. Making the substitutions $x \to X$ in Eq. (10.36) leads to

$$\sum_{n=0}^{l}(-1)^n\binom{l}{n}\sum_{k=0}^{n}\binom{n}{k}\frac{(-1)^k}{2^k k!}\mathrm{H}_k(X)=\frac{1}{2^l l!}\mathrm{H}_l(X). \tag{10.37}$$

So, in order to prove Eq. (10.36), we only need to prove Eq. (10.37). Noting that

$$\sum_{k=0}^{n}\binom{n}{k}\frac{(-1)^k}{2^k k!}\mathrm{H}_k(X)=\sum_{k=0}^{n}\binom{n}{k}\frac{(-1)^k}{k!}:X^k:\;=:\mathrm{L}_n(X):, \tag{10.38}$$

and using the summation formula of Laguerre polynomials in normal ordering in Eq. (10.32), and the operator summation formula

$$\sum_{n=0}^{\infty}(-1)^n\binom{l}{n}\mathrm{L}_n(X)=\frac{X^l}{l!}, \tag{10.39}$$

the left side of Eq. (10.37) becomes

$$\begin{aligned}&\sum_{n=0}^{l}(-1)^n\binom{l}{n}\sum_{k=0}^{n}\binom{n}{k}\frac{(-1)^k}{k!}:X^k:\\&=\sum_{n=0}^{l}(-1)^n\binom{l}{n}:\mathrm{L}_n(X):\\&=:\frac{X^l}{l!}:\;=\frac{1}{2^l l!}\mathrm{H}_l(X),\end{aligned} \tag{10.40}$$

thus the proof of Eq. (10.37) is completed. The above discussion also indicates that Laguerre polynomials can be derived from Hermite polynomials.

10.2 Hermite Polynomial Excited Quantum States of Light Fields

10.2.1 Coherent States as Input States

Operating the operator Hermite polynomials $\mathrm{H}_m\left(ga^\dagger\right)$ as a function of the creation operator $a^\dagger$ on the coherent states, we get

$$|\psi\rangle=\mathfrak{N}^{-\frac{1}{2}}\mathrm{H}_m\left(ga^\dagger\right)D(\alpha)|0\rangle=\mathfrak{N}^{-\frac{1}{2}}\mathrm{H}_m\left(ga^\dagger\right)|\alpha\rangle, \tag{10.41}$$

where $D(\alpha)=\mathrm{e}^{\alpha a^\dagger-\alpha^* a}$ is the displaced operator [5], g is real, $\mathfrak{N}$ is an undetermined normalization factor, so we call the states $|\psi\rangle$ Hermite polynomial excited coherent states.

For calculating the factor $\mathfrak{N}$, we need to know the normal ordering product of the operator $\mathrm{H}_n(fa)\mathrm{H}_m(ga^\dagger)$. For this, we need to introduce the generating function for $\mathrm{H}_n(fa)\mathrm{H}_m(ga^\dagger)$, which is of the form:

$$\begin{aligned}
&\sum_{n,m=0}^{\infty}\frac{t^n s^m}{n!m!}\mathrm{H}_n(fa)\,\mathrm{H}_m(ga^\dagger)\\
&=\mathrm{e}^{-t^2+2tfa}\mathrm{e}^{-s^2+2sga^\dagger}=\mathrm{e}^{4ftsg}:\mathrm{e}^{-t^2+2tfa-s^2+2sga^\dagger}:\\
&=\sum_{k=0}^{\infty}\frac{(4ftsg)^k}{k!}:\mathrm{e}^{-t^2+2tfa-s^2+2sga^\dagger}:\\
&=\sum_{k=0}^{\infty}\frac{1}{k!}:\left(\frac{\partial}{\partial a^\dagger}\right)^k\left(\frac{\partial}{\partial a}\right)^k\mathrm{e}^{-t^2+2tfa-s^2+2sga^\dagger}:\\
&=\sum_{k=0}^{\infty}\frac{(fg)^k}{k!}:\left(\frac{\partial}{g\partial a^\dagger}\right)^k\left(\frac{\partial}{f\partial a}\right)^k\sum_{n,m=0}^{\infty}\frac{t^n s^m}{n!m!}\mathrm{H}_m(ga^\dagger)\,\mathrm{H}_n(fa):.
\end{aligned} \tag{10.42}$$

Using the recurrence relation of Hermite polynomials, that is

$$\frac{\mathrm{d}^l}{\mathrm{d}x^l}\mathrm{H}_n(x)=\frac{2^l n!}{(n-l)!}\mathrm{H}_{n-l}(x), \tag{10.43}$$

Equation (10.42) can be expressed as

$$\begin{aligned}
&\sum_{n,m=0}^{\infty}\frac{t^n s^m}{n!m!}\mathrm{H}_n(fa)\,\mathrm{H}_m(ga^\dagger)\\
&=\sum_{n,m=0}^{\infty}\frac{t^n s^m}{n!m!}\sum_{k=0}^{\min(n,m)}k!\,(4fg)^k\\
&\quad\times\binom{n}{k}\binom{m}{k}:\mathrm{H}_{m-k}(ga^\dagger)\,\mathrm{H}_{n-k}(fa):,
\end{aligned} \tag{10.44}$$

which leads to

$$\begin{aligned}
\mathrm{H}_n(fa)\,\mathrm{H}_m(ga^\dagger)=&\sum_{k=0}^{\min(n,m)}k!\,(4fg)^k\binom{n}{k}\\
&\times\binom{m}{k}\mathrm{H}_{m-k}(ga^\dagger)\,\mathrm{H}_{n-k}(fa).
\end{aligned} \tag{10.45}$$

Using Eq. (10.45), we can calculate the normalization of the state $|\psi\rangle$ as

$$1=\mathfrak{N}^{-1}\langle\alpha|\,\mathrm{H}_m(g^*a)\,\mathrm{H}_m(ga^\dagger)\,|\alpha\rangle$$

$$= \mathfrak{N}^{-1} \sum_{k=0}^{m} \frac{\left(4\left|g\right|^2\right)^k (m!)^2}{k!\left[(m-k)!\right]^2} \left|\mathrm{H}_{m-k}\left(g\alpha^*\right)\right|^2, \tag{10.46}$$

it follows that the normalization factor $\mathfrak{N}$ reads

$$\mathfrak{N} = \sum_{k=0}^{m} \frac{\left(4\left|g\right|^2\right)^k (m!)^2}{k!\left[(m-k)!\right]^2} \left|\mathrm{H}_{m-k}\left(g\alpha^*\right)\right|^2. \tag{10.47}$$

On the other hand, using the completeness relation of coherent states in Eq. (3.103) and the P-representation, that is $\rho = \pi^{-1} \int \mathrm{d}^2 z \mathrm{P}(z) \left|z\right\rangle \left\langle z\right|$, we therefore have

$$\begin{aligned} &\mathrm{H}_n(fa)\,\mathrm{H}_m\left(ga^\dagger\right) \\ &= \int \frac{\mathrm{d}^2 z}{\pi} \mathrm{H}_n(fz) \left|z\right\rangle \left\langle z\right| \mathrm{H}_m\left(gz^*\right) \\ &= \int \frac{\mathrm{d}^2 z}{\pi} :\mathrm{H}_n(fz)\,\mathrm{H}_m\left(gz^*\right) \exp\left[-\left(z^* - a^\dagger\right)(z-a)\right]:. \end{aligned} \tag{10.48}$$

Considering that Eqs. (10.45) and (10.48) are within the normal ordering symbol $: \ :$, thus making the following substitutions $a^\dagger \to \lambda$ and $a \to \sigma$ in Eqs. (10.45) and (10.48), we naturally obtain a new integration formula

$$\begin{aligned} &\int \frac{\mathrm{d}^2 z}{\pi} \mathrm{H}_n(fz)\,\mathrm{H}_m\left(gz^*\right) \exp\left[-\left(z^* - \lambda\right)(z-\sigma)\right] \\ &= \sum_{k=0}^{\min(n,m)} k!\,(4fg)^k \binom{n}{k} \binom{m}{k} \mathrm{H}_{n-k}(f\sigma)\,\mathrm{H}_{m-k}(g\lambda). \end{aligned} \tag{10.49}$$

Next, we consider the normal ordering product of the operator $\mathrm{e}^{\beta a}\mathrm{H}_n(fa)\mathrm{H}_m\left(ga^\dagger\right)\mathrm{e}^{\gamma a^\dagger}$. Using the same method as obtaining Eq. (10.48), we have

$$\begin{aligned} &\mathrm{e}^{\beta a}\mathrm{H}_n(fa)\,\mathrm{H}_m\left(ga^\dagger\right)\mathrm{e}^{\gamma a^\dagger} \\ &= \int \frac{\mathrm{d}^2 z}{\pi} \mathrm{e}^{\beta z}\mathrm{H}_n(fz) \left|z\right\rangle \left\langle z\right| \mathrm{H}_m\left(gz^*\right)\mathrm{e}^{\gamma z^*} \\ &= \int \frac{\mathrm{d}^2 z}{\pi} :\mathrm{H}_n(fz)\,\mathrm{H}_m\left(gz^*\right) \mathrm{e}^{-|z|^2 + z\left(a^\dagger+\beta\right) + z^*(a+\gamma) - a^\dagger a}: \\ &= \int \frac{\mathrm{d}^2 z}{\pi} :\mathrm{H}_n(fz)\,\mathrm{H}_m\left(gz^*\right) \exp\{-\left[\left(z^* - \left(a^\dagger + \beta\right)\right)\right] \\ &\times \left[z - (a+\gamma)\right] + \beta a + \gamma a^\dagger + \beta\gamma\}: \\ &= \sum_{k=0}^{\min(n,m)} k!\,(4fg)^k \binom{n}{k} \binom{m}{k} :\mathrm{H}_{m-k}\left[g\left(a^\dagger + \beta\right)\right] \end{aligned}$$

$$\times \mathrm{H}_{n-k}\left[f\left(a+\gamma\right)\right]\mathrm{e}^{\beta a+\gamma a^{\dagger}+\beta\gamma}:$$

$$= \mathrm{e}^{\alpha a^{\dagger}} \sum_{k=0}^{\min(n,m)} k!\,(4fg)^{k} \binom{n}{k}\binom{m}{k}$$

$$\times \mathrm{H}_{m-k}\left[g\left(a^{\dagger}+\beta\right)\right]\mathrm{H}_{n-k}\left[f\left(a+\gamma\right)\right]\mathrm{e}^{\beta\gamma}. \tag{10.50}$$

Combining Eqs. (10.45) and (10.46), some new identities can be obtained. For examples, letting $g = \mathrm{i}/\sqrt{2}$ and $\alpha = 0$, thus we have

$$\langle 0|\,\mathrm{H}_m\left(\frac{-\mathrm{i}a}{\sqrt{2}}\right)\mathrm{H}_m\left(\frac{\mathrm{i}a^{\dagger}}{\sqrt{2}}\right)|0\rangle = \sum_{k=0}^{m}\frac{(m!)^2}{k!\,[(m-k)!]^2}\left|\mathrm{H}_{m-k}\left(0\right)\right|^2. \tag{10.51}$$

Using the operator identity

$$X^n = (2\mathrm{i})^{-n} : \mathrm{H}_n\left(\mathrm{i}X\right):, \quad X = (a+a^{\dagger})/\sqrt{2}, \tag{10.52}$$

which can be obtain by comparing the same power in the following equation

$$\begin{aligned}\sum_{n=0}^{\infty}\frac{(2t)^n}{n!}X^n &= \mathrm{e}^{2tX} = \mathrm{e}^{\sqrt{2}t\left(a+a^{\dagger}\right)} \\ &= :\mathrm{e}^{2(-\mathrm{i}t)\mathrm{i}X-(-\mathrm{i}t)^2}: \\ &= \sum_{n=0}^{\infty}\frac{(-\mathrm{i}t)^n}{n!}:\mathrm{H}_n\left(\mathrm{i}X\right):,\end{aligned} \tag{10.53}$$

and noting that

$$X^m\,|0\rangle = (2\mathrm{i})^{-m} : \mathrm{H}_m\left(\mathrm{i}X\right): \,|0\rangle = (2\mathrm{i})^{-m}\,\mathrm{H}_m\left(\frac{\mathrm{i}a^{\dagger}}{\sqrt{2}}\right)|0\rangle\,, \tag{10.54}$$

so comparing Eqs. (10.51) and (10.54) leads to

$$\begin{aligned}\langle 0|X^{2m}\,|0\rangle &= \langle 0|X^m X^m\,|0\rangle \\ &= 4^{-m}\,\langle 0|\,\mathrm{H}_m\left(\frac{-\mathrm{i}a}{\sqrt{2}}\right)\mathrm{H}_m\left(\frac{\mathrm{i}a^{\dagger}}{\sqrt{2}}\right)|0\rangle \\ &= 4^{-m}\sum_{k=0}^{m}\frac{(m!)^2}{k!\,[(m-k)!]^2}\left|\mathrm{H}_{m-k}\left(0\right)\right|^2.\end{aligned} \tag{10.55}$$

On the other hand, using the operator identity

$$X^{2n} =: n!\mathrm{L}_n^{-1/2}\left(-X^2\right):, \tag{10.56}$$

we get

$$\langle 0|X^{2m}|0\rangle = \langle 0|:m!\mathrm{L}_m^{-1/2}\left(-X^2\right):|0\rangle = m!\mathrm{L}_m^{-1/2}(0). \tag{10.57}$$

So, comparing Eqs. (10.55) and (10.57), we give

$$(4)^{-m}\sum_{k=0}^{m}\frac{(m!)^2}{k!\left[(m-k)!\right]^2}\left|\mathrm{H}_{m-k}(0)\right|^2 = m!\mathrm{L}_m^{-1/2}(0). \tag{10.58}$$

Besides, the normal ordering product of the operator X^{2n} is expressed as

$$X^{2n} = (2\mathrm{i})^{-2n}:\mathrm{H}_{2n}(\mathrm{i}X):, \tag{10.59}$$

thus comparing Eqs. (10.56) and (10.59) leads to

$$(2\mathrm{i})^{-2n}:\mathrm{H}_{2n}(\mathrm{i}X):\; =: n!\mathrm{L}_n^{-1/2}\left(-X^2\right):. \tag{10.60}$$

Similarly, both sides of Eq. (10.60) are within the normal ordering, so we obtain the new identity

$$(2\mathrm{i})^{-2n}\mathrm{H}_{2n}(\mathrm{i}x) = n!\mathrm{L}_n^{-1/2}\left(-x^2\right) \tag{10.61}$$

10.2.2 Squeezed States as Input States

Usually, a two-mode squeezed vacuum state is expressed as [6]

$$S_2(r)|00\rangle = \exp\left[r\left(ab - a^\dagger b^\dagger\right)\right]|00\rangle = \mathrm{sech}r\exp\left(a^\dagger b^\dagger\tanh r\right)|00\rangle, \tag{10.62}$$

its corresponding integration form reads

$$S_2(r)|00\rangle = \frac{1}{\sinh r}\int\frac{\mathrm{d}^2z}{\pi}\exp\left(-\frac{|z|^2}{\tanh r} + za^\dagger + z^*b^\dagger\right)|00\rangle. \tag{10.63}$$

If the two-mode squeezed state is used as the input state for the two input ports of the beam splitter, the output state is

$$\begin{aligned}RS_2(r)|00\rangle &= \frac{1}{\sinh r}\int\frac{\mathrm{d}^2z}{\pi}\mathrm{e}^{-\frac{|z|^2}{\tanh r}}R\mathrm{e}^{za^\dagger + z^*b^\dagger}R^{-1}R|00\rangle\\ &= \frac{1}{\sinh r}\int\frac{\mathrm{d}^2z}{\pi}\mathrm{e}^{-\frac{|z|^2}{\tanh r}}\exp\left[b^\dagger\left(z^*\cos\theta - z\sin\theta\right)\right.\\ &\quad\left.+a^\dagger\left(z\cos\theta + z^*\sin\theta\right)\right]|00\rangle,\end{aligned} \tag{10.64}$$

where we have used the transformation relation of the beam splitter operator $R = \exp[\theta(a^\dagger b - ab^\dagger)]$, i.e.,

$$Ra^\dagger R^{-1} = a^\dagger \cos\theta - b^\dagger \sin\theta, \quad Rb^\dagger R^{-1} = b^\dagger \cos\theta + a^\dagger \sin\theta. \tag{10.65}$$

If the coherent state ${}_b\langle\alpha|$ is measured in the b-mode, thus the output state at the other port is a displaced thermal state, that is

$$\begin{aligned}
&{}_b\langle\alpha| RS_2(r)|00\rangle \\
&= \frac{1}{\sinh r}\, {}_b\langle\alpha| \int \frac{d^2z}{\pi} e^{-\frac{|z|^2}{\tanh r}} \exp[b^\dagger(z^*\cos\theta - z\sin\theta) \\
&+a^\dagger(z\cos\theta + z^*\sin\theta)]\,|0\rangle_a|0\rangle_b \\
&= \frac{1}{\sinh r}\, {}_b\langle 0| \int \frac{d^2z}{\pi} e^{-\frac{|z|^2}{\tanh r}} e^{\alpha^* b} \exp[b^\dagger(z^*\cos\theta - z\sin\theta) \\
&+a^\dagger(z\cos\theta + z^*\sin\theta)]\,|0\rangle_a|0\rangle_b \\
&= \frac{1}{\sinh r} \int \frac{d^2z}{\pi} e^{-\frac{|z|^2}{\tanh r}} e^{\alpha^*(z^*\cos\theta - z\sin\theta)} \exp\left[a^\dagger(z\cos\theta + z^*\sin\theta)\right]|0\rangle_a \\
&= \frac{1}{\sinh r} \int \frac{d^2z}{\pi} \exp\left[-\frac{|z|^2}{\tanh r} + z\left(a^\dagger\cos\theta - \alpha^*\sin\theta\right)\right. \\
&\left.+z^*\left(\alpha^*\cos\theta + a^\dagger\sin\theta\right)\right]|0\rangle_a \\
&= \mathrm{sech} r \exp\left[\tanh r\left(a^\dagger\cos\theta - \alpha^*\sin\theta\right)\left(\alpha^*\cos\theta + a^\dagger\sin\theta\right)\right]|0\rangle_a \\
&= \mathrm{sech} r \exp\left[\left(\frac{a^{\dagger 2}}{2} - \frac{\alpha^{*2}}{2}\right)\tanh r\sin 2\theta + \alpha^* a^\dagger\cos 2\theta\tanh r\right]|0\rangle_a.
\end{aligned} \tag{10.66}$$

If the number state ${}_b\langle m|$ is measured in the b-mode, the output quantum state is

$$\begin{aligned}
&{}_b\langle m| RS_2(r)|00\rangle \\
&= {}_b\langle 0| \frac{b^m}{\sqrt{m!}\sinh r} \int \frac{d^2z}{\pi} e^{-\frac{|z|^2}{\tanh r}} \\
&\times \exp\left[b^\dagger(z^*\cos\theta - z\sin\theta) + a^\dagger(z\cos\theta + z^*\sin\theta)\right]|0\rangle_a|0\rangle_b,
\end{aligned} \tag{10.67}$$

where

$$\begin{aligned}
&{}_b\langle 0| b^m e^{b^\dagger(z^*\cos\theta - z\sin\theta)}|0\rangle_b \\
&= \left(\frac{\partial}{\partial f}\right)^m {}_b\langle 0| e^{fb} e^{b^\dagger(z^*\cos\theta - z\sin\theta)}|0\rangle_b\Big|_{f=0} \\
&= \left(\frac{\partial}{\partial f}\right)^m \exp\left[f(z^*\cos\theta - z\sin\theta)\right]\Big|_{f=0}.
\end{aligned} \tag{10.68}$$

Substituting Eq. (10.68) into Eq. (10.67) leads to

$$\begin{aligned}
&{}_b\langle m|\, RS_2(r)\,|00\rangle \\
&= \frac{1}{\sqrt{m!}\sinh r}\left(\frac{\partial}{\partial f}\right)^m \int \frac{d^2 z}{\pi} e^{-\frac{|z|^2}{\tanh r}} \exp[a^\dagger \left(z\cos\theta + z^*\sin\theta\right) \\
&\quad + f\left(z^*\cos\theta - z\sin\theta\right)]\,|0\rangle_a\big|_{f=0} \\
&= \frac{1}{\sqrt{m!}\cosh r}\left(\frac{\partial}{\partial f}\right)^m \exp[\tanh r(a^\dagger\cos\theta \\
&\quad - f\sin\theta)\left(f\cos\theta + a^\dagger\sin\theta\right)]\,|0\rangle_a\big|_{f=0} \\
&= \frac{1}{\sqrt{m!}\cosh r}\left(\frac{\partial}{\partial f}\right)^m \exp\left[\left(\frac{a^{\dagger 2}}{2} - \frac{f^2}{2}\right)\tanh r\sin 2\theta \right. \\
&\quad \left. + f a^\dagger \cos 2\theta \tanh r\right]|0\rangle_a\big|_{f=0} \\
&= \frac{\left(\sqrt{\frac{\tanh r\sin 2\theta}{2}}\right)^m}{\sqrt{m!}\cosh r} e^{\frac{a^{\dagger 2}}{2}\tanh r\sin 2\theta}\left(\frac{\partial}{\partial f\sqrt{\frac{\tanh r\sin 2\theta}{2}}}\right)^m \\
&\quad \exp\left[-\left(f\sqrt{\frac{\tanh r\sin 2\theta}{2}}\right)^2 \right. \\
&\quad \left. + 2\left(f\sqrt{\frac{\tanh r\sin 2\theta}{2}}\right) a^\dagger \frac{\cos 2\theta\tanh r}{\sqrt{2\tanh r\sin 2\theta}}\right]|0\rangle_a\Big|_{f=0} \\
&= \frac{1}{\sqrt{m!}\cosh r}\left(\sqrt{\frac{\tanh r\sin 2\theta}{2}}\right)^m \mathrm{H}_m\left(a^\dagger\frac{\sqrt{\tanh r\cos 2\theta}}{\sqrt{2\tan 2\theta}}\right) \\
&\quad \times e^{\frac{a^{\dagger 2}}{2}\tanh r\sin 2\theta}\,|0\rangle_a\,.
\end{aligned} \tag{10.69}$$

In the last step in Eq. (10.69), we have used the generating function of the single-mode Hermite polynomial, i.e.,

$$\mathrm{H}_m(x) = \left(\frac{\partial}{\partial t}\right)^m e^{-t^2+2tx}\Big|_{t=0}. \tag{10.70}$$

Equation (10.69) shows that $e^{\frac{a^{\dagger 2}}{2}\tanh r\sin 2\theta}\,|0\rangle_a$ refers to a single-mode squeezed vacuum with squeezing parameter $\tanh r\sin(2\theta)$, which is smaller than the original squeezing parameter $\tanh r$ in the input squeezed vacuum, so $\mathrm{H}_m\left(a^\dagger\sqrt{\frac{\tanh r\cos 2\theta}{2\tan 2\theta}}\right)e^{\frac{a^{\dagger 2}}{2}\tanh r\sin 2\theta}\,|0\rangle_a$ is actually a Hermite polynomial excited squeezd vacuum.

When $\theta = \pi/4$, corresponding to the symmetrical beam splitter, Eq. (10.69) becomes a single-mode squeezed vacuum, that is

$$
\begin{aligned}
&{}_b\langle m|\, RS_2(r)\,|00\rangle \\
&= \frac{1}{\sqrt{m!}\cosh r}\left(\frac{\partial}{\partial f}\right)^m \exp\left[\left(\frac{a^{\dagger 2}}{2}-\frac{f^2}{2}\right)\tanh r\right]|0\rangle_a\,|_{f=0} \\
&= \frac{1}{\sqrt{m!}\cosh r}\left(\frac{\partial}{\partial f}\right)^m \exp\left(-\frac{f^2}{2}\tanh r+\frac{a^{\dagger 2}}{2}\tanh r\right)|0\rangle_a\Big|_{f=0} \\
&\rightarrow \left[\frac{1}{\sqrt{\cosh r}}\exp\left(\frac{a^{\dagger 2}}{2}\tanh r\right)|0\rangle_a\right].
\end{aligned} \tag{10.71}
$$

This is because when the squeezed vacuum state passes through a symmetrical beam splitter, the output state is no longer an entangled state, but a direct product state of two single-mode squeezed states. Therefore, measuring one output mode will not have any impact on the output mode at the other port. Or, in order to generate the Hermite polynomial excited squeezed states, the beam splitters used are required to be asymmetric.

As is well known, coherent states are related to Bargmann spaces through the basis functions:

$$
\frac{z^n}{\sqrt{n!}} = \langle z\,|n\rangle\,, \tag{10.72}
$$

where $|n\rangle = \frac{a^{\dagger m}}{\sqrt{m!}}\,|0\rangle$ is a Fock state. The basis functions can form an orthogonal complete function space, that is

$$
\int \frac{d^2 z}{\pi}\frac{z^n z^{*m}}{\sqrt{n!m!}}e^{-|z|^2} = \delta_{n,m}, \tag{10.73}
$$

where $e^{-|z|^2}$ is an integral weighted function. Using Eq. (10.73), we obtain

$$
\begin{aligned}
&\int \frac{d^2 z}{\pi}\,(z+i\sigma)^m\,(z^*+i\lambda)^n\,e^{-|z|^2} \\
&= \sum_{l,k=0}^{\infty}\binom{n}{l}\binom{m}{k}(i\lambda)^{n-l}\,(i\sigma)^{m-k}\int \frac{d^2 z}{\pi} z^{*l} z^k e^{-|z|^2} \\
&= \sum_{l,k=0}^{\infty}\binom{n}{l}\binom{m}{k}(i\lambda)^{n-l}\,(i\sigma)^{m-k}\sqrt{l!k!}\delta_{l,k} \\
&= i^{m+n}\sum_{l=0}^{\min(m,n)}\frac{(-1)^l m!n!}{l!(m-l)!(n-l)!}\sigma^{m-l}\lambda^{n-l}.
\end{aligned} \tag{10.74}
$$

Further, using the definition of two-variable Hermite polynomials $H_{m,n}(\xi,\xi^*)$, that is

$$
H_{m,n}(\xi,\xi^*) = \sum_{l=0}^{\min(m,n)}\frac{(-1)^l m!n!}{l!(m-l)!(n-l)!}\xi^{m-l}\xi^{*n-l}, \tag{10.75}
$$

so Eq. (10.74) becomes

$$\int \frac{d^2 z}{\pi} (z + i\sigma)^m \left(z^* + i\lambda\right)^n e^{-|z|^2} = i^{m+n} H_{m,n}(\sigma, \lambda) \tag{10.76}$$

or

$$\int \frac{d^2 z}{\pi} z^m z^{*n} e^{-(z^*-i\lambda)(z-i\sigma)} = i^{m+n} H_{m,n}(\sigma, \lambda). \tag{10.77}$$

Obviously, Eqs. (10.76) and (10.77) are the integral transformations from the Bargmann space to the new function space. Using Eqs. (3.103) and (10.77), we have

$$\begin{aligned} a^n a^{\dagger m} &= \int \frac{d^2 z}{\pi} a^n |z\rangle \langle z| a^{\dagger m} \\ &= \int \frac{d^2 z}{\pi} z^n z^{*m} : \exp\left[-\left(z^* - a^\dagger\right)(z - a)\right] : \\ &= (-i)^{m+n} : H_{m,n}\left(ia^\dagger, ia\right) :, \end{aligned} \tag{10.78}$$

which shows that, using the integral transform relation in Eq. (10.77), the anti-normal ordering operator $a^n a^{\dagger m}$ can be concisely and effectively rewritten as its normal ordering form. On the other hand, summing up the operator $a^n a^{\dagger m}$ leads to the result

$$\begin{aligned} \sum_{n,m=0}^{\infty} \frac{\tau^n t^m}{n!m!} a^n a^{\dagger m} &= e^{\tau a} e^{t a^\dagger} = : \exp\left(\tau a + t a^\dagger + \tau t\right) : \\ &= : \exp\left[(-i\tau)(ia) + (-it)\left(ia^\dagger\right) - (-i\tau)(-it)\right] :, \end{aligned} \tag{10.79}$$

it follows from Eq. (10.78) that

$$\sum_{n,m=0}^{\infty} \frac{\tau^n t^m}{n!m!} a^n a^{\dagger m} = \sum_{n,m=0}^{\infty} \frac{(-i\tau)^n (-it)^m}{n!m!} : H_{m,n}\left(ia^\dagger, ia\right) : . \tag{10.80}$$

Comparing Eqs. (10.79) and (10.80), we can obtain the generating function for $H_{n,m}(x, y)$ as

$$\sum_{n,m=0}^{\infty} \frac{t^n \tau^m}{n!m!} H_{n,m}(x, y) = \exp(tx + \tau y - t\tau). \tag{10.81}$$

Noting that the operators $a^\dagger, b^\dagger$ are commutative, so Eq. (10.81) becomes

$$\sum_{n,m=0}^{\infty} \frac{a^{\dagger n} b^{\dagger m}}{n!m!} H_{n,m}(\xi, \xi^*) = \exp\left(a^\dagger \xi + b^\dagger \xi^* - a^\dagger b^\dagger\right). \tag{10.82}$$

Comparing Eqs. (10.72) and (10.81), we can find that the polynomials $\mathrm{H}_{n,m}(\xi, \xi^*)$ are the bases of two-mode function space (orthogonal and complete). Therefore, when the basis function $\frac{z^n}{\sqrt{n!}}$ is extended to $\frac{1}{\sqrt{n!m!}}\mathrm{H}_{n,m}(\xi, \xi^*)$, it naturally accompanies the extension from coherent states to entangled states.

Noting that

$$\mathrm{H}_{n,m}(x, y) = \frac{\partial^{n+m}}{\partial t^n \partial \tau^m} \exp(tx + \tau y - t\tau)\,|_{t,\tau=0}, \tag{10.83}$$

thus the generating function of the single-variable Hermite polynomial in Eq. (10.70) are clearly the degeneracy of $\mathrm{H}_{n,m}(x, y)$ in Eq. (10.83).

Using the Baker-Hausdorff operator formula and Eq. (10.81), we can obtain the anti-normal ordering product of the operator $\mathrm{e}^{ta^\dagger}\mathrm{e}^{\tau a}$ as

$$\begin{aligned} \mathrm{e}^{ta^\dagger}\mathrm{e}^{\tau a} &= \vdots \exp\left(\tau a + ta^\dagger - \tau t\right) \vdots \\ &= \sum_{n,m=0}^{\infty} \frac{t^n \tau^m}{n!m!} \vdots \mathrm{H}_{n,m}\left(a^\dagger, a\right) \vdots \\ &= \sum_{n,m=0}^{\infty} \frac{t^n \tau^m}{n!m!} a^{\dagger n} a^m, \end{aligned} \tag{10.84}$$

which leads to the anti-normal ordering product of $a^{\dagger n}a^m$, i.e.,

$$a^{\dagger n}a^m = \vdots \mathrm{H}_{n,m}\left(a^\dagger, a\right) \vdots . \tag{10.85}$$

As an application, using Eq. (10.85) and the completeness relation of coherent states, we get

$$\begin{aligned} \vdots \mathrm{H}_{m,n}\left(a^\dagger, a\right) \vdots &= \int \frac{\mathrm{d}^2 z}{\pi} \vdots \mathrm{H}_{m,n}\left(a^\dagger, a\right) \vdots\, |z\rangle \langle z| \\ &=: \int \frac{\mathrm{d}^2 z}{\pi} \mathrm{H}_{m,n}\left(z^*, z\right) \exp\left[-\left(z^* - a^\dagger\right)(z - a)\right] : \\ &= a^{\dagger m} a^n, \end{aligned} \tag{10.86}$$

since the last two steps in Eq. (10.86) are within the normal ordering symbol : : , the operators $a^\dagger$ and a are replaced by the c numbers λ and σ, we thus obtain a new mathematical integral, that is

$$\int \frac{\mathrm{d}^2 z}{\pi} \mathrm{H}_{m,n}\left(z^*, z\right) \exp\left[-\left(z^* - \lambda\right)(z - \sigma)\right] = \lambda^m \sigma^n, \tag{10.87}$$

which is actually the inverse relation of Eq. (10.77).

10.3 Generalized Binomial Theorem

The binomial theorem refers to

$$(y+z)^n = \sum_{l=0}^{n} \binom{n}{l} y^l z^{n-l}, \tag{10.88}$$

when replacing y^l with a two-variable Hermite polynomial $\mathrm{H}_{n-l,l}(x,y)$ in Eq. (10.88), what results will be obtained?

We want to prove that:

$$\sum_{l=0}^{n} \binom{n}{l} z^{n-l} \mathrm{H}_{n-l,l}(x,y) = \sqrt{z^n}\mathrm{H}_n\left(\frac{zx+y}{2\sqrt{z}}\right). \tag{10.89}$$

To achieve this goal, we first replace $\mathrm{H}_{n-l,l}(x,y)$ with the anti-normal ordering operator $\vdots \mathrm{H}_{n-l,l}\left(a,a^\dagger\right)\vdots$ on the left side of Eq. (10.89), and then use Eq. (2.78) to calculate

$$\sum_{l=0}^{n} \binom{n}{l} z^{n-l} \vdots \mathrm{H}_{n-l,l}\left(a,a^\dagger\right)\vdots = \sum_{l=0}^{n} \binom{n}{l} z^{n-l} : a^{\dagger l} a^{n-l} : \; =: \left(a^\dagger + za\right)^n : . \tag{10.90}$$

Further, constructing the following summation, that is

$$\begin{aligned}\sum_{l=0}^{n} \frac{t^n}{n!} : (a^\dagger + za)^n : \; &= \; : \mathrm{e}^{t\left(a^\dagger + za\right)} : \; = \mathrm{e}^{ta^\dagger}\mathrm{e}^{tza} \\ &= \vdots \mathrm{e}^{tza+ta^\dagger} \vdots \mathrm{e}^{[ta^\dagger, tza]} = \vdots \mathrm{e}^{tza+ta^\dagger} \vdots \mathrm{e}^{-t^2 z} \\ &= \vdots \mathrm{e}^{2\left(\frac{\sqrt{z}a + a^\dagger/\sqrt{z}}{2}\right)(t\sqrt{z})} \vdots \mathrm{e}^{-(t\sqrt{z})^2} \\ &= \sum_{n=0}^{\infty} \frac{t^n \sqrt{z^n}}{n!} \vdots \mathrm{H}_n\left(\frac{\sqrt{z}a + a^\dagger/\sqrt{z}}{2}\right) \vdots \end{aligned} \tag{10.91}$$

and comparing the same power of t^n on both sides of Eq. (10.91), we have

$$: \left(a^\dagger + za\right)^n : \; = \sqrt{z^n} \vdots \mathrm{H}_n\left(\frac{\sqrt{z}a + a^\dagger/\sqrt{z}}{2}\right) \vdots . \tag{10.92}$$

Combining Eqs. (10.90) and (10.92), we give

$$\sum_{l=0}^{n} \binom{n}{l} z^{n-l} \vdots \mathrm{H}_{n-l,l}\left(a,a^\dagger\right)\vdots = \sqrt{z^n} \vdots \mathrm{H}_n\left(\frac{\sqrt{z}a + a^\dagger/\sqrt{z}}{2}\right) \vdots . \tag{10.93}$$

Noting that both sides of Eq. (10.93) are within anti-normal ordering, and the operators a and $a^{\dagger}$ are commutative in the symbol $\vdots\,\vdots$, so we obtain the generalized binomial theorem as

$$\sum_{l=0}^{n}\binom{n}{l}z^{n-l}\mathrm{H}_{n-l,l}\left(x,y\right)=\sqrt{z^{n}}\mathrm{H}_{n}\left(\frac{zx+y}{2\sqrt{z}}\right). \tag{10.94}$$

On the other hand, using the formula in Eq. (2.76), thus Eq. (10.92) becomes

$$\begin{aligned}:\left(a^{\dagger}+za\right)^{n}: &=\int\frac{\mathrm{d}^{2}\beta}{\pi}\vdots\left\langle-\beta\right|:\left(a^{\dagger}+za\right)^{n}:\left|\beta\right\rangle\mathrm{e}^{-|\beta|^{2}+\beta^{*}a-\beta a^{\dagger}+a^{\dagger}a}\vdots\\ &=\int\frac{\mathrm{d}^{2}\beta}{\pi}\vdots\left(-\beta^{*}+z\beta\right)^{n}\mathrm{e}^{-|\beta|^{2}+\beta^{*}a-\beta a^{\dagger}+a^{\dagger}a}\vdots\\ &=\sqrt{z^{n}}\vdots\mathrm{H}_{n}\left(\frac{\sqrt{z}a+a^{\dagger}/\sqrt{z}}{2}\right)\vdots,\end{aligned} \tag{10.95}$$

which leads to a new integral formula

$$\int\frac{\mathrm{d}^{2}\beta}{\pi}\left(z\beta-\beta^{*}\right)^{n}\mathrm{e}^{-|\beta|^{2}+\beta^{*}\lambda-\beta\lambda^{*}}=\sqrt{z^{n}}\mathrm{H}_{n}\left(\frac{\sqrt{z}\lambda+\lambda^{*}/\sqrt{z}}{2}\right)\mathrm{e}^{-|\lambda|^{2}}. \tag{10.96}$$

Now we derive a more general summation formula [7, 8], i.e.,

$$\sum_{l=0}^{n}\binom{n}{l}z^{n-l}q^{l}\mathrm{H}_{l,m}\left(x,y\right)=q^{n}\mathrm{H}_{n,m}\left(\frac{z}{q}+x,y\right). \tag{10.97}$$

According to the operator Hermite polynomial method, we consider

$$\begin{aligned}\sum_{l=0}^{n}\binom{n}{l}z^{n-l}q^{l}\vdots\mathrm{H}_{l,m}\left(a,a^{\dagger}\right)\vdots &=\sum_{l=0}^{n}\binom{n}{l}z^{n-l}q^{l}:a^{\dagger m}a^{l}:\\ &=:a^{\dagger m}\left(z+qa\right)^{n}:,\end{aligned} \tag{10.98}$$

and construct the following summation

$$\begin{aligned}&\sum_{m,n=0}^{\infty}\frac{s^{m}t^{n}}{m!n!}:a^{\dagger m}\left(z+qa\right)^{n}:\\ &=\mathrm{e}^{sa^{\dagger}}\mathrm{e}^{(z+qa)t}=\vdots\mathrm{e}^{\left(\frac{z}{q}+a\right)tq}\mathrm{e}^{sa^{\dagger}}\mathrm{e}^{-stq}\vdots\\ &=\sum_{n,m=0}^{\infty}\frac{s^{m}\left(tq\right)^{n}}{m!n!}\vdots\mathrm{H}_{n,m}\left(\frac{z}{q}+a,a^{\dagger}\right)\vdots.\end{aligned} \tag{10.99}$$

Comparing the same power of $s^m t^n$ on both sides of Eq. (10.99), we obtain

$$: a^{\dagger m} (z + qa)^n : \; = q^n \vdots \mathrm{H}_{n,m} \left(\frac{z}{q} + a, a^\dagger \right) \vdots . \tag{10.100}$$

Combining Eqs. (2.29) and (10.98) leads to

$$\sum_{l=0}^{n} \binom{n}{l} z^{n-l} q^l \vdots \mathrm{H}_{l,m} \left(a, a^\dagger\right) \vdots = q^n \vdots \mathrm{H}_{n,m} \left(\frac{z}{q} + a, a^\dagger \right) \vdots , \tag{10.101}$$

which leads to the following formula

$$\sum_{l=0}^{n} \binom{n}{l} z^{n-l} q^l \mathrm{H}_{l,k} (x, y) = q^n \mathrm{H}_{n,k} \left(\frac{z}{q} + x, y \right). \tag{10.102}$$

Especially, for $q = 1$, Eq. (10.102) becomes

$$\sum_{l=0}^{n} \binom{n}{l} t^{n-l} \mathrm{H}_{l,k} (x, y) = \mathrm{H}_{n,k} (t + x, y) . \tag{10.103}$$

Interestingly, comparing Eq. (10.103) with the binomial theorem of single-variable Hermite polynomial $\mathrm{H}_l (x)$, that is

$$\sum_{l=0}^{n} \binom{n}{l} z^{n-l} \mathrm{H}_l (x) = \mathrm{H}_n (z + x) , \tag{10.104}$$

which can deepen the impression. Also, Eq. (10.103) can be extended to the binomial theorem of two-variable Hermite polynomials:

$$\sum_{r=0}^{l} \sum_{q=0}^{k} \binom{l}{r} \binom{k}{q} \mathrm{H}_{r,q} (x, y) \, t^{l-r} s^{k-q} = \mathrm{H}_{r,q} (x + t, y + s) . \tag{10.105}$$

In fact, the operator Hermite polynomial method can not only derive new binomial formulas, but also help to find new relationships of special functions. For example, noting that the operator identity

$$\begin{aligned} &\sum_{l=0}^{n} \binom{n}{l} z^{n-l} q^l \vdots \mathrm{H}_{l+k,m} \left(a, a^\dagger\right) \vdots \\ &= \sum_{l=0}^{n} \binom{n}{l} z^{n-l} q^l : a^{\dagger m} a^{l+k} : \\ &= \; : a^{\dagger m} (z + qa)^n a^k : , \end{aligned} \tag{10.106}$$

thus we construct the following summation:

$$
\begin{aligned}
&\sum_{m,n=0}^{\infty} \frac{s^m t^n}{m!n!} : a^{\dagger m} (z+qa)^n a^k : \\
&= e^{sa^\dagger} e^{(z+qa)t} a^k \\
&= (a-s)^k \vdots e^{\left(\frac{z}{q}+a\right)tq} e^{sa^\dagger} e^{-stq} \vdots \\
&= \sum_{j=0}^{k} \binom{k}{j} a^{k-j} s^j \sum_{n,m=0}^{\infty} \frac{s^{m'} (tq)^n}{m'!n!} \vdots \mathrm{H}_{n,m'} \left(\frac{z}{q}+a, a^\dagger\right) \vdots, \qquad (10.107)
\end{aligned}
$$

which leads to the result

$$
\begin{aligned}
&: a^{\dagger m} (z+qa)^n a^k : \\
&= q^n \sum_{j=0}^{k} \binom{k}{j} a^{k-j} \vdots \mathrm{H}_{n,m-j} \left(\frac{z}{q}+a, a^\dagger\right) \vdots \frac{m!}{(m-j)!} \\
&= q^n \sum_{j=0}^{k} \frac{k!m!}{(m-j)!(k-j)!j!} a^{k-j} \vdots \mathrm{H}_{n,m-j} \left(\frac{z}{q}+a, a^\dagger\right) \vdots. \qquad (10.108)
\end{aligned}
$$

Combining Eqs. (10.106) and (10.108), we obtain

$$
\begin{aligned}
&\sum_{l=0}^{n} \binom{n}{l} z^{n-l} q^l \vdots \mathrm{H}_{l+k,m} \left(a, a^\dagger\right) \vdots \\
&= q^n \vdots \sum_{j=0}^{k} \frac{k!m!}{(m-j)!(k-j)!j!} a^{k-j} \vdots \mathrm{H}_{n,m-j} \left(\frac{z}{q}+a, a^\dagger\right) \vdots. \qquad (10.109)
\end{aligned}
$$

Similar to obtaining Eq. (10.94), we get

$$
\begin{aligned}
&\sum_{l=0}^{n} \binom{n}{l} z^{n-l} q^l \mathrm{H}_{l+k,m} (x, y) \\
&= q^n \sum_{j=0}^{k} \frac{k!m!}{(m-j)!(k-j)!j!} x^{k-j} \mathrm{H}_{n,m-j} \left(\frac{z}{q}+x, y\right). \qquad (10.110)
\end{aligned}
$$

Especially, for $q = 1$, $z = 0$, $n = l$, thus the new relation of two-variable Hermite polynomials reads

$$
\sum_{j=0}^{k} \frac{k!m!}{j!(m-j)!(k-j)!} x^{k-j} \mathrm{H}_{l,m-j} (x, y) = \mathrm{H}_{l+k,m} (x, y). \qquad (10.111)
$$

10.4 New Generating Functions

Using the operator identity $\mathrm{H}_n(X) = 2^n : X^n :$, we can easily derive some new generating functions for Hermite polynomials, they will occasionally play a role in studying the physical properties of new light fields [9, 10]. For example, in terms of

$$\begin{aligned}\sum_{k=0}^{\infty} \frac{z^k}{k!} \mathrm{H}_{n+k}(X) &= 2^n \sum_{k=0}^{\infty} \frac{2^k z^k}{k!} : X^{n+k} : \ = 2^n : \mathrm{e}^{2zX} X^n : \\ &= 2^n \mathrm{e}^{\sqrt{2}za^\dagger} : X^n : \mathrm{e}^{\sqrt{2}za} \\ &= \mathrm{e}^{\sqrt{2}za^\dagger} \mathrm{H}_n(X) \mathrm{e}^{\sqrt{2}za} \\ &= \mathrm{e}^{\sqrt{2}za^\dagger} \mathrm{e}^{\sqrt{2}za} \mathrm{e}^{-\sqrt{2}za} \mathrm{H}_n\left(\frac{a^\dagger + a}{\sqrt{2}}\right) \mathrm{e}^{\sqrt{2}za} \\ &= \mathrm{e}^{\sqrt{2}za^\dagger} \mathrm{e}^{\sqrt{2}za} \mathrm{H}_n\left(\frac{a^\dagger + a - \sqrt{2}z}{\sqrt{2}}\right) \\ &= \mathrm{e}^{2zX - z^2} \mathrm{H}_n(X - z), \end{aligned} \tag{10.112}$$

we can directly obtain the generating function for a new Hermite polynomial, i.e.,

$$\sum_{k=0} \frac{z^k}{k!} \mathrm{H}_{n+k}(x) = \mathrm{H}_n(x - z) \mathrm{e}^{2zx - z^2}. \tag{10.113}$$

It follows from Eq. (10.113) that

$$\begin{aligned}\sum_{n=0}^{\infty} \frac{t^n}{2^n n!} \mathrm{H}_n(X) \mathrm{H}_{n+l}(y) &= \sum_{n=0}^{\infty} \frac{t^n : X^n :}{n!} \mathrm{H}_{n+l}(y) \\ &= : \mathrm{H}_l(y - tX) \mathrm{e}^{2tyX - t^2 X^2} : . \end{aligned} \tag{10.114}$$

Further, noting

$$S \equiv \sum_{l=0}^{\infty} \frac{z^l}{l!} \sum_{n=0}^{\infty} \frac{t^n}{2^n n!} \mathrm{H}_n(X) \mathrm{H}_{n+l}(y), \tag{10.115}$$

inserting Eq. (10.9) into Eq. (10.115) leads to

$$\begin{aligned} S &= \sum_{l=0}^{\infty} \frac{z^l}{l!} : \mathrm{H}_l(y - tX) \mathrm{e}^{2ytX - t^2 X^2} : \\ &= : \mathrm{e}^{2z(y - tX) - z^2} \mathrm{e}^{2ytX - t^2 X^2} : \\ &= : \mathrm{e}^{2(y - z)tX - t^2 X^2} : \mathrm{e}^{2zy - z^2}. \end{aligned} \tag{10.116}$$

It follows from Eq. (1.78) that

$$:\mathrm{e}^{2(y-z)tX-t^2X^2}: = \frac{1}{\sqrt{1-t^2}}\exp\left[\frac{2t\,(y-z)\,X - t^2X^2 - t^2\,(y-z)^2}{1-t^2}\right], \tag{10.117}$$

so Eq. (10.116) becomes

$$\begin{aligned} S &= \frac{1}{\sqrt{1-t^2}}\exp\left[\frac{2t\,(y-z)\,X - t^2X^2 - t^2\,(y-z)^2}{1-t^2} + 2zy - z^2\right] \\ &= \exp\left(\frac{2tyX - t^2X^2 - t^2y^2}{1-t^2}\right)\exp\left[f\,(z)\right], \end{aligned} \tag{10.118}$$

where

$$f\,(z) \equiv \frac{2Xtz - 2yz + z^2}{t^2-1}. \tag{10.119}$$

According to Eq. (10.1), we have

$$\begin{aligned} \exp\left[f\,(z)\right] &= \exp\left[-\left(\frac{z}{\sqrt{1-t^2}}\right)^2 + 2\frac{y-Xt}{\sqrt{1-t^2}}\frac{z}{\sqrt{1-t^2}}\right] \\ &= \sum_{l=0}^{\infty}\frac{z^l}{\sqrt{\left(1-t^2\right)^l l!}}\mathrm{H}_l\left(\frac{y-Xt}{\sqrt{1-t^2}}\right), \end{aligned} \tag{10.120}$$

so Eq. (10.115) becomes

$$\begin{aligned} S &= \sum_{l=0}^{\infty}\frac{z^l}{l!}\sum_{n=0}^{\infty}\frac{t^n}{2^n n!}\mathrm{H}_n\,(X)\,\mathrm{H}_{n+l}\,(y) \\ &= \frac{1}{\sqrt{1-t^2}}\sum_{l=0}^{\infty}\frac{z^l}{\sqrt{\left(1-t^2\right)^l l!}}\mathrm{H}_l\left(\frac{y-Xt}{\sqrt{1-t^2}}\right) \\ &\quad\times\exp\left(\frac{2tyX - t^2X^2 - t^2y^2}{1-t^2}\right). \end{aligned} \tag{10.121}$$

Comparing the coefficient of z^l on both sides of Eq. (10.121), we get

$$\begin{aligned} \sum_{n=0}^{\infty}\frac{t^n}{2^n n!}\mathrm{H}_n\,(x)\,\mathrm{H}_{n+l}\,(y) &= \frac{1}{\sqrt{\left(1-t^2\right)^{l+1}}}\mathrm{H}_l\left(\frac{y-xt}{\sqrt{1-t^2}}\right) \\ &\quad\times\exp\left(\frac{2tyx - t^2x^2 - t^2y^2}{1-t^2}\right). \end{aligned} \tag{10.122}$$

Especially, for $l = 0$, Eq. (10.122) reduces to

$$\sum_{n=0}^{\infty} \frac{t^n}{2^n n!} \mathrm{H}_n(x)\, \mathrm{H}_n(y) = \frac{1}{\sqrt{1-t^2}} \exp\left(\frac{2tyx - t^2x^2 - t^2y^2}{1-t^2}\right). \tag{10.123}$$

Next, we want to calculate a new summation form of Hermite polynomials, i.e.,

$$\sum_{n,m=0}^{\infty} \frac{s^m t^n}{m!n!} \mathrm{H}_{m+l,n+k}(x, y) =? \tag{10.124}$$

For this, using the operator Hermite polynomial method and Eq. (2.78), we obtain

$$\begin{aligned}
&\sum_{n,m=0}^{\infty} \frac{s^m t^n}{m!n!} \vdots \mathrm{H}_{m+l,n+k}(a^\dagger, a) \vdots \\
&= \sum_{n,m=0}^{\infty} \frac{s^m t^n}{m!n!} a^{\dagger m+l} a^{n+k} = a^{\dagger l} : \mathrm{e}^{sa^\dagger + ta} : a^k \\
&= \mathrm{e}^{-st} a^{\dagger l} \mathrm{e}^{ta} \mathrm{e}^{sa^\dagger} a^k \\
&= \mathrm{e}^{-st} \mathrm{e}^{ta} \left(a^\dagger - t\right)^l (a-s)^k \mathrm{e}^{sa^\dagger}.
\end{aligned} \tag{10.125}$$

Also, using the binomial theorem and Eq. (2.78) to obtain

$$\begin{aligned}
\left(a^\dagger - t\right)^l (a-s)^k &= \sum_{r,q=0}^{\infty} \binom{l}{r} a^{\dagger r} (-t)^{l-r} \binom{k}{q} a^q (-s)^{k-q} \\
&= \sum_{r,q=0}^{\infty} \binom{l}{r}\binom{k}{q} \vdots \mathrm{H}_{r,q}\left(a^\dagger, a\right) \vdots (-t)^{l-r} (-s)^{k-q} \\
&= \vdots \mathrm{H}_{l,k}\left(a^\dagger - t, a - s\right) \vdots.
\end{aligned} \tag{10.126}$$

Substituting Eqs. (10.125) and (10.126), we obtain the following operator identity

$$\sum_{n,m=0}^{\infty} \frac{s^m t^n}{m!n!} \vdots \mathrm{H}_{m+l,n+k}(a^\dagger, a) \vdots = \vdots \mathrm{e}^{-st} \mathrm{e}^{ta} \mathrm{H}_{l,k}\left(a^\dagger - t, a - s\right) \mathrm{e}^{sa^\dagger} \vdots, \tag{10.127}$$

which leads to the new generating function for two-variable Hermite polynomial $\mathrm{H}_{m+l,n+k}(x, y)$, i.e.,

$$\sum_{n,m=0}^{\infty} \frac{s^m t^n}{m!n!} \mathrm{H}_{m+l,n+k}(x, y) = \mathrm{e}^{-st+sx+ty} \mathrm{H}_{l,k}(x - t, y - s). \tag{10.128}$$

For $l = k = 0$, Eq. (10.128) reduces to Eq. (10.24), as expected.

As an application of Eq. (10.128), according to the two-variable Hermite polynomials, we expand the two-mode entangled states $|\xi\rangle$ as

$$
\begin{aligned}
|\xi\rangle &= \exp\left(-\frac{1}{2}|\xi|^2 + a^\dagger\xi + \xi^* b^\dagger - a^\dagger b^\dagger\right)|00\rangle \\
&= \mathrm{e}^{-|\xi|^2/2}\sum_{m,n=0}^{\infty}\frac{a^{\dagger m}b^{\dagger n}}{m!n!}\mathrm{H}_{m,n}(\xi,\xi^*)\,|00\rangle\,,
\end{aligned}
\tag{10.129}
$$

where $\left[b, b^\dagger\right] = 1$, and the states $|\xi\rangle$ yield the following eigen-equations:

$$
\left(a + b^\dagger\right)|\xi\rangle = \xi\,|\xi\rangle\,,\quad \left(a^\dagger + b\right)|\xi\rangle = \xi^*\,|\xi\rangle\,. \tag{10.130}
$$

Using Eq. (10.130) and the following identity

$$
\begin{aligned}
&\mathrm{H}_{l,k}\left(\xi - b^\dagger, \xi^* - a^\dagger\right)|\xi\rangle \\
&= \mathrm{e}^{-\frac{1}{2}|\xi|^2}\mathrm{e}^{a^\dagger\xi + \xi^* b^\dagger - b^\dagger a^\dagger} \\
&\quad\times\mathrm{H}_{l,k}\left(\xi - b^\dagger, \xi^* - a^\dagger\right)|00\rangle\,,
\end{aligned}
\tag{10.131}
$$

we therefore have

$$
\mathrm{H}_{l,k}(a,b)\,|\xi\rangle = \mathrm{H}_{l,k}\left(\xi - b^\dagger, \xi^* - a^\dagger\right)|\xi\rangle\,. \tag{10.132}
$$

Using Eq. (10.128), we obtain

$$
\begin{aligned}
&\mathrm{e}^{a^\dagger\xi + \xi^* b^\dagger - b^\dagger a^\dagger}\mathrm{H}_{l,k}\left(\xi - b^\dagger, \xi^* - a^\dagger\right)|00\rangle \\
&= \sum_{n,m=0}^{\infty}\frac{a^{\dagger m}b^{\dagger n}}{m!n!}\mathrm{H}_{m+l,n+k}(\xi,\xi^*)\,|00\rangle \\
&= \sum_{n,m=0}^{\infty}\frac{\mathrm{H}_{m+l,n+k}(\xi,\xi^*)}{\sqrt{m!n!}}\,|m,n\rangle\,,
\end{aligned}
\tag{10.133}
$$

where $|m,n\rangle = a^{\dagger m}b^{\dagger n}\,|00\rangle\,/\sqrt{m!n!}$. Combining the above four equations, we get

$$
\mathrm{H}_{l,k}(a,b)\,|\xi\rangle = \mathrm{e}^{-|\xi|^2/2}\sum_{n,m=0}^{\infty}\frac{\mathrm{H}_{m+l,n+k}(\xi,\xi^*)}{\sqrt{m!n!}}\,|m,n\rangle\,. \tag{10.134}
$$

Therefore, the wave function of the state $\mathrm{H}_{l,k}(a,b)\,|\xi\rangle$ in Fock space can be directly given by Eq. (10.134), that is

$$
\langle m,n|\,\mathrm{H}_{l,k}(a,b)\,|\xi\rangle = \mathrm{e}^{-\frac{1}{2}|\xi|^2}\frac{\mathrm{H}_{m+l,n+k}(\xi,\xi^*)}{\sqrt{m!n!}}. \tag{10.135}
$$

Above, we use the operator Hermite polynomial method to derive some new binomial identities and generalized generating functions involving two-variable Hermite polynomials, and present some of their applications. This method will play a role in finding new quantum light field states.

10.5 New Special Function 1

Using the generating function for the two-variable Hermite polynomial in Eq. (10.83), and replacing x and y with two commutative coordinate operators X and Y, respectively, i.e., $x \to X$, $y \to Y$, thus we have the following operator identities [11]

$$\mathrm{e}^{-ts+tX+sY} = \sum_{n,m=0}^{\infty} \frac{s^n t^m}{n!m!} \mathrm{H}_{n,m}(X, Y) \tag{10.136}$$

and

$$\begin{aligned} \mathrm{H}_{n,m}(X, Y) &= \left.\frac{\partial^{n+m}}{\partial s^n \partial t^m} \mathrm{e}^{-ts+tX+sY}\right|_{t,s=0} \\ &= \sum_{l=0}^{\min(m,n)} \frac{n!m!(-1)^l}{l!(n-l)!(m-l)!} X^{n-l} Y^{m-l}. \end{aligned} \tag{10.137}$$

Substituting Eq. (10.59) into Eq. (10.137) leads to

$$\begin{aligned} \left.\frac{\partial^{n+m}}{\partial s^n \partial t^m} \mathrm{e}^{-ts+tX+sY}\right|_{t=s=0} &= (\mathrm{i}2)^{-n-m} \sum_{l=0}^{\min(m,n)} \frac{4^l n!m!}{l!(n-l)!(m-l)!} \\ &\quad \times : \mathrm{H}_{n-l}(\mathrm{i}X) \mathrm{H}_{m-l}(\mathrm{i}Y) : . \end{aligned} \tag{10.138}$$

On the other hand, noting

$$\mathrm{e}^{-ts+tX+sY} =: \exp\left(sX + \frac{s^2}{4} + tY + \frac{t^2}{4} - st\right) :, \tag{10.139}$$

so using Eqs. (10.138) and (10.139), we obtain

$$\begin{aligned} &\frac{\partial^{n+m}}{\partial s^n \partial t^m} : \exp\left(sX + \frac{s^2}{4} + tY + \frac{t^2}{4} - st\right) : |_{t,s=0} \\ &= (2\mathrm{i})^{-n-m} \sum_{l=0}^{\min(m,n)} \frac{4^l n!m!}{l!(n-l)!(m-l)!} : \mathrm{H}_{n-l}(\mathrm{i}X) \mathrm{H}_{m-l}(\mathrm{i}Y) : . \end{aligned} \tag{10.140}$$

Noting that both sides of Eq. (10.140) are in normal ordering, we thus have

$$\exp\left(sx+ty-st+\frac{s^2}{4}+\frac{t^2}{4}\right)=\sum_{n,m=0}^{\infty}\frac{s^n t^m}{n!m!}\mathfrak{F}_{n,m}(x,y), \tag{10.141}$$

where

$$\mathfrak{F}_{n,m}(x,y)=(\mathrm{i}2)^{-n-m}\sum_{l=0}^{\min(m,n)}\frac{4^l n!m!}{l!(n-l)!(m-l)!}\mathrm{H}_{n-l}(\mathrm{i}x)\,\mathrm{H}_{m-l}(\mathrm{i}y) \tag{10.142}$$

is a new special function with the generating function $\exp\left(sx+ty-st+\frac{s^2}{4}+\frac{t^2}{4}\right)$.

Using the differential relation of single-variable Hermite polynomials:

$$\mathrm{H}_n'(\mathrm{i}x)=2\mathrm{i}n\mathrm{H}_{n-1}(\mathrm{i}x), \tag{10.143}$$

we differentiate the function $\mathfrak{F}_{n,m}(x,y)$ in Eq. (10.142) and obtain the identities

$$\begin{aligned}
&\frac{\partial}{\partial x}\mathfrak{F}_{n,m}(x,y)\\
&=(2\mathrm{i})^{-n-m}\sum_{l=0}^{\min(m,n)}\frac{4^l n!m!}{l!(n-l)!(m-l)!}2\mathrm{i}(n-l)\,\mathrm{H}_{n-l-1}(\mathrm{i}x)\,\mathrm{H}_{m-l}(\mathrm{i}y)\\
&=n(2\mathrm{i})^{-n-m+1}\sum_{l=0}^{\min(m,n)}\frac{4^l(n-1)!m!}{l!(n-l-1)!(m-l)!}\mathrm{H}_{n-l-1}(\mathrm{i}x)\,\mathrm{H}_{m-l}(\mathrm{i}y)\\
&=n\mathfrak{F}_{n-1,m}(x,y)
\end{aligned} \tag{10.144}$$

and

$$\begin{aligned}
\frac{\partial}{\partial y}\mathfrak{F}_{n,m}(x,y)&=(2\mathrm{i})^{-n-m}\sum_{l=0}^{\min(m,n)}\frac{4^l n!m!}{l!(n-l)!(m-l)!}\mathrm{H}_{n-l}(\mathrm{i}x)\,\mathrm{H}_{m-l}'(\mathrm{i}y)\\
&=m\mathfrak{F}_{n,m-1}(x,y).
\end{aligned} \tag{10.145}$$

Therefore, we have

$$\frac{\partial^2}{\partial x\partial y}\mathfrak{F}_{n,m}(x,y)=nm\mathfrak{F}_{n-1,m-1}(x,y), \tag{10.146}$$

which is the differentiate relation of the new special function $\mathfrak{F}_{n,m}(x,y)$.

Firstly, using the new special function $\mathfrak{F}_{n,m}(x,y)$ to find some operator identities and integration formulas. Using Eqs. (10.137), (10.139) and (10.141), we obtain the new operator identity:

$$\mathrm{H}_{n,m}(X,Y) =: \mathfrak{F}_{n,m}(X,Y):. \tag{10.147}$$

On the other hand, it follows from Eq. (10.141) that

$$\begin{aligned}\sum_{n,m=0}^{\infty}\frac{s^n t^m}{n!m!}\mathfrak{F}_{n,m}(\mathrm{i}X,\mathrm{i}Y) &= \exp\left(\mathrm{i}sX+\mathrm{i}tY-st+\frac{s^2}{4}+\frac{t^2}{4}\right)\\ &=: \exp(\mathrm{i}sX+\mathrm{i}tY-st):\\ &= \sum_{n,m=0}^{\infty}\frac{s^n t^m}{n!m!}: \mathrm{H}_{n,m}(\mathrm{i}X,\mathrm{i}Y):. \end{aligned}\tag{10.148}$$

Comparing Eqs. (10.147) and (10.148) leads to another new operator identity

$$\mathfrak{F}_{n,m}(\mathrm{i}X,\mathrm{i}Y) =: \mathrm{H}_{n,m}(\mathrm{i}X,\mathrm{i}Y):. \tag{10.149}$$

Using the completeness relation of coordinate states, i.e.,

$$\int_{-\infty}^{\infty}\mathrm{d}x\,|x\rangle\langle x| = \int_{-\infty}^{\infty}\frac{\mathrm{d}x}{\sqrt{\pi}}: \mathrm{e}^{-(x-X)^2}: = 1, \tag{10.150}$$

where $|x\rangle$ is the eigen-state of coordinate operator, we thus have

$$\mathrm{H}_{n,m}(X,Y) = \iint_{-\infty}^{\infty}\frac{\mathrm{d}x\mathrm{d}y}{\pi}: \mathrm{e}^{-(x-X)^2-(y-Y)^2}: \mathrm{H}_{n,m}(x,y) \tag{10.151}$$

$$= : \mathfrak{F}_{n,m}(X,Y):. \tag{10.152}$$

Noting that both sides of Eq. (10.152) are in normal ordering, so we obtain a new integration formula

$$\frac{1}{\pi}\iint_{-\infty}^{\infty}\mathrm{d}x'\mathrm{d}y'\mathrm{e}^{-(x'-x)^2-(y'-y)^2}\mathrm{H}_{n,m}(x',y') = \mathfrak{F}_{n,m}(x,y). \tag{10.153}$$

On the other hand, using Eqs. (10.150) and (10.57) to obtain

$$\begin{aligned}\mathfrak{F}_{n,m}(\mathrm{i}X,\mathrm{i}Y) &= \iint_{-\infty}^{\infty}\mathrm{d}x\mathrm{d}y\,\mathfrak{F}_{n,m}(\mathrm{i}x,\mathrm{i}y)\,|x,y\rangle\langle x,y|\\ &= \frac{1}{\pi}\iint_{-\infty}^{\infty}\mathrm{d}x\mathrm{d}y: \mathrm{e}^{-(x-X)^2-(y-Y)^2}: \mathfrak{F}_{n,m}(\mathrm{i}x,\mathrm{i}y)\\ &=: \mathrm{H}_{n,m}(\mathrm{i}X,\mathrm{i}Y):, \end{aligned}\tag{10.154}$$

which means another new integration formula:

$$\frac{1}{\pi}\iint_{-\infty}^{\infty} dx'dy' e^{-(x'-x)^2-(y'-y)^2}\mathfrak{F}_{n,m}(ix, iy) = H_{n,m}(ix, iy). \tag{10.155}$$

Secondly, we present some applications of the new special function $\mathfrak{F}_{n,m}(x, y)$. Using two-variable operator Hermite polynomials in Eqs. (10.142) and (10.152), we can obtain the two-variable Hermite polynomial excited vacuum, i.e.,

$$\begin{aligned} H_{n,m}(X, Y)|00\rangle &= \frac{1}{(2i)^{n+m}}\sum_{l=0}^{\min(m,n)} \frac{4^l n!m!}{l!(n-l)!(m-l)!} \\ &\times H_{n-l}\left(\frac{ia^\dagger}{\sqrt{2}}\right) H_{m-l}\left(\frac{ib^\dagger}{\sqrt{2}}\right)|00\rangle, \end{aligned} \tag{10.156}$$

which is a two-mode entangled state. Besides, noting the commutative relation $\left[\left(a+b^\dagger\right), \left(b+a^\dagger\right)\right] = 0$, thus using Eq. (10.141), we have

$$e^{s(a+b^\dagger)+t(b+a^\dagger)-st+\frac{s^2}{4}+\frac{t^2}{4}} = \sum_{n,m=0}^{\infty} \frac{s^n t^m}{n!m!}\mathfrak{F}_{n,m}\left(a+b^\dagger, b+a^\dagger\right). \tag{10.157}$$

Further, using Baker-Hausdorff formula, the left of Eq. (10.157) can be written as

$$\begin{aligned} &\exp\left[s\left(a+b^\dagger\right)+t\left(b+a^\dagger\right)-st+\frac{s^2}{4}+\frac{t^2}{4}\right] \\ &=: \exp\left[s\left(a+b^\dagger\right)+t\left(b+a^\dagger\right)-\left(\frac{-is}{2}\right)^2-\left(\frac{-it}{2}\right)^2\right]: \\ &= \sum_{n,m=0}^{\infty} \frac{\left(\frac{-is}{2}\right)^n\left(\frac{-it}{2}\right)^m}{n!m!} : H_n\left[i\left(a+b^\dagger\right)\right] H_m\left[i\left(b+a^\dagger\right)\right]:. \end{aligned} \tag{10.158}$$

Comparing Eqs. (10.157) and (10.158) leads to the following operator identity:

$$\mathfrak{F}_{n,m}\left(a+b^\dagger, b+a^\dagger\right) = \left(\frac{-i}{2}\right)^{m+n} : H_n\left[i\left(a+b^\dagger\right)\right] H_m\left[i\left(b+a^\dagger\right)\right]:. \tag{10.159}$$

Acting the operator $\mathfrak{F}_{n,m}\left(a+b^\dagger, b+a^\dagger\right)$ on a two-mode vacuum, we obtain

$$\mathfrak{F}_{n,m}\left(a+b^\dagger, b+a^\dagger\right)|00\rangle = \left(\frac{-i}{2}\right)^{m+n} H_n\left(ib^\dagger\right) H_m\left(ia^\dagger\right)|00\rangle, \tag{10.160}$$

which is a Hermite polynomial excited vacuum state.

On the other hand, using Eq. (10.141), we also obtian

$$
\begin{aligned}
&: \exp\left[s\left(a+b^{\dagger}\right)+t\left(b+a^{\dagger}\right)-st+\frac{s^2}{4}+\frac{t^2}{4}\right]: \\
&=\sum_{n,m=0}^{\infty}\frac{s^n t^m}{n!m!}: \mathfrak{F}_{n,m}\left(a+b^{\dagger},b+a^{\dagger}\right):.
\end{aligned} \tag{10.161}
$$

Using the Baker-Hausdorff formula and Eq. (10.158), thus the left side of Eq. (10.161) is equivalent to

$$
\begin{aligned}
&: \exp\left[s\left(a+b^{\dagger}\right)+t\left(b+a^{\dagger}\right)-st+\frac{s^2}{4}+\frac{t^2}{4}\right]: \\
&=\exp\left[s\left(a+b^{\dagger}\right)+t\left(b+a^{\dagger}\right)+\frac{s^2}{4}+\frac{t^2}{4}\right] \\
&=\sum_{n,m=0}^{\infty}\frac{\left(\frac{-is}{2}\right)^n\left(\frac{-it}{2}\right)^m}{n!m!}\mathrm{H}_n\left[\mathrm{i}\left(a+b^{\dagger}\right)\right]\mathrm{H}_m\left[\mathrm{i}\left(b+a^{\dagger}\right)\right].
\end{aligned} \tag{10.162}
$$

Comparing Eqs. (10.161) and (10.162) leads to the following new operator identity

$$
\mathrm{H}_n\left[\mathrm{i}\left(a+b^{\dagger}\right)\right]\mathrm{H}_m\left[\mathrm{i}\left(b+a^{\dagger}\right)\right]=(2\mathrm{i})^{m+n}:\mathfrak{F}_{n,m}\left(a+b^{\dagger},b+a^{\dagger}\right):. \tag{10.163}
$$

Operating Eq. (10.163) on a two-mode vacuum, we can obtain the new special function excited vacuum state, i.e.,

$$
\mathrm{H}_n\left[\mathrm{i}\left(a+b^{\dagger}\right)\right]\mathrm{H}_m\left[\mathrm{i}\left(b+a^{\dagger}\right)\right]|00\rangle=(2\mathrm{i})^{m+n}\mathfrak{F}_{n,m}\left(b^{\dagger},a^{\dagger}\right)|00\rangle. \tag{10.164}
$$

By virtue of the completeness relation of two-mode entangled states $|\xi\rangle$, that is

$$
\int\frac{\mathrm{d}^2\xi}{\pi}|\xi\rangle\langle\xi|=\int\frac{\mathrm{d}^2\xi}{\pi}:\mathrm{e}^{-[\xi-(a+b^{\dagger})][\xi^*-(a^{\dagger}+b)]}:=1, \tag{10.165}
$$

where the entangled state $|\xi\rangle$ is

$$
|\xi\rangle=\exp\left(-\frac{|\xi|^2}{2}+a^{\dagger}\xi+b^{\dagger}\xi^*-a^{\dagger}b^{\dagger}\right)|00\rangle \tag{10.166}
$$

yielding the eigen-equation:

$$
\left(a+b^{\dagger}\right)|\xi\rangle=\xi|\xi\rangle,\ \left(b+a^{\dagger}\right)|\xi\rangle=\xi^*|\xi\rangle, \tag{10.167}
$$

and Eq. (10.163), we therefore obtain

$$\begin{aligned}
&\mathrm{H}_n\left[\mathrm{i}\left(a+b^{\dagger}\right)\right]\mathrm{H}_m\left[\mathrm{i}\left(b+a^{\dagger}\right)\right] \\
&=\mathrm{H}_n\left[\mathrm{i}\left(a+b^{\dagger}\right)\right]\mathrm{H}_m\left[\mathrm{i}\left(b+a^{\dagger}\right)\right]\int\frac{\mathrm{d}^2\xi}{\pi}\left|\xi\right\rangle\left\langle\xi\right| \\
&=\int\frac{\mathrm{d}^2\xi}{\pi}\mathrm{H}_n\left(\mathrm{i}\xi\right)\mathrm{H}_m\left(\mathrm{i}\xi^*\right):\mathrm{e}^{-[\xi-(a+b^{\dagger})][\xi^*-(a^{\dagger}+b)]}: \\
&=(2\mathrm{i})^{m+n}:\mathfrak{F}_{n,m}\left(a+b^{\dagger},b+a^{\dagger}\right):. \qquad (10.168)
\end{aligned}$$

It follows from Eq. (10.168) that the new integral formula is obtained as

$$\int\frac{\mathrm{d}^2\xi}{\pi}\mathrm{H}_n\left(\mathrm{i}\xi\right)\mathrm{H}_m\left(\mathrm{i}\xi^*\right)\mathrm{e}^{-[\xi-\lambda][\xi^*-\lambda^*]}=(2\mathrm{i})^{m+n}\mathfrak{F}_{n,m}\left(\lambda,\lambda^*\right). \qquad (10.169)$$

As an application of the integral formula in (10.169), we consider the normalization of the Hermite polynomial excited state $\mathrm{H}_m\left(\mathrm{i}a^{\dagger}\right)|0\rangle$. Using the completeness relation of coherent states in Eqs. (3.103) and (10.169), we obtain

$$\begin{aligned}
\mathrm{H}_m\left(-\mathrm{i}a\right)\mathrm{H}_m\left(\mathrm{i}a^{\dagger}\right)&=(-1)^m\,\mathrm{H}_m\left(\mathrm{i}a\right)\mathrm{H}_m\left(\mathrm{i}a^{\dagger}\right) \\
&=(-1)^m\int\frac{\mathrm{d}^2z}{\pi}\mathrm{H}_m\left(\mathrm{i}z\right)|z\rangle\langle z|\,\mathrm{H}_m\left(\mathrm{i}z^*\right) \\
&=(-1)^m\int\frac{\mathrm{d}^2z}{\pi}\mathrm{H}_n\left(\mathrm{i}z\right)\mathrm{H}_m\left(\mathrm{i}z^*\right):\mathrm{e}^{-\left(z^*-a^{\dagger}\right)(z-a)}: \\
&=4^m:\mathfrak{F}_{m,m}\left(a^{\dagger},a\right):. \qquad (10.170)
\end{aligned}$$

Therefore, the normalization factor of the state $\mathrm{H}_m\left(\mathrm{i}a^{\dagger}\right)|0\rangle$ reads

$$\langle 0|\,\mathrm{H}_m\left(-\mathrm{i}a\right)\mathrm{H}_m\left(\mathrm{i}a^{\dagger}\right)|0\rangle=4^m\mathfrak{F}_{m,m}\left(0,0\right). \qquad (10.171)$$

Above, we have used the operator Hermite polynomial method and the IWOP method to find a new class of special functions that are closely related to single- and two-variable Hermite polynomials.

10.6 New Special Function 2

In the above section, we introduced a new special function $\mathfrak{F}_{n,m}\left(x,y\right)$, its generating function and simple applications. In fact, the function $\mathfrak{F}_{n,m}\left(x,y\right)$ can be obtained via replacing $x^{n-l}y^{m-l}$ by $\mathrm{H}_{n-l}\left(\mathrm{i}x\right)\mathrm{H}_{m-l}\left(\mathrm{i}y\right)$ in the definition of two-variable Hermite polynomials [12], i.e.,

$$\begin{aligned}\mathrm{H}_{n,m}(x,y) &= \frac{\partial^{n+m}}{\partial s^n \partial t^m}\mathrm{e}^{-ts+tx+sy}|_{t,s=0} \\ &= \sum_{l=0}^{\min(m,n)} \frac{n!m!(-1)^l}{l!(n-l)!(m-l)!}x^{n-l}y^{m-l}.\end{aligned} \tag{10.172}$$

This special function $\mathfrak{F}_{n,m}(x,y)$ appears in the calculations related to quantum optical theory.

Another interesting question is: if a more general single-variable Hermite polynomial is used instead of x^{n-l} in Eq. (10.172), for example, $x^{n-l} \to \mathrm{H}_{n-l}(fx)$, that is

$$\begin{aligned}&\sum_{l=0}^{\min(m,n)} \frac{n!m!(-1)^l}{l!(n-l)!(m-l)!}x^{n-l}y^{m-l} \\ &\to \sum_{l=0}^{\min(m,n)} \frac{n!m!(-1)^l}{l!(n-l)!(m-l)!}\mathrm{H}_{n-l}(fx)y^{m-l} \equiv \mathfrak{G}_{n,m}(fx,y),\end{aligned} \tag{10.173}$$

thus is the new special function $\mathfrak{G}_{n,m}(x,y)$ introduced in this way a meaningful special function? If yes, what is its generating function? What are its specific applications?

Firstly, let's discuss under what physical conditions such a special function $\mathfrak{G}_{n,m}(x,y)$ may occur. For example, when calculating the photon number distribution of the Hermite polynomial excited state $\mathrm{H}_n\left(\sqrt{g}a^\dagger\right)|0\rangle$, it is necessary to calculate

$$\langle l|\mathrm{H}_n\left(\sqrt{g}a^\dagger\right)|0\rangle = \frac{1}{\sqrt{l!}}\langle 0|a^l\mathrm{H}_n\left(\sqrt{g}a^\dagger\right)|0\rangle, \tag{10.174}$$

where $\langle l| = \frac{1}{\sqrt{l!}}\langle 0|a^l$ is a number state. Using the integral formula

$$\int_{-\infty}^{\infty}\frac{\mathrm{d}x}{\sqrt{\pi}}\mathrm{H}_n(fx)\mathrm{e}^{-(x-y)^2} = \left(1-f^2\right)^{n/2}\mathrm{H}_n\left(fy/\sqrt{1-f^2}\right) \tag{10.175}$$

and the IWOP method, we obtain

$$\begin{aligned}\mathrm{H}_n(fX) &= \mathrm{H}_n(fX)\int_{-\infty}^{\infty}\mathrm{d}x\,|x\rangle\langle x| \\ &= \int_{-\infty}^{\infty}\frac{\mathrm{d}x}{\sqrt{\pi}}\mathrm{H}_n(fx) : \mathrm{e}^{-(x-X)^2} : \\ &= \left(1-f^2\right)^{n/2} : \mathrm{H}_n\left(fX/\sqrt{1-f^2}\right) :,\end{aligned} \tag{10.176}$$

where we have used the completeness relation of coordinate states, i.e., $\int_{-\infty}^{\infty}\mathrm{d}x\,|x\rangle\langle x| = 1$. Letting

$$f=\sqrt{\frac{2g}{1+2g}},$$

thus we obtain

$$\begin{aligned}\mathrm{H}_n\left(\sqrt{\frac{2g}{1+2g}}X\right)&=\left(\frac{1}{1+2g}\right)^{n/2}:\mathrm{H}_n\left(\sqrt{2g}X\right):\\&=\left(\frac{1}{1+2g}\right)^{n/2}:\mathrm{H}_n\left[\sqrt{g}\left(a+a^{\dagger}\right)\right]:.\end{aligned}\tag{10.177}$$

Therefore, acting the right side of Eq. (10.177) on vacuum $|0\rangle$, and acting the left of Eq. (10.177) on number state $\langle l|$, we get

$$\langle l|\,\mathrm{H}_n\left(\sqrt{g}a^{\dagger}\right)|0\rangle=(1+2g)^{n/2}\frac{1}{\sqrt{l!}}\langle 0|\,a^l\mathrm{H}_n\left(\sqrt{\frac{2g}{1+2g}}X\right)|0\rangle\,.\tag{10.178}$$

It can be seen that when calculating the matrix element $\langle l|\mathrm{H}_n\left(\sqrt{g}a^{\dagger}\right)|0\rangle$, the operator $a^l\mathrm{H}_n\left(fX\right)$ has appeared, and its Hermite conjugate operator is $\mathrm{H}_n\left(fX\right)a^{\dagger l}$.

Next, we obtain the normal ordering product of the operator $\mathrm{H}_n\left(fX\right)a^{\dagger l}$. Using the operator identity

$$\mathrm{e}^{fX}a^{\dagger}\mathrm{e}^{-fX}=a^{\dagger}+\left[f\frac{a+a^{\dagger}}{\sqrt{2}},a^{\dagger}\right]=a^{\dagger}+\frac{f}{\sqrt{2}}\tag{10.179}$$

to obtain

$$\begin{aligned}\mathrm{H}_n\left(fX\right)a^{\dagger l}&=\frac{\mathrm{d}^n}{\mathrm{d}t^n}\mathrm{e}^{-t^2+2tfX}\bigg|_{t=0}a^{\dagger l}\\&=\frac{\mathrm{d}^n}{\mathrm{d}t^n}\left(a^{\dagger}+\sqrt{2}tf\right)^l\mathrm{e}^{-t^2+2tfX}\bigg|_{t=0}\\&=\sum_{k=0}^{n}\binom{n}{k}\frac{\mathrm{d}^k}{\mathrm{d}t^k}\left(a^{\dagger}+\sqrt{2}tf\right)^l\frac{\mathrm{d}^{n-k}}{\mathrm{d}t^{n-k}}\mathrm{e}^{-t^2+2tfX}\bigg|_{t=0},\end{aligned}\tag{10.180}$$

where

$$\begin{aligned}\frac{\mathrm{d}^k}{\mathrm{d}t^k}\left(a^{\dagger}+\sqrt{2}tf\right)^l\bigg|_{t=0}&=\frac{\mathrm{d}^k}{\mathrm{d}t^k}\sum_{j=0}^{l}\binom{l}{j}a^{\dagger l-j}\left(\sqrt{2}tf\right)^j\bigg|_{t=0}\\&=\sum_{j=0}^{l}\binom{l}{j}a^{\dagger l-j}\left(\sqrt{2}f\right)^j\delta_{kj}k!\\&=\left(\sqrt{2}f\right)^k k!a^{\dagger l-k}\binom{l}{k}.\end{aligned}\tag{10.181}$$

Therefore, the normal ordering product of the operator $\mathrm{H}_n(fX)\,a^{\dagger l}$ is

$$\begin{aligned}
\mathrm{H}_n(fX)\,a^{\dagger l} &= \sum_{k=0}^{n}\binom{n}{k}\left(\sqrt{2}f\right)^k k!a^{\dagger l-k}\binom{l}{k}\mathrm{H}_{n-k}(fX) \\
&= \sum_{k=0}^{n}\frac{n!l!\left(\sqrt{2}f\right)^k}{(l-k)!k!\,(n-k)!}a^{\dagger l-k}\mathrm{H}_{n-k}(fX) \\
&= \left(-\sqrt{2}f\right)^l\left(1-f^2\right)^{n/2}\sum_{k=0}^{\min(l,n)}\frac{n!l!\,(-1)^k\left(\frac{-1}{\sqrt{2}f}\right)^{l-k}}{(l-k)!k!\,(n-k)!} \\
&\quad\times{:}\,a^{\dagger l-k}\mathrm{H}_{n-k}\left(fX/\sqrt{1-f^2}\right){:}\,. \qquad (10.182)
\end{aligned}$$

In terms of the definition of the special function $\Omega_{n,m}(x,y)$ in Eq. (10.173), we thus obtain

$$\mathrm{H}_n(fX)\,a^{\dagger l} = \left(-\sqrt{2}f\right)^l\left(1-f^2\right)^{n/2}{:}\,\Omega_{n,l}\left(\frac{fX}{\sqrt{1-f^2}},\frac{-1}{\sqrt{2}f}a^\dagger\right){:}\,. \qquad (10.183)$$

It can be seen that when calculating the photon number distributions of the Hermite polynomial excited states, we can encounter the new special function $\Omega_{n,m}(x,y)$.

Similarly, we obtain the anti-normal ordering product of the operator $\mathrm{H}_n(X)\,a^l$, that is

$$\begin{aligned}
\mathrm{H}_n(X)\,a^l &= \frac{\mathrm{d}^n}{\mathrm{d}t^n}\,\mathrm{e}^{-t^2+2tX}\Big|_{t=0}a^l = \frac{\mathrm{d}^n}{\mathrm{d}t^n}\left(a-\sqrt{2}t\right)^l\mathrm{e}^{-t^2+2tX}|_{t=0} \\
&= \sum_{k=0}^{n}\binom{n}{k}\frac{\mathrm{d}^k}{\mathrm{d}t^k}\left(a-\sqrt{2}t\right)^l\frac{\mathrm{d}^{n-k}}{\mathrm{d}t^{n-k}}\mathrm{e}^{-t^2+2tX}|_{t=0} \\
&= \sum_{k=0}^{n}\frac{n!l!\left(-\sqrt{2}\right)^k}{(l-k)!k!\,(n-k)!}a^{l-k}\mathrm{H}_{n-k}(X)\,. \qquad (10.184)
\end{aligned}$$

Further, inserting the anti-normal ordering product of the operator $\mathrm{H}_n(X)$, i.e.,

$$\mathrm{H}_n(X) = 2^{n/2}\vdots\mathrm{H}_n\left(X/\sqrt{2}\right)\vdots \qquad (10.185)$$

into Eq. (10.184) and using the definition of the special function $\Omega_{n,m}(x,y)$ in Eq. (10.173), we give

$$
\begin{aligned}
\mathrm{H}_n(X)\,a^l &= 2^{n/2}\sum_{k=0}^{n}\frac{n!l!\,(-1)^k}{(l-k)!k!\,(n-k)!}a^{l-k}\,\vdots\mathrm{H}_{n-k}\left(X/\sqrt{2}\right)\vdots \\
&\equiv \vdots\Omega_{n,l}\left(a, X/\sqrt{2}\right)\vdots.
\end{aligned}
\tag{10.186}
$$

Finally, we want to investigate the generating function of $\Omega_{n,m}(x,y)$. Using the operator Hermite polynomial method to calculate the summation

$$
\begin{aligned}
\sum_{n,l=0}^{\infty}\frac{t^n\tau^l}{n!l!}\mathrm{H}_n(X)\,a^l &= \mathrm{e}^{-t^2+2tX}\mathrm{e}^{\tau a} = \mathrm{e}^{-t^2+\sqrt{2}t\left(a+a^{\dagger}\right)}\mathrm{e}^{\tau a} \\
&= \mathrm{e}^{-t^2+\sqrt{2}ta}\mathrm{e}^{\sqrt{2}ta^{\dagger}}\mathrm{e}^{\tau a}\mathrm{e}^{-t^2} \\
&= \mathrm{e}^{-2t^2-\sqrt{2}t\tau}\mathrm{e}^{\tau a+\sqrt{2}ta}\mathrm{e}^{\sqrt{2}ta^{\dagger}},
\end{aligned}
\tag{10.187}
$$

and substituting Eq. (10.186) into Eq. (10.187), thus we obtain

$$
\sum_{n,l=0}^{\infty}\frac{t^n\tau^l}{n!l!}\vdots\Omega_{n,l}\left(a, X/\sqrt{2}\right)\vdots = \mathrm{e}^{-2t^2-\sqrt{2}t\tau}\vdots\mathrm{e}^{(\tau a+2tX)}\vdots.
\tag{10.188}
$$

Since both sides of Eq. (10.188) are in anti-normal ordering, thus we make the substitutions $X/\sqrt{2}\to x$, $a\to y$, $\sqrt{2}t\to t$ in Eq. (10.188) and obtain the generating function for $\Omega_{n,l}$ as

$$
\sum_{n,l=0}^{\infty}\frac{t^n\tau^l}{n!l!}\Omega_{n,l}(x,y) = \mathrm{e}^{-2t^2-\sqrt{2}t\tau}\mathrm{e}^{\left(\tau x+2\sqrt{2}ty\right)}
\tag{10.189}
$$

or

$$
\Omega_{n,l}(x,y) = \frac{\partial^{n+l}}{\partial t^n\partial\tau^l}\mathrm{e}^{-2t^2-\sqrt{2}t\tau}\,\mathrm{e}^{\left(\tau x+2\sqrt{2}ty\right)}\Big|_{t=\tau=0}.
\tag{10.190}
$$

In short, using the operator Hermite polynomial method and the IWOP method, we propose another new special function $\Omega_{n,l}(x,y)$, which includes the power series and Hermite polynomials. It may appear for the case of calculating physical quantities in quantum theory.

References

1. Fan HY, Fan Y (2002) New bosonic operator ordering identities gained by the entangled state representation and two-variable Hermite polynomials. Commun Theor Phys 38(3):297–300
2. Fan HY, Yuan HC, Jiang NQ (2011) New identities about operator Hermite polynomials and their related integration formulas. Sci China Phys Mech Astron 54(12):2145–2149

3. Zhou J, Fan HY, Song J (2013) Deriving new identities of Hermite-polynomial operators and Laguerre-polynomial operators by using representation theory and operator-ordering technique. Optik 124(10):911–914
4. Meng XG, Wang JS, Liang BL (2023) Entangled state representations in quantum optics. Springer, Singapore
5. Fan HY (2004) Entangled states, squeezed states gained via the route of developing Dirac's symbolic method and their applications. Int J Mod Phys B 18(10–11):1387–1455
6. Xu XX, Yuan HC, Fan HY (2015) Generating Hermite polynomial excited squeezed states by means of conditional measurements on a beam splitter. J Opt Soc Am B 32(6):1146–1154
7. Wang JS, Meng XG, Fan HY (2019) Time evolution of angular momentum coherent state derived by virtue of entangled state representation and a new binomial theorem. Chin Phys B 28(10):100301
8. Meng XG, Liu JM, Wang JS, Fan HY (2019) New generalized binomial theorems involving two-variable Hermite polynomials via quantum optics approach and their applications. Eur Phys J D 73(2):32–37
9. Zhan DH, Fan HY (2014) Some new generating function formulae of the two-variable Hermite polynomials and their application in quantum optics. Chin Phys B 23(12):120301
10. Fan HY, Zhang PF, Wang Z (2015) Generating function of product of bivariate Hermite polynomials and their applications in studying quantum optical states. Chin Phys B 24(5):050303
11. Chen F, Fan HY (2016) New useful special function in quantum optics theory. Chin Phys B 25(8):080303
12. Fan HY, Wan ZL, Wu Z, Zhang PF (2015) A new kind of special function and its application. Chin Phys B 24(10):100302

Chapter 11
Evolutions of Quantum States in a Laser Channel

This chapter investigates the evolutions of the chaotic state and the negative binomial state in a laser channel. The laser channel can be described by the following master equation [1–4]:

$$\begin{aligned}\frac{\mathrm{d}\rho(t)}{\mathrm{d}t} &= g\left[2a^{\dagger}\rho(t)\,a - aa^{\dagger}\rho(t) - \rho(t)\,aa^{\dagger}\right] \\ &\quad + \kappa\left[2a\rho(t)\,a^{\dagger} - a^{\dagger}a\rho(t) - \rho(t)\,a^{\dagger}a\right], \end{aligned} \tag{11.1}$$

where $\rho(t)$ is the density operator of the system at time t, and g, κ are, respectively, gain and loss. For $g = 0$, Eq. (11.1) reduces to the master equation for the amplitude decay channel, that is

$$\frac{\mathrm{d}\rho(t)}{\mathrm{d}t} = \kappa\left[2a\rho(t)\,a^{\dagger} - a^{\dagger}a\rho(t) - \rho(t)\,a^{\dagger}a\right], \tag{11.2}$$

while for $\kappa = 0$, Eq. (11.1) becomes the master equation with only gain, i.e.,

$$\frac{\mathrm{d}\rho(t)}{\mathrm{d}t} = g\left[2a^{\dagger}\rho(t)\,a - aa^{\dagger}\rho(t) - \rho(t)\,aa^{\dagger}\right]. \tag{11.3}$$

11.1 Initial Chaotic State Case

The main concerns here are as follows: If the initial quantum state is the chaotic state with the density operator $\rho(0) = \left(1 - \mathrm{e}^{f}\right)\mathrm{e}^{fa^{\dagger}a}$, what final state will the system evolve into? And we derive the differential equation satisfied by the parameters in the density operator $\rho(t)$, solve the corresponding differential equation and find an exact solution for $\rho(t)$ to clarify the evolution law of the initial chaotic state.

H. Fan and L. Hu, *Dissipation and Diffusion of Quantum Light Field*,
https://doi.org/10.1007/978-981-95-1528-8_11

Assuming that the initial state is the chaotic state $\rho(0)$, there is reason to believe that the state $\rho(t)$ remains chaotic at time t during the gain and decay process (Bose distribution in average photon numbers still holds) [5]. In this process, there is neither squeezing nor nonlinear effect, so the final state can be set as

$$\rho(t) = \left(1 - e^{f'}\right) e^{f' a^\dagger a}, \tag{11.4}$$

where $f' = f'(t)$ is an undetermined time function. Substituting Eq. (11.4) into the left side of Eq. (11.1) leads to

$$\begin{aligned}\frac{d}{dt}\left(1 - e^{f'}\right) e^{f' a^\dagger a} &= e^{f' a^\dagger a}\frac{d}{dt}\left(1 - e^{f'}\right) + \left(1 - e^{f'}\right)\frac{d}{dt} e^{f' a^\dagger a} \\ &= \left[\left(1 - e^{f'}\right) a^\dagger a - e^{f'}\right] e^{f' a^\dagger a}\frac{df'}{dt}.\end{aligned} \tag{11.5}$$

Substituting Eq. (11.4) into the right side of Eq. (11.1), we have

$$\begin{aligned}&\text{The right side of Eq. (11.1)} \\ &= 2\left(1 - e^{f'}\right)\left\{g\left[a^\dagger a e^{-f'} - a a^\dagger\right] + \kappa\left[a a^\dagger e^{f'} - a^\dagger a\right]\right\} e^{f' a^\dagger a} \\ &= 2\left(1 - e^{f'}\right)\left\{a^\dagger a\left[g\left(e^{-f'} - 1\right) + \kappa\left(e^{f'} - 1\right)\right]' + \kappa e^{f'} - g\right\} e^{f' a^\dagger a}.\end{aligned} \tag{11.6}$$

Combining Eqs. (11.5) and (11.6), and comparing the same items $a^\dagger a e^{f' a^\dagger a}$ or $e^{f' a^\dagger a}$, we obtain two new differential equations:

$$\frac{df'}{dt} = 2\left[g\left(e^{-f'} - 1\right) + \kappa\left(e^{f'} - 1\right)\right], \tag{11.7}$$

and

$$2\left(1 - e^{f'}\right)\left(\kappa e^{f'} - g\right) = -e^{f'}\frac{df'}{dt}. \tag{11.8}$$

The above two equations are the same, so only one of them needs to be solved. Using Eq. (11.8) to obtain

$$\int \frac{df'}{g\left(e^{-f'} - 1\right) + \kappa\left(e^{f'} - 1\right)} = 2\int dt + c, \tag{11.9}$$

where c is an integral constant. Using the integration formula

$$\int \frac{dx}{g\left(e^{-x} - 1\right) + \kappa\left(e^{x} - 1\right)} = \frac{\ln\left[\left(\kappa e^{x} - g\right)/\left(1 - e^{x}\right)\right]}{g - \kappa} + c', \tag{11.10}$$

we obtain the result of Eq. (11.9) as

$$\ln\left[\left(\kappa e^{f'}-g\right)/\left(1-e^{f'}\right)\right]=2\left(g-\kappa\right)t+c'', \tag{11.11}$$

where c'' is an new integral constant. When $t=0$, $f'\to f$, so the integral constant c'' reads

$$c''=\ln\frac{\kappa e^{f}-g}{1-e^{f}} \tag{11.12}$$

and there exists the identity

$$\frac{\kappa e^{f'}-g}{1-e^{f'}}=\frac{\kappa e^{f}-g}{1-e^{f}}e^{2(g-\kappa)t}. \tag{11.13}$$

Substituting Eqs. (11.2) and (11.3) into Eq. (11.11) leads to the result

$$f'=\ln\frac{g\left(1-e^{f}\right)+\left(\kappa e^{f}-g\right)e^{2(g-\kappa)t}}{\kappa\left(1-e^{f}\right)+\left(\kappa e^{f}-g\right)e^{2(g-\kappa)t}}. \tag{11.14}$$

So, in the channel with gain and decay, the density operator $\rho\left(0\right)$ of the chaotic state evolves into

$$\rho\left(t\right)=\frac{\left(\kappa-g\right)\left(1-e^{f}\right)}{\kappa\left(1-e^{f}\right)+\left(\kappa e^{f}-g\right)e^{2(g-\kappa)t}}\left[\frac{g\left(1-e^{f}\right)+\left(\kappa e^{f}-g\right)e^{2(g-\kappa)t}}{\kappa\left(1-e^{f}\right)+\left(\kappa e^{f}-g\right)e^{2(g-\kappa)t}}\right]^{a^{\dagger}a}. \tag{11.15}$$

Especially, when $\kappa=0$, no loss case, Eq. (11.14) becomes

$$f'=\ln[(e^{f}-1)e^{-2gt}+1], \tag{11.16}$$

thus the density operator in the channel with only gain at time t is represented as

$$\begin{aligned}\rho\left(t\right)&=\left(1-e^{f'}\right)e^{f'a^{\dagger}a}\\&=e^{-2gt}\left(1-e^{f}\right)e^{a^{\dagger}a\ln[1-e^{-2gt}\left(1-e^{f}\right)]}\\&=e^{-2gt}\left(1-e^{f}\right)\left[1-e^{-2gt}\left(1-e^{f}\right)\right]^{a^{\dagger}a}.\end{aligned} \tag{11.17}$$

While for $g=0$, no gain case, we similarly obtain

$$f'=\ln\frac{1}{1+\left(e^{-f}-1\right)e^{2t\kappa}} \tag{11.18}$$

and

$$\rho\left(t\right)=\frac{\left(e^{-f}-1\right)e^{2\kappa t}}{\left(e^{-f}-1\right)e^{2\kappa t}+1}\left[\frac{1}{\left(e^{-f}-1\right)e^{2\kappa t}+1}\right]^{a^{\dagger}a}. \tag{11.19}$$

To verify the correctness of Eq. (11.14), substituting Eq. (11.14) into the right side of Eq. (11.7), we have

$$\begin{aligned}
&2\left[g\left(\mathrm{e}^{-f'}-1\right)+\kappa\left(\mathrm{e}^{f'}-1\right)\right] \\
&=\frac{2\left(\mathrm{e}^{f}-1\right)(g-\kappa)^{2}\left(\kappa\mathrm{e}^{f}-g\right)\mathrm{e}^{2(g-\kappa)t}}{\left[g\left(1-\mathrm{e}^{f}\right)+\left(\kappa\mathrm{e}^{f}-g\right)\mathrm{e}^{2(g-\kappa)t}\right]\left[\kappa\left(1-\mathrm{e}^{f}\right)+\left(\kappa\mathrm{e}^{f}-g\right)\mathrm{e}^{2(g-\kappa)t}\right]}.
\end{aligned} \tag{11.20}$$

On the other hand, substituting Eq. (11.14) into the left side of Eq. (11.7) leads to

$$\begin{aligned}
\frac{\mathrm{d}f'}{\mathrm{d}t}&=\frac{\kappa\left(1-\mathrm{e}^{f}\right)+\left(\kappa\mathrm{e}^{f}-g\right)\mathrm{e}^{2(g-\kappa)t}}{g\left(1-\mathrm{e}^{f}\right)+\left(\kappa\mathrm{e}^{f}-g\right)\mathrm{e}^{2(g-\kappa)t}}\frac{\mathrm{d}}{\mathrm{d}t}\frac{g\left(1-\mathrm{e}^{f}\right)+\left(\kappa\mathrm{e}^{f}-g\right)\mathrm{e}^{2(g-\kappa)t}}{\kappa\left(1-\mathrm{e}^{f}\right)+\left(\kappa\mathrm{e}^{f}-g\right)\mathrm{e}^{2(g-\kappa)t}} \\
&=\frac{2\left(\mathrm{e}^{f}-1\right)(g-\kappa)^{2}\left(\kappa\mathrm{e}^{f}-g\right)^{2(g-\kappa)t}}{\left[g\left(1-\mathrm{e}^{f}\right)+\left(\kappa\mathrm{e}^{f}-g\right)\mathrm{e}^{2(g-\kappa)t}\right]\left[\kappa\left(1-\mathrm{e}^{f}\right)+\left(\kappa\mathrm{e}^{f}-g\right)\mathrm{e}^{2(g-\kappa)t}\right]}.
\end{aligned} \tag{11.21}$$

Obviously, Eqs. (11.20) and (11.21) are equal, so Eq. (11.14) is just the formal solution to Eq. (11.7).

Thus, we have solved the master equation of the chaotic state in the channel with gain and decay in a direct and concise method. This method is different from the motheds in other papers that require to know the Kraus operator-sum representation of the evolved density operator.

11.2 Initial Negative Binomial State Case

Now, we first present the operator-sum solution for the master equation describing the laser channel. For this, introducing the operator $\tilde{a}$ of fictitious mode, $\left[\tilde{a},\tilde{a}^{\dagger}\right]=1$, and the two-mode entangled states [6, 7]

$$|\eta\rangle=\exp\left(-\frac{1}{2}|\eta|^{2}+\eta a^{\dagger}-\eta^{*}\tilde{a}^{\dagger}+a^{\dagger}\tilde{a}^{\dagger}\right)\left|0\tilde{0}\right\rangle, \tag{11.22}$$

which obey the following orthogonal completeness, i.e.,

$$\left\langle\eta'\right|\eta\rangle=\pi\delta\left(\eta'-\eta\right)\delta\left(\eta'^{*}-\eta^{*}\right) \tag{11.23}$$

$$\int\frac{\mathrm{d}^{2}\eta}{\pi}|\eta\rangle\langle\eta|=1. \tag{11.24}$$

Also, the squeezed transformation $|\eta\rangle\rightarrow\frac{1}{\mu}|\eta/\mu\rangle$ is actually obtained by the following two-mode squeezed operator

$$S \equiv \int \frac{\mathrm{d}^2\eta}{\pi\mu}|\eta/\mu\rangle\langle\eta| = \exp\left[\lambda\left(a^\dagger\widetilde{a}^\dagger - a\widetilde{a}\right)\right], \quad \mu = \mathrm{e}^\lambda. \tag{11.25}$$

Using Eqs. (11.23) and (11.25) to obtain

$$S|\eta\rangle = \frac{1}{\mu}|\eta/\mu\rangle. \tag{11.26}$$

Operating both sides of Eq. (11.1) on the state $|I\rangle = |\eta = 0\rangle$, and denoting $|\rho(t)\rangle \equiv \rho(t)|I\rangle$, we thus have

$$\begin{aligned}\frac{\mathrm{d}}{\mathrm{d}t}|\rho(t)\rangle = [g\left(2a^\dagger\rho(t)a - aa^\dagger\rho(t) - \rho(t)aa^\dagger\right) \\ +\kappa\left(2a\rho(t)a^\dagger - a^\dagger a\rho(t) - \rho(t)a^\dagger a\right)]|I\rangle.\end{aligned} \tag{11.27}$$

Using Eqs. (6.8) and (6.9), we obtain the formal solution to Eq. (11.27) as

$$|\rho(t)\rangle = \exp\left[gt\left(2a^\dagger\tilde{a}^\dagger - aa^\dagger - \tilde{a}\tilde{a}^\dagger\right) + \kappa t\left(2a\tilde{a} - a^\dagger a - \tilde{a}^\dagger\tilde{a}\right)\right]|\rho(0)\rangle, \tag{11.28}$$

where $|\rho(0)\rangle \equiv \rho(0)|I\rangle$, $\rho(0)$ is the initial density operator. Therefore, $\rho(t)$ can be expressed as the infinite dimensional operator-sum representation [similar to deriving Eq. (6.21)], i.e.,

$$\begin{aligned}\rho(t) &= \sum_{n,j=0}^{\infty} T_3\frac{\kappa^n g^j T_1^{n+j}}{n!j!T_2^{2j}}\mathrm{e}^{a^\dagger a\ln T_2}a^{\dagger j}a^n\rho(0)a^{\dagger n}a^j\mathrm{e}^{a^\dagger a\ln T_2} \\ &= \sum_{n,j=0}^{\infty} M_{n,j}\rho(0)M_{n,j}^\dagger,\end{aligned} \tag{11.29}$$

where

$$\begin{aligned}T_1 &= \frac{1-\mathrm{e}^{-2(\kappa-g)t}}{\kappa - g\mathrm{e}^{-2t(\kappa-g)}}, \; T_2 = \frac{(\kappa-g)\,\mathrm{e}^{-(\kappa-g)t}}{\kappa - g\mathrm{e}^{-2t(\kappa-g)}} \\ T_3 &= \frac{\kappa-g}{\kappa - g\mathrm{e}^{-2t(\kappa-g)}} = 1 - gT_1\end{aligned} \tag{11.30}$$

and

$$M_{n,j} = \sqrt{\frac{T_3\kappa^n g^j}{n!j!T_2^{2j}}}T_1^{n+j}\mathrm{e}^{a^\dagger a\ln T_2}a^{+j}a^n. \tag{11.31}$$

Also, we can prove

$$\sum_{n,j=0}^{\infty} M_{n,j}^\dagger M_{n,j} = 1. \tag{11.32}$$

Therefore, $\rho(t)$ in Eq. (11.29) is exactly a density operator, and obeys the relation

$$\mathtt{tr}\rho(t) = \mathtt{tr}\left(\sum_{n,j=0}^{\infty} M_{n,j}\rho(0) M_{n,j}^{\dagger}\right) = \mathtt{tr}\rho(0). \tag{11.33}$$

Therefore, for any given initial density operator $\rho(0)$, the density operator $\rho(t)$ at time t can be directly obtained from Eq. (11.29). It can be seen from the above results that the entangled state representation provides us with a beautiful way to obtain the Kraus operator-sum representation of the density operator $\rho(t)$.

Next, we investigate the evolution of the negative binomial state $\rho_\gamma(0)$ [also see in Eq. (7.17)] in the laser channel [8]. For convenience, writting the negative binomial state $\rho_\gamma(0)$ again as

$$\rho_\gamma(0) = \frac{1}{s!(n_c)^s} a^s \gamma e^{a^\dagger a \ln(1-\gamma)} a^{\dagger s}, \; n_c = \frac{1-\gamma}{\gamma}. \tag{11.34}$$

Substituting Eq. (11.34) into Eq. (11.29) leads to

$$\begin{aligned}
\rho(t) &= \frac{T_3\gamma}{s!n_c^s} \sum_{n,j=0}^{\infty} \frac{\kappa^n g^j}{n!j!T_2^{2j}} T_1^{n+j} e^{a^\dagger a \ln T_2} \\
&\times a^{\dagger j} a^{n+s} e^{a^\dagger a \ln(1-\gamma)} a^{\dagger n+s} a^j e^{a^\dagger a \ln T_2} \\
&= \frac{T_3\gamma}{s!n_c^s} \sum_{j=0}^{\infty} \frac{g^j T_1^j}{j!T_2^{2j}} \sum_{n=0}^{\infty} \frac{(\kappa T_1)^n}{n!} \left(a^\dagger T_2\right)^j e^{a^\dagger a \ln T_2} \\
&\times a^{n+s} e^{a^\dagger a \ln(1-\gamma)} a^{\dagger n+s} e^{a^\dagger a \ln T_2} (aT_2)^j \\
&= \frac{T_3\gamma}{s!n_c^s T_2^{2s}} \sum_{j=0}^{\infty} \frac{g^j T_1^j}{j!T_2^{2j}} \left(a^\dagger T_2\right)^j \sum_{n=0}^{\infty} \frac{\left(\kappa T_1/T_2^2\right)^n}{n!} \\
&\times a^{n+s} e^{a\dagger a \ln\left[(1-\gamma)T_2^2\right]} a^{\dagger n+s} (aT_2)^j \\
&= \frac{T_3\gamma}{s!n_c^s T_2^{2s}} \sum_{j=0}^{\infty} \frac{g^j T_1^j}{j!T_2^{2j}} \left(a^\dagger T_2\right)^j \sum_{n=0}^{\infty} \frac{\left(\kappa T_1/T_2^2\right)^n (n+s)!}{n!} \\
&\times \left[(1-\gamma) T_2^2\right]^{n+s} : e^{\left[(1-\gamma)T_2^2-1\right]a\dagger a} \mathrm{L}_{n+s}\left[a^\dagger a (\gamma-1) T_2^2\right] : (aT_2)^j.
\end{aligned} \tag{11.35}$$

To further calculate Eq. (11.35), we need to introduce the new generating function for the Laguerre polynomial, that is

$$\sum_{n=0}^{\infty} \frac{(n+s)!(-\lambda)^n}{n!s!} \mathrm{L}_{n+s}(z) = (1+\lambda)^{-s-1} e^{\frac{\lambda z}{1+\lambda}} \mathrm{L}_s\left(\frac{z}{1+\lambda}\right). \tag{11.36}$$

Equation (11.36) is proven as follows. For this, we first calculate the summation

$$\sum_{n=0}^{\infty} \frac{\lambda^n}{n!} \mathrm{H}_{n+m,n+s}(x, y), \tag{11.37}$$

where

$$\mathrm{H}_{m,s}(x, y) = \sum_{n=0}^{\min(m,s)} \frac{m!s!(-1)^n}{n!(m-n)!(s-n)!} x^{m-n} y^{s-n} \tag{11.38}$$

is a two-variable Hermite polynomial. Here, we do not directly calculate (11.37), while consider the summation

$$\sum_{n=0}^{\infty} \frac{\lambda^n}{n!} (-\mathrm{i})^{m+s+2n} \mathrm{H}_{n+m,n+s}\left(\mathrm{i}a^\dagger, \mathrm{i}a\right). \tag{11.39}$$

Using the operator identity

$$a^s a^{\dagger m} = (-\mathrm{i})^{m+s} : \mathrm{H}_{m,s}\left(\mathrm{i}a^\dagger, \mathrm{i}a\right) :, \tag{11.40}$$

we have

$$\begin{aligned} \sum_{n=0}^{\infty} \frac{\lambda^n}{n!} (-\mathrm{i})^{m+s+2n} \mathrm{H}_{n+m,n+s}\left(\mathrm{i}a^\dagger, \mathrm{i}a\right) &= \sum_{n=0}^{\infty} \frac{\lambda^n}{n!} a^{n+s} a^{\dagger n+m} \\ &= a^s \vdots \mathrm{e}^{\lambda a a^\dagger} \vdots a^{\dagger m}. \end{aligned} \tag{11.41}$$

Further, using the IWOP method and the completeness relation:

$$\int \frac{\mathrm{d}^2 z}{\pi} |z\rangle \langle z| = \int \frac{\mathrm{d}^2 z}{\pi} : \exp\left(-|z|^2 + z a^\dagger + z^* a - a^\dagger a\right) : = 1, \tag{11.42}$$

where $|z\rangle = \exp\left(-|z|^2/2 + z a^\dagger\right)|0\rangle$ is a coherent state, thus Eq. (11.41) becomes

$$\begin{aligned} a^s \vdots \mathrm{e}^{\lambda a a^\dagger} \vdots a^{\dagger m} &= \int \frac{\mathrm{d}^2 z}{\pi} z^s \mathrm{e}^{\lambda |z|^2} z^{*m} |z\rangle \langle z| \\ &= \int \frac{\mathrm{d}^2 z}{\pi} z^s z^{*m} |z\rangle \langle z| : \mathrm{e}^{-(1-\lambda)|z|^2 + z a^\dagger + z^* a - a^\dagger a} : \\ &= (-\mathrm{i})^{m+s} (1-\lambda)^{-\frac{(s+m)}{2} - 1} : \mathrm{e}^{\frac{\lambda a^\dagger a}{1-\lambda}} \mathrm{H}_{m,s}\left(\frac{\mathrm{i}a^\dagger}{\sqrt{1-\lambda}}, \frac{\mathrm{i}a}{\sqrt{1-\lambda}}\right) :. \end{aligned} \tag{11.43}$$

Comparing Eqs. (11.41) and (11.43), we obtain the operator identity:

$$\sum_{n=0}^{\infty}\frac{(-\lambda)^n}{n!}:\mathrm{H}_{n+m,n+s}\left(\mathrm{i}a^{\dagger},\mathrm{i}a\right):$$
$$=(1-\lambda)^{-\frac{(s+m)}{2}-1}:\mathrm{e}^{\frac{\lambda a^{\dagger}a}{1-\lambda}}\mathrm{H}_{m,s}\left(\frac{\mathrm{i}a^{\dagger}}{\sqrt{1-\lambda}},\frac{\mathrm{i}a}{\sqrt{1-\lambda}}\right):. \tag{11.44}$$

Noting that both sides of Eq. (11.44) are in normal ordering, thus the operators $a^{\dagger}$, a are equivalent to the classical c numbers, thus making the substitutions $x\rightarrow \mathrm{i}a^{\dagger}$ and $y\rightarrow \mathrm{i}a$ in Eq. (11.44), we have

$$\sum_{n=0}^{\infty}\frac{\lambda^n}{n!}\mathrm{H}_{n+m,n+s}\left(x,y\right)=(1+\lambda)^{-\frac{(s+m)}{2}-1}\mathrm{e}^{\frac{\lambda xy}{1+\lambda}}\mathrm{H}_{m,s}\left(\frac{x}{\sqrt{1+\lambda}},\frac{y}{\sqrt{1+\lambda}}\right), \tag{11.45}$$

which is the new generating function for the two-variable Hermite polynomial $\mathrm{H}_{n+m,n+s}\left(x,y\right)$. Especially, when $m=s$, Eq. (11.45) becomes

$$\sum_{n=0}^{\infty}\frac{\lambda^n}{n!}\mathrm{H}_{n+s,n+s}\left(x,y\right)=(1+\lambda)^{-s-1}\mathrm{e}^{\frac{\lambda xy}{1+\lambda}}\mathrm{H}_{s,s}\left(\frac{x}{\sqrt{1+\lambda}},\frac{y}{\sqrt{1+\lambda}}\right). \tag{11.46}$$

Further, using the relations between two-variable Hermite polynomials and Laguerre polynomials, i.e.,

$$\mathrm{L}_s\left(xy\right)=\frac{(-1)^s}{s!}\mathrm{H}_{s,s}\left(x,y\right), \tag{11.47}$$

thus Eq. (11.46) becomes Eq. (11.36).

Using Eq. (11.36) and noting $n_c=(1-\gamma)/\gamma$, and summing over n in Eq. (11.35), we obtain

$$\rho(t)=T_3\frac{\gamma}{n_c^sT_2^{2s}}\sum_{j=0}^{\infty}\frac{g^jT_1^j}{j!T_2^{2j}}:\left(a^{\dagger}aT_2^2\right)^j\left[(1-\gamma)\,T_2^2\right]^s\mathrm{e}^{\left[(1-\gamma)T_2^2-1\right]a^{\dagger}a}$$
$$\times\left[1-\kappa T_1\left(1-\gamma\right)\right]^{-s-1}\mathrm{e}^{\frac{\kappa T_1(1-\gamma)^2T_2^2}{1-\kappa T_1(1-\gamma)}a^{\dagger}a}\mathrm{L}_s\left[\frac{a^{\dagger}a\left(\gamma-1\right)T_2^2}{1-\kappa T_1\left(1-\gamma\right)}\right]:$$
$$=G:\gamma'^{s+1}\mathrm{e}^{\left(gT_1-\gamma'\right)a^{\dagger}a}\mathrm{L}_s\left[a^{\dagger}a\left(\gamma'-1\right)\right]:, \tag{11.48}$$

where

$$G\equiv\left[\frac{\gamma}{1-\left(\kappa T_1+T_2^2\right)\left(1-\gamma\right)}\right]^{s+1}\left(1-gT_1\right) \tag{11.49}$$

and

$$\gamma'=1-\frac{T_2^2\left(1-\gamma\right)}{1-\kappa T_1\left(1-\gamma\right)}. \tag{11.50}$$

To better understand the physical meaning of the density operator $\rho(t)$, using Eq. (11.48) to obtain the identity

$$\begin{aligned}\rho(t) &= G\sum_{n=0}^{\infty}\frac{(gT_1)^n}{n!}a^{\dagger n}:\gamma'^{s+1}e^{-\gamma' a^\dagger a}L_s\left[a^\dagger a\left(\gamma'-1\right)\right]:a^n\\ &= G\sum_{n=0}^{\infty}\frac{(gT_1)^n}{n!}a^{\dagger n}\rho_{\gamma'}a^n,\end{aligned} \tag{11.51}$$

where

$$\rho_{\gamma'} = \gamma'^{s+1}:e^{-\gamma' a^\dagger a}L_s\left[\left(\gamma'-1\right)a^\dagger a\right]:. \tag{11.52}$$

Obviously, $\rho_{\gamma'}$ is a new negative binomial state since its parameter γ' is different from the parameter γ of the initial negative binomial state in Eq. (11.34). Therefore, after passing through the gain-decay channel, the initial negative binomial state becomes the superposition of photon-added negative binomial states.

Using the completeness relation of coherent states and Eq. (11.48), we can prove $\mathrm{tr}\rho(t) = 1$, i.e.,

$$\begin{aligned}\mathrm{tr}\rho(t) &= G\gamma'^{s+1}\int\frac{d^2z}{\pi}\langle z|:e^{(gT_1-\gamma')a^\dagger a}L_s\left[a^\dagger a\left(\gamma'-1\right)\right]:|z\rangle\\ &= G\gamma'^{s+1}\sum_{l=0}^{s}\frac{\left(1-\gamma'\right)^l s!}{(l!)^2(s-l)!}\int\frac{d^2z}{\pi}|z|^{2l}e^{(gT_1-\gamma')|z|^2}\\ &= G\gamma'^{s+1}\sum_{l=0}^{s}\frac{\left(1-\gamma'\right)^l s!}{(l!)^2(s-l)!}\frac{l!}{(\gamma'-gT_1)^{l+1}}\\ &= \left(\frac{\gamma}{1-\left(\kappa T_1+T_2^2\right)(1-\gamma)}\right)^{s+1}\frac{\gamma'^{s+1}(1-gT_1)}{\gamma'-gT_1}\left(\frac{1-\gamma'}{\gamma'-gT_1}+1\right)^s\\ &= 1.\end{aligned} \tag{11.53}$$

Besides, we calculate the average photon number of the state $\rho(t)$ in Eq. (11.51). In the initial negative binomial state in Eq. (11.34), the average photon number reads

$$\begin{aligned}\langle N\rangle_0 &= \mathrm{tr}\left[a^\dagger a\rho_\gamma(0)\right]\\ &= \mathrm{tr}\left(\sum_{n=0}^{\infty}\frac{(n+s)!}{n!s!}\gamma^{s+1}(1-\gamma)^n n|n\rangle\langle n|\right)\\ &= \sum_{n=1}^{\infty}\frac{(n+s)!}{(n-1)!s!}\gamma^{s+1}(1-\gamma)^n\\ &= \frac{(1+s)(1-\gamma)}{\gamma}.\end{aligned} \tag{11.54}$$

Using Eq. (11.48) and the operator identity $\left[a, :f\left(a, a^{\dagger}\right):\right] = \frac{\partial}{\partial a^{\dagger}} : f\left(a, a^{\dagger}\right):$, the average photon number of the state in Eq. (11.51) is calculated as

$$
\begin{aligned}
\langle N\rangle_t &= \mathrm{tr}\left(a^{\dagger} a \rho(t)\right) \\
&= G\gamma'^{s+1} \int \frac{d^2 z}{\pi} \langle z| a^{\dagger} a : e^{\left(gT_1-\gamma'\right) a^{\dagger} a} \mathrm{L}_s\left[a^{\dagger} a\left(\gamma'-1\right)\right] : |z\rangle \\
&= G\gamma'^{s+1} \int \frac{d^2 z}{\pi} \langle z| a^{\dagger} : e^{\left(gT_1-\gamma'\right) a^{\dagger} a} \mathrm{L}_s\left[a^{\dagger} a\left(\gamma'-1\right)\right] : a|z\rangle \\
&+ G\gamma'^{s+1} \int \frac{d^2 z}{\pi} \langle z| a^{\dagger} : \frac{\partial}{\partial a^{\dagger}}\left\{e^{\left(gT_1-\gamma'\right) a^{\dagger} a} \mathrm{L}_s\left[a^{\dagger} a\left(\gamma'-1\right)\right]\right\} : |z\rangle \\
&= G\gamma'^{s+1}\left\{\left(1+gT_1-\gamma'\right) \int \frac{d^2 z}{\pi}|z|^2 e^{\left(gT_1-\gamma'\right)|z|^2} \mathrm{L}_s\left[|z|^2\left(\gamma'-1\right)\right]\right. \\
&\left. + \int \frac{d^2 z}{\pi} e^{\left(gT_1-\gamma'\right)|z|^2} \sum_{l=0}^{s} \frac{l|z|^{2l}\left(1-\gamma'\right)^l s!}{(l!)^2(s-l)!}\right\} \\
&= G\gamma'^{s+1} \sum_{l=0}^{s} \frac{\left(1-\gamma'\right)^l s!}{l!(s-l)!} \frac{\left(gT_1+1-\gamma'\right)(l+1)+l\left(\gamma'-gT_1\right)}{\left(\gamma'-gT_1\right)^{l+2}} \\
&= \frac{(1+s)(1-\gamma)}{\gamma} e^{-2(\kappa-g)t}.
\end{aligned} \tag{11.55}
$$

Comparing Eqs. (11.54) and (11.55), we have

$$
\langle N\rangle_t = \langle N\rangle_0 \, e^{-2(\kappa-g)t}. \tag{11.56}
$$

It can be seen that the average number of photons at any given time t alway decays or gains exponentially, which depends on the decay and gain coefficients κ, g.

So far, we have discussed how a negative binomial state evolves in a laser channel. By using the newly derived Laguerre polynomial's generating function formula, we found that the final state can be precisely expressed as the superposition of photon-added negative binomial states and the average number of photons of the final state increases or decreases exponentially, which depends on the decay and gain coefficients.

References

1. Scully MO, Zubairy MS (1997) Quantum optics. Cambridge University Press, Cambridge
2. Chen JH, Fan HY (2013) Entropy evolution law in a laser process. Ann Phys 334:272–279
3. Meng XG, Liang BL, Liu JM, Zhou XC (2025) Entropy evolution law of general quadratic state in a laser channel. Front Phys 20(2):022205

4. Wan ZL (2025) The entropy variation and quantum statistical behaviors of a general linear state in a laser channel. Chin J Phys 94:215–225
5. Jiang NQ, Fan HY, Hu LY (2011) Evolution of the chaotic field in the laser process: evolution law of density operator and photon number decay. J Phys A Math Theor 44(19):195302
6. Meng XG, Wang JS, Liang BL (2023) Entangled state representations in quantum optics. Springer, Singapore
7. Fan HY (2004) Entangled states, squeezed states gained via the route of developing Dirac's symbolic method and their applications. Int J Mod Phys B 18(10–11):1387–1455
8. Da C, Chen QF, Fan HY (2014) Evolution law of the negative binomial state in laser channel and its photon-number decay formula. Int J Theor Phys 53(12):4372–4380

FSC
www.fsc.org
MIX
Papier aus verantwortungsvollen Quellen
Paper from responsible sources
FSC® C105338

If you have any concerns about our products,
you can contact us on
ProductSafety@springernature.com

In case Publisher is established outside the EU,
the EU authorized representative is:
Springer Nature Customer Service Center GmbH
Europaplatz 3, 69115 Heidelberg, Germany

Printed by Libri Plureos GmbH
in Hamburg, Germany